D0141858

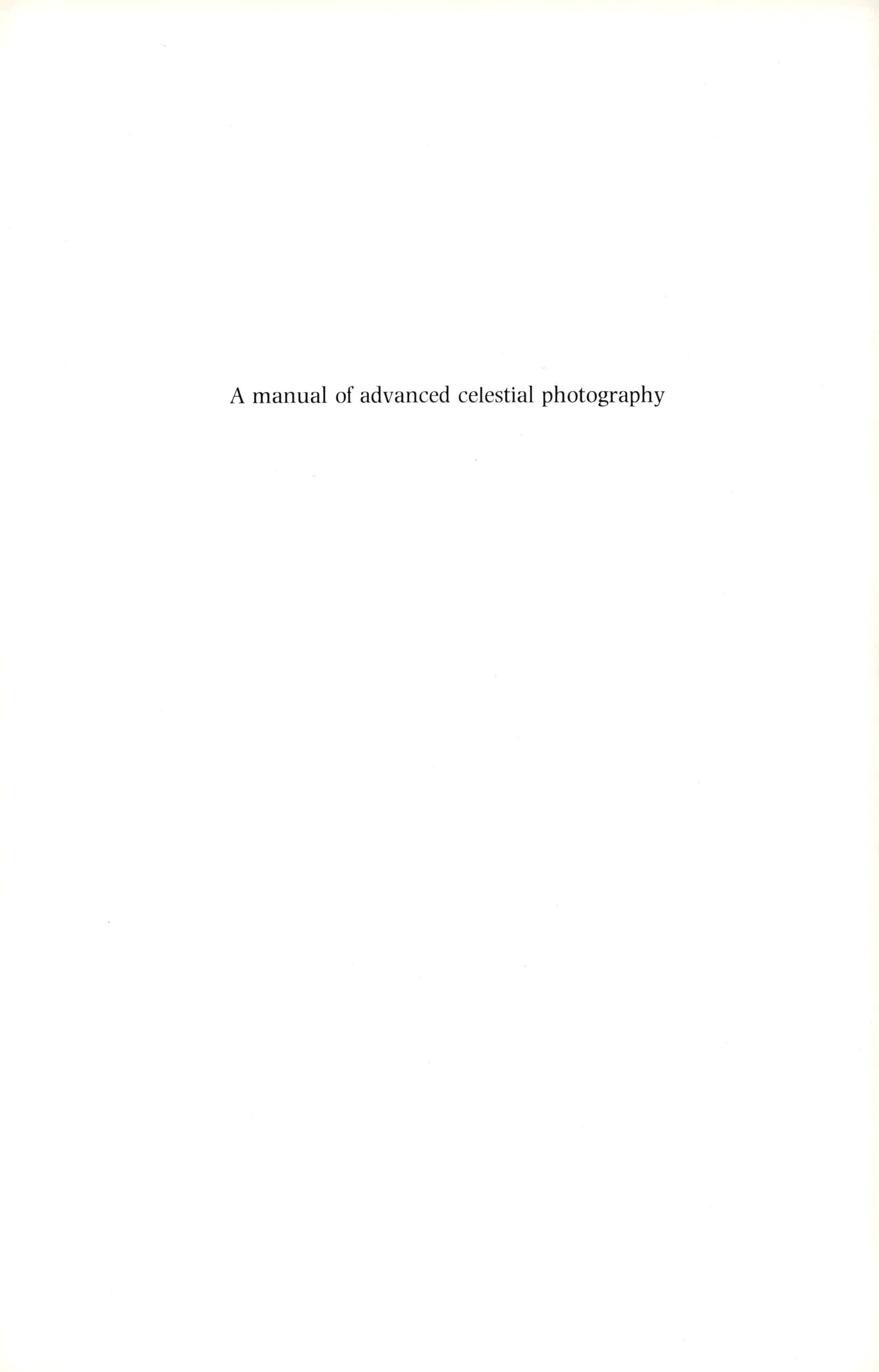

A manual of advanced celestial photography

Concrete is heavy and iron is hard
but the grass will prevail

Edward Abbey

A manual of advanced celestial photography

BRAD D. WALLIS

ROBERT W. PROVIN

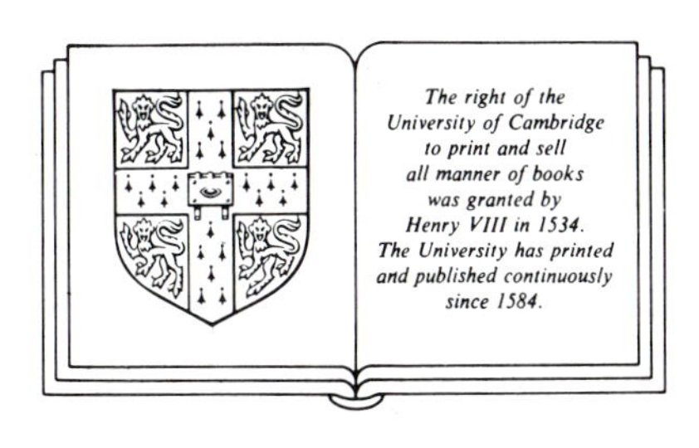

CAMBRIDGE UNIVERSITY PRESS

Cambridge

New York New Rochelle Melbourne Sydney

Published by the Press Syndicate of the University of Cambridge
The Pitt Building, Trumpington Street, Cambridge CB2 1RP
32 East 57th Street, New York, NY 10022, USA
10 Stamford Road, Oakleigh, Melbourne 3166, Australia

First published 1988

Printed in Great Britain by Ebenezer Baylis & Son Ltd, Worcester

British Library cataloguing in publication data

Wallis, Brad D.
A manual of advanced celestial photography.
1. Astronomical photography
I. Title II. Provin, Robert W.
778.9'9523 QB121

Library of Congress cataloguing in publication data

Wallis, Brad D.
A manual of advanced celestial photography.
Bibliography:
Includes index.
1. Astronomical photography. I. Provin, Robert W.
II. Title.
QB121.W25 1988 522'.63 86–12946

ISBN 0 521 25553 8

DEDICATION

We are the brothers of the boulders, the cousins of the clouds.
Harlow Shapley

We would like to dedicate this book to all those who came before us:

To Ernest D. Wallis, photographer, who taught us much of the photographic techniques and 'tricks' that have been of so much value in our work. Who also taught us the value of excellence and taught us never to settle for mediocrity when true excellence could be obtained. Whose knowledge, assistance, and direction allowed us to overcome many of the photographic problems that faced us as we explored the realm of astronomical photography. Who died before we could finish this work.

To William A. Provin who always points to and insists on the best path, even if it be the most difficult;

To Wm ('Bill') C. Miller who helped and encouraged us in several of our advanced techniques and who taught us the need for careful controls and quantitative measurements. Who took the first astronomical color photographs with the Palomar 200 inch telescope and who worked to establish rigorous photographic controls in observatories around the world. Who we will also miss;

To E.S. King whose numerous published works have proved invaluable to us and who laid much of the groundwork that has made astrophotography what it is today;

To Alan McClure, Everett Kreimer, and Orien Ernest whose photographs inspired us and showed us that there really was more to astrophotography;

We would also like to dedicate this book to all those that will follow us, we hope that this work can help avoid many of the problems that we have had to face and solve over the years. Our primary role over the last two decades has been to learn and apply professional techniques and processes in the area of amateur astrophotography and to convey our knowledge to the many amateurs who enjoy this blend of astronomy and photography. If we can teach a portion of what we have learned, our goal will have been achieved.

Contents

Preface – A dialogue

The location is basically unimportant, but for the sake of choosing a stimulating environment that is still pure and largely untouched by the ravage of today's civilization we will hold this dialogue in the plateaus of southeast Utah on the edge of a 2000 foot cliff that overlooks the lands where the dinosaurs reigned some 150 million years ago. Here one is surrounded by a fierce, primitive, even hostile beauty. The mind is free to soar with the eagles and is unfettered by the traditions that normally hold it prisoner in today's commercially directed world. Here, where the great lizards roamed, bred and finally died we shall sip beer and discuss our reasons for writing this work while watching a large vulture patiently circle overhead waiting for some animal's universe to come to an end.

Robert: Here it is, *the* contract. We finally have a firm commitment to get all of our work out where it will be available to a whole lot of people.

Brad: Yes, you most certainly did it! It's what we have both been wanting, but you sure picked a publisher! I hope that we can produce the work we want to. There is so much to say and we're going to undoubtedly forget some important points as we go, not to mention all the things that we still haven't had time to learn in detail.

Robert: That has always been the problem: so much to learn and so little time! We have been doing this over 20 years and still have just scratched the surface. We always wanted to do a book but have never felt that we knew enough to do a good job. Finally it dawned we were never going to know all we wanted to know and it was time to get what we have learned into a book that would be available to just about anyone.

Brad: I will definitely drink to that! If anyone should write a book on astronomical photography it should have been Bill Miller, but it would have been less oriented at the amateur world. We keep writing articles, and presenting papers, but we can not reach enough people that way! We keep getting inquiries about the 1977 *Sky and Telescope* series and letters asking for copies of all our articles, but that only hits a small group spread around the world. We just can't get to enough people without putting the information out through a large publisher.

Robert: I really want people to know that the reason behind all of this is to try and give them a helping hand, to help avoid many of the pitfalls we struggled through because there was no real reference work on the topic when we started. There still is no manifesto to get one over the beginning hurdles and on into the more challenging areas of the field. If we can't teach some of what we have learned, much of our efforts will have been wasted.

Brad: I sure wish we had known all that we know now! It would have saved so much time! Hopefully, we will be able to expose enough of the myths that continue to hinder progress in the amateur field that we will get people to stop listening to the arm-chair experts that inhabit so many local astronomy clubs. Then, with a larger group of amateurs getting out of the habits and practices of astrophotography of the 1950s, the rate of progress should increase dramatically. After all, it has taken professionals over 50 years to finally adopt a technique as simple as contrast masking, yet it is so simple that anyone can apply it with great success.

Robert: It would seem that a more educated group of amateurs should also help to improve considerably the quality of commercial products for astrophotography. That should be of great benefit considering some of the inferior products that have been foisted on the market by some rather deceptive marketing practices. It sure is hot out here, it's time for another beer!

Brad: We have already seen the effects of a number of well designed products on the quality of the photos that are being turned out. The 'low profile focuser' allows people to get photos that show less vignetting than was evident 10 years ago, and the off-axis guiders that are now available for small instruments have allowed everyone to get out and photograph without having to fight flexure (until a comet comes rolling through!).

Robert: That's all true, but it seems to me that people are now using these accessories without really knowing the reason for using them. For instance, what good is a low profie focuser without a proper size diagonal? What use is a scope that has been used exclusively with an off-axis guider when one wants to photograph a comet.

Brad: Exactly! That means that it is important that we try to explain the '*why's*' behind any rules or equations or techniques that we discuss. If people know the logical basis for something, they can use that to reason forward to some new conclusions that may be useful in an entirely different problem. With just a set of rules and no logical foundation,it is impossible to make much progress without first stepping backwards in search of the logic and we have seen that virtually no one is willing to take those steps. For some reason, all but the most recent popular literature is ignored.

Robert: The problem with that attitude of ignoring everything but the latest and most popular articles is that the amateur literature just seems to keep repeating the same concepts over and over with just minor variations. How many articles deal with making mirrors or building observatories and how many deal with how to test film speeds accurately? If we can manage to create the tome that we would like to, we will be able to put countless tidbits of information into a single source that will be available for everyone for years to come! I hope we can do it right!

Brad: It is scary when you, or we, really get down to it! We have to do this the way we really want to do it and if we blow it, we fail! This is going to put everything on the line: Either we all win or, if we blow it, everyone loses!

Robert: If that isn't frightening enough, we have a couple of other problems to take on! It seems that when we try to let people know of some technique we have found that is helpful in getting better photos, or that can help to bypass some nasty problem that often ruins hours of work, we often end up being accused of telling people that their work is inferior and are told we are too negative.

Brad: I've noticed! The times when that has not happened have been when we put the information into large articles instead of doing a presentation at some conference. I do not understand it. It seems like the offer of help is often seen as an accusatory finger. Besides that, we are often accused of having such a negative attitude because of the fact that we keep pushing to make things better and keep pushing to get people to do better work. Why is the desire to improve on existing conditions seen as a negative attitude? I wish someone could explain the logic behind that attitude!

Robert: I don't know, but it keeps coming back at us from so many different directions! There is a core of highly skilled astrophotographers out there that are in total agreement with our philosophy, but may others just want to keep on doing the same sort of work that has been done for the past 20 years, while others have achieved great popularity by campaigning to get people to forget the theory and technology and just go out and shoot regardless of how poor the final results are. I just do not see the point of these attitudes. If you have no intention of doing the work well, there is no point in doing the work!

Brad: Those same people then claim that they are simply taking the photos for themselves whenever pushed on the point. However, as my photographer/teacher friend, Al Weber, will point out to his students who take that tact, 'If you are just taking the photos for yourself, why are you spending so much effort seeing that they are seen by so many others?' I am particularly puzzled by the negative attitude that we supposedly have. If you were doing good work in grade school and one of your instructors came up to talk and told you that you should keep on doing better and better work and that you should set your sights on college, would you say that he was being negative? There are so many things that amateurs could be doing with the equipment, skills, and time that they have yet many seem to just want to take the same photos over and over again.

Robert: The group that seems to accept anything that comes out of the camera without any sort of discrimination or sense of aesthetics have created a considerable back slide with their revolt against theory and technology. It is a sort of 'socialized astrophotography'. By lowering the standards and teaching it is 'OK' to turn out second rate work, not only do they get their own work circulated as 'astrophotography for fun', but they seduce many others into this 'laid-back' approach. It is fine to encourage beginners, and to enjoy the work, but to make no attempt at progress as one works is to invite the intellectual stagnation that has already stifled this field for over 20 years!

Brad: Time for more beer! Yes, here's to continued moral and intellectual decay! Sacrifice quality photos because they require slightly more time and effort. Sacrifice quality optics because there is a larger market for cheap optics or because everyone would rather have a tube that is only a foot long! The few master machinists around that make high precision drives for telescopes can barely stay in business because people can not understand why one drive should cost $250 when another can be bought for a mere $10. Unfortunately, this growing trend of maximum compromise seems to pervade more than just astrophotography! Just imagine what we would have if Michelangelo had sculpted David from chalk because of the high cost of marble!

Robert: It sure would be a dull world without the few artists who take the time and effort to make something just a little better before thinking that it is close to being finished! Yes indeed! Here's to moral and intellectual decay! It takes just a little more effort to produce a markedly better result. There are so many projects that can even be carried out in darkroom away from the telescope. A little reading now and then can add so much perspective to the field and a little writing now and then can help others to avoid the pitfalls that have made this latest task so challenging.

Brad: Speaking of writing, we have about 125 000 words to crank out before we run out of the space that we have been allowed for this effort, and we had better finish off this beer, get out of the hot sun, say goodbye to our vulture friends and get to work.

Robert: I'll drink to that! Say, what do you suppose that sign over there means? 'Hayduke's Road . . . Take The Other'.

Brad: Probably just some sunbaked fool who stayed out here too long and decided to give up civilization for some things of real value: beauty, peace, tranquility, and principles.

Robert: It's amazing what can happen to the mind when exposed to the baking sun, clean air and the desert solitaire for too long. How about another beer!

To complete our work, we must journey away from this unspoiled site down the long twisted road to what is euphemistically called 'civilization'. A sign post up ahead reads 'Los Angeles City Limits', some poor demented sunbaked prophet has slapped a large, vividly colored bumper sticker over it that reads 'out of order'. Perhaps there is some hope for the future!

> Growth for the sake of growth
> is the ideology of the cancer cell!
> Edward Abbey

Acknowledgements

Over the many years that we have worked together, and in the last four years of creating this work, we have received help from numerous sources. While some are expressly mentioned elsewhere, many have provided help and information that can not be expressly cited in the text. We would like to use this opportunity to offer our sincerest thanks to all that helped to make this volume as complete as it is. Among those that have contributed we must list the following persons and establishments:

Bill Miller
Penny Oldenberger
Al Weber
Martin Germano
Ron Potter
Edwin Hirsch
Dick Nelson
Bob Little
Ruthanne Greenwood-Zuellig
Richard Berry
Rhea Goodwin
Lonne Lane

Steve Walther
Joan Gantz
Jack Newton
Jack Marling
Jim Rouse
Lee Coombs
Fred Larsen
Orien Ernest
Peggy Michaels
Dennis di Cicco

Mt Wilson and Las Campanas Observatories Library
Eastman Kodak Research Library
Astronomy Magazine
Sky and Telescope

Finally, we would like to thank the following individuals for their indirect but, nonetheless, essential contributions.

Edward Abbey
Friedrich Neitsche
Arthur Schopenhauer
François Marie Arouet (aka Voltaire)

Introduction

Our purpose in creating *A Manual of Advanced Celestial Photography* is to fill a long standing void in the literature on astronomical photography. The popular books currently available (circa 1985) on the subject are largely out of date and are universally limited in scope and elementary in their coverage and, while much fine advanced work can be found in periodicals, few of these works address the theories and principles that form the foundation of fundamental photographic methods and techniques. The contemporary literature is full of beautifully illustrated up-to-date tests of the most recent films, developers, and hypering systems yet these tests are of very limited scientific value since few practicing amateurs possess even a rudimentary understanding of sensitometry or of the mechanics and chemistry of photographic materials. It is our hope to shed some light on these subjects in an attempt to narrow the gap that exists between amateur and professional and to thereby make techniques that have long been employed by professionals in both photography and in astronomy readily available to the amateur community.

With this goal in mind, we have deliberately omitted much introductory material that is readily available in other sources. The material that we do cover can largely be divided into two broad categories: the first deals primarily with optical and mechanical systems and their optimal application to astrophotography while the second deals with the mechanics, processes and testing of photographic materials for use in astromonical applications. Wherever necessary these distinct areas are treated together, but in general they have been covered in separate sections. Although some of the material that is presented under the first category can be found elsewhere in the amateur literature it is here concentrated into a single source and has been augmented with information from professional sources and heavily tempered by over 40 years of the authors' accumulated field experience. It is our desire that the readers of this book will benefit by the many years of testing and trial and error work that was happily spent at the telescope and in the darkroom moving towards the realization of this work.

The second category, dealing with the photographic process, is perhaps the most important in that areas which have been previously unexplored by the amateur astrophotographer are now examined. It is in these sections of the book where we have striven to describe and analyze the tools that will enable those interested to carry out meaningful work, to advance their own knowledge, and perhaps astrophotography in general. A conscious effort has been made to avoid references to specific products in order to circumvent the inevitable dating that results from the constant rapid advances we are now experiencing in photographic technology. Instead, we have chosen to concentrate on process and methodology. The sections on black-and-white and color photography, darkroom techniques and, in particular, sensitometry should provide the interested amateur with the basic knowledge needed to derive precise conclusions about many of the photographic processes and to use these to maximum effect. More important however, is the fact that these methods produce more exact quantitative results that can be demonstrated repeatedly and thus be of more value to all working in the field.

Amateur astronomers have established a long tradition of close cooperation with professionals and many valuable contributions to the science have been made by the 'backyard' observer or photographer. A fine example of this kind of cooperation is the *American Astromomical Society Photo-Bulletin* which serves as a forum where professional and advanced amateur astrophotographers alike can discuss the latest advances in the field. However, while a number of amateurs have contributed to the *AAS Photo-Bulletin*, there is a definite need for greater scientific accuracy on their part. The *Photo-Bulletin's* editor Alex G. Smith stated the problem very well with 'A plea for scientific sensitometry' which appeared in the 1980 No.2 issue. In his 'plea', Smith '. . . urges all astrophotographers seriously interested in . . . experimenting with emulsions to think long and hard about performing quantitative laboratory test of a reasonably traditional nature. It is, after all, only in this way that we can all communicate effectively with each other!' Of course this idea can be applied to all areas of astrophotography and it is to this end and in this spirit that we dedicate this work.

Finally, in any endeavor of this magnitude, a point is reached where the authors' experience and research are no longer equal to the task. We are also certain that we have left out numerous details that probably should have been included for the benefit of our readers. In an effort to fill some of these gaps, a collection of papers by widely recognized practitioners has been included as the last chapter of the book. This work represents the current 'state of the art' for the topics included and goes well beyond the authors' resources in these selected fields. For a variety of reasons we could not include all the topics that we would have liked to include. As time, space, and energy ran out we were forced to draw the line and call a halt. Perhaps another volume is in order, or a second edition some years down the line. In the meantime, we hope that our work will be of use and we welcome any comments, suggestions, corrections, or debates that readers may wish to relate. Best wishes, clear skies, and good Seeing to all! ! !

. . . is it really so difficult simply to accept everything that one has been brought up on and that has gradually struck deep roots – what is considered truth in the circle of one's relatives and of many good men, and what, moreover, really comforts and elevates man? Is that more difficult than to strike new paths, fighting the habitual, experiencing the insecurity of independence and the frequent wavering of one's feelings and even one's conscience, proceeding often without any consolation, but ever with the eternal goal of the true, the beautiful, and the good? . . . Do we after all seek rest, peace, and pleasure in our inquiries? No, only truth–even if it be the most abhorrent and ugly . . . Faith offers the least support for a proof of objective truth. Here the ways of men part: if you wish to strive for peace of soul and pleasure, then believe. If you wish to be a devotee of truth, then inquire . . .

Letter to his sister, 1865
Friedrich Nietzsche

1

Facts do not cease to exist because they are ignored.
Aldous Huxley

A brief history of astronomical photography

Light itself would seem to be the origin of all astronomers' problems. Those mysterious objects out there are emitting light and the astronomer's goal is to learn all that he can about 'out there'. When light is all you have to work with, a means needs to be devised to capture the details presented by that light.

For many years, the astronomer's only recourse was to draw the images that he could view through his telescope. This methodology has one innate drawback: **you can only draw the details that you can see**. That seems quite obvious and, when one is studying details on distant planetary surfaces, it does not present a major drawback because such studies are basically limited by the stability of the Earth's atmosphere. However, when the objects under scrutiny include stars, nebulae and galaxies this becomes a major problem because of the faintness of these distant objects.

For over two centuries after the initial invention of the telescope, the only means at the astronomer's disposal for seeing fainter and fainter objects was to increase the size of the telescope that was being used to study his subjects. The problem presented by this solution is obvious from the data presented in Table 1.1. The aperture of the telescope must increase dramatically in order to reach farther and farther out into the visible universe. The technology needed to create such massive telescope objectives grew slowly by comparison with the astronomer's needs. Galileo's telescope was built around 1609 and had an aperture of about 1 inch. The first reflecting telescope was built by Newton in 1668 and it possessed a

Table 1.1. *Limiting visual magnitude as a function of aperture*

Aperture (inches)	Limiting visual magnitude (approx.)
14	15
22	16
35	17
55	18
87	19
138	20
219	21

speculum primary mirror of only 1 inch aperture. However, the invention of the reflecting telescope allowed larger and larger instruments to be built and the size syndrome spiralled upwards after this breakthrough. The largest behemoth of the mid-1880s would appear to be Lord Rosse's 72 inch reflector which surpassed the 48 inch instrument built by Sir William Herschel. These beasts were so massive that it was impossible to construct a conventional mounting for their use. Both of these instruments were mounted like meridian telescopes allowing observations of any particular object to be carried out for only a few minutes each evening while the object crossed the meridian.

The astronomer was the absolute slave of this size syndrome until 1838 when a novel chemical process was discovered which utilized silver plates (actually a silver film deposited on plates made from copper), iodine crystals, and mercury vapor. This process was known as the Daguerreotype, after its inventor, but John Herschel gave it the name by which it has been known ever since: **photography**. The Daguerreotype process was turned towards the sky less than a year after its invention was announced when John W. Draper made the first photograph of the moon in March of 1840. In 1842, G.A. Majocchi succeeded in photographing the crescent sun just before totality, but failed to record any trace of the corona during totality. The disk of the sun was first photographed in April of 1845 by Foucault and Fizeau at the Paris Observatory using the Daguerreotype process. However, photography was a long way away from becoming a useful tool for the astronomer.

From the late 1840s until the late 1850s, photography was but a toy to the astronomer owing to the extremely slow emulsion speeds. To give an idea of the speed of a Daguerreotype plate, a nominal mid-day portrait in bright sunlight with a lens of about f/8 required an exposure lasting 5 or 6 min. This period was also marked by a rapid series of improvements in the photographic process. The Daguerreotype was first replaced by the wet collodion process which was, in turn, replaced by the dry collodion process. Because of the extremely slow speed of the Daguerreotype, the one area of astronomy where photography could be employed was in the study of the sun. Outside of this one application, the only advantage offered by the photographic process was the permanence of the 'observational' record.

Table 1.2. *Approximate relative sensitivities of successive photographic processes*

Period	Process	Sensitivity
1839–1851	Daguerreotype	0.1–10
1851–1879	Wet collodion	50–100
1879–1887	First silver bromide emulsions	1 000–2 000
1887–1939	Modern silver bromide emulsions	2 000–10 000
1939–1959	Fastest pre-sensitized emulsions	10 000–30 000
Present	Fastest pre-sensitized emulsions	30 000–100 000

To illustrate the problem that astronomers faced in attempting to apply photography to astronomical studies, consider this: on 17 July 1850, John A. Whipple made the very first star photograph of the star Vega, and the exposure required to record an image using an improved Daguerreotype process was 100 s! The general consensus of opinion after the first few attempts to photograph stars and planets was that the photographic process was too slow and that the telescope drives available were inadequate for this application.

Even though the development of stellar photography was temporarily arrested by the slow emulsion speeds of the time, the merger of photography and astronomy continued. In 1851 a successful Daguerreotype was obtained of the solar corona by Berkowski and Busch which also showed the solar prominences which were totally misunderstood at that time. At that same eclipse, P. Secchi secured photographs of the partial phases that proved that the phenomenon of solar-limb darkening was real. The role of photography for regular recording of the solar disk was pressed by John Herschel until the Royal Society officially established such a program in 1855. Solar photographs taken by Reade in 1854 were of such quality that the sun's mottled appearance was very clearly shown. In the case of solar photography, one of the main problems slowing its development was the difficulty of obtaining exposures that were short enough without severely jarring the telescope! In addition to the realm of solar photography, lunar photography continued to make advances. In 1852, Da la Rue used the new wet collodion process to make some very good lunar photographs. During the next decade, numerous workers contributed to the field of lunar photography until Ellery made a series of lunar photographs in 1873 which remained unsurpassed until the 100 inch Hooker telescope was used for this purpose.

From around 1870 to the turn of the century, photographic technology gradually improved. In addition, a number of studies began to show some of the real advantages of the photographic process. In 1878, visual and photographic records of the solar eclipse were collected and published which demonstrated the unreliability of observations drawn from memory of such events. Throughout the period that the use of photographs was proving its value to astronomy, the accuracy of the photographic record was continually being questioned. The literature of the period is full of articles showing that astrometric measures of star positions were at least as accurate as those from numerous measurements produced by highly trained observers. After repeatedly being proved the dimensional stability of the photographic plate was finally accepted as was the ability to make accurate photometric measurements; the articles 'proving' that photographic measures equalled or exceeded visual measures slowly disappeared from the literature.

One of the prominent figures in this 'battle' to firmly establish the credentials of astronomical photography was G.P. Bond. It was Bond who turned John Whipple's attention to the pursuit of astronomical photography (Whipple was a prominent studio Daguerreotypist of the day) and it was Bond who propounded the numerous advantages and conveniences of the photographic process to his contemporaries. In addition, Bond conducted test after test of the accuracy of the photographic record and first demonstrated the dimensional stability of the photographic emulsion. Bond also recognized the importance of 'good seeing' to the quality of astronomical photographs and made many photographic records of the effects of seeing on star images. His studies led him to recommend strongly that observatories of the future not be located at the convenient site of some university, but that they be situated in locations where the seeing and transparency was of exceptional quality.

One of the biggest improvements from the standpoint of early astronomical photography came in 1871 with the introduction of the first plates that could be exposed in a dry state. This represented a major improvement in convenience for the astronomers since, prior to the introduction of the dry collodion plate, it was necessary for all photographers to apply a freshly mixed collodion solution to a glass plate and expose that plate while it was still wet. For most types of 'normal' photography this was simply an inconvenience; for Wm H. Jackson and his large 14 × 17 inch camera photographing in the American southwest desert from his horsedrawn wagon–studio, this was more than a *minor* inconvenience. However, for the astronomer, the inconvenience was the least of his worries. Astronomical exposures could be 1–3 h in length and, as the plate dried, its sensitivity decreased significantly. Furthermore, since the plate was most probably tilted from horizontal, the sensitivity of the emulsion would vary non-uniformly across the plate as time progressed and as the plate slowly dried. The dry plate was a gift well-received by all photographers and the sensitivity of the dry plate improved steadily after its introduction.

In September of 1880, Henry Draper made the first successful photograph of a nebula: his 51 min exposure of the central region of the Orion nebula with an 11 inch Clark refractor showed considerably less than the human eye can pick out with a 6 inch instrument (A.A. Common had made an unsuccessful attempt in January of that year). In 1882, Draper made another exposure of the Orion nebula of 137 min which captured far more detail but which still fell short of visual capabilities. On 28 February 1883, A.A. Common used the photographic process on the Orion nebula using a 36 inch reflector which he had personally designed and built (which was later to be purchased by Lick Observatory and become known as the Crossley Reflector) to obtain the first photograph recording details and stars which could not be visually detected (Fig. 1.1). This represented a major step in the evolution of this application and the potentials offered by this process were readily apparent.

It is also interesting to note that these early astrophotographers had to invent their own equipment in order to pursue advances in this field. Draper invented a governor for his clock drive so as to control the rate of the driving clock. Common invented a double slide-plate holder and guiding head that is very similar to those still in use today in major observatories. Common also adopted a mounting design for his later 60 inch telescope using a fork mount that is now almost a standard for photographic instruments (Common's 36 inch telescope used a rather unorthodox variation of the fork mount which proved somewhat unwieldy).

In 1859, Temple had claimed a visual discovery of a new nebula surrounding the star Merope in the Pleiades. From that time, astronomers had argued over whether or not this nebulosity actually existed or was merely a figment of the imaginations of the few astronomers who reported sighting the nebula. On 16 November 1885, Paul and Prosper Henry made an exposure of the Pleiades which not only ended the debate over the Merope nebulosity, but showed the presence of clouds of nebulosity where none had ever been recorded. The Henry brothers had previously been engaged in visually preparing a detailed atlas of the stars along the ecliptic in order to aid in the identification of asteroids. They saw that photography would make the old, laborious, visual method of searching for asteroids obsolete since an asteroid would record on the plate as a small trail instead

Fig. 1.1. This photograph of the Orion nebula was taken by Ainslee Common in 1883 using the 36 inch reflector of his own design. This 37 min exposure, made on a dry gelatin emulsion plate, was the very first astronomical photograph to show stars fainter than the eye could perceive through the same telescope.

of as a sharp star image; they set out to construct a camera for that purpose. Their work in this and other areas helped to establish firmly the potential of photography for astronomy. Their lens design was adopted by the first Astro-Photographic Congress in 1886 as a model for all photographic instruments of the time (their design was a singlet, while Pickering had recommended adoption of a doublet design as the standard).

The photographs by A.A. Common and the Henry brothers greatly impressed E.E. Barnard of the Lick Observatory who quickly recognized the potential of astronomical photography for the discovery of nebulae and the study of the structure of the Milky Way. Barnard conceived the project of conducting a photographic survey of the skies and secured a 6 inch aperture lens of about 30 inches focus which had been constructed and figured in

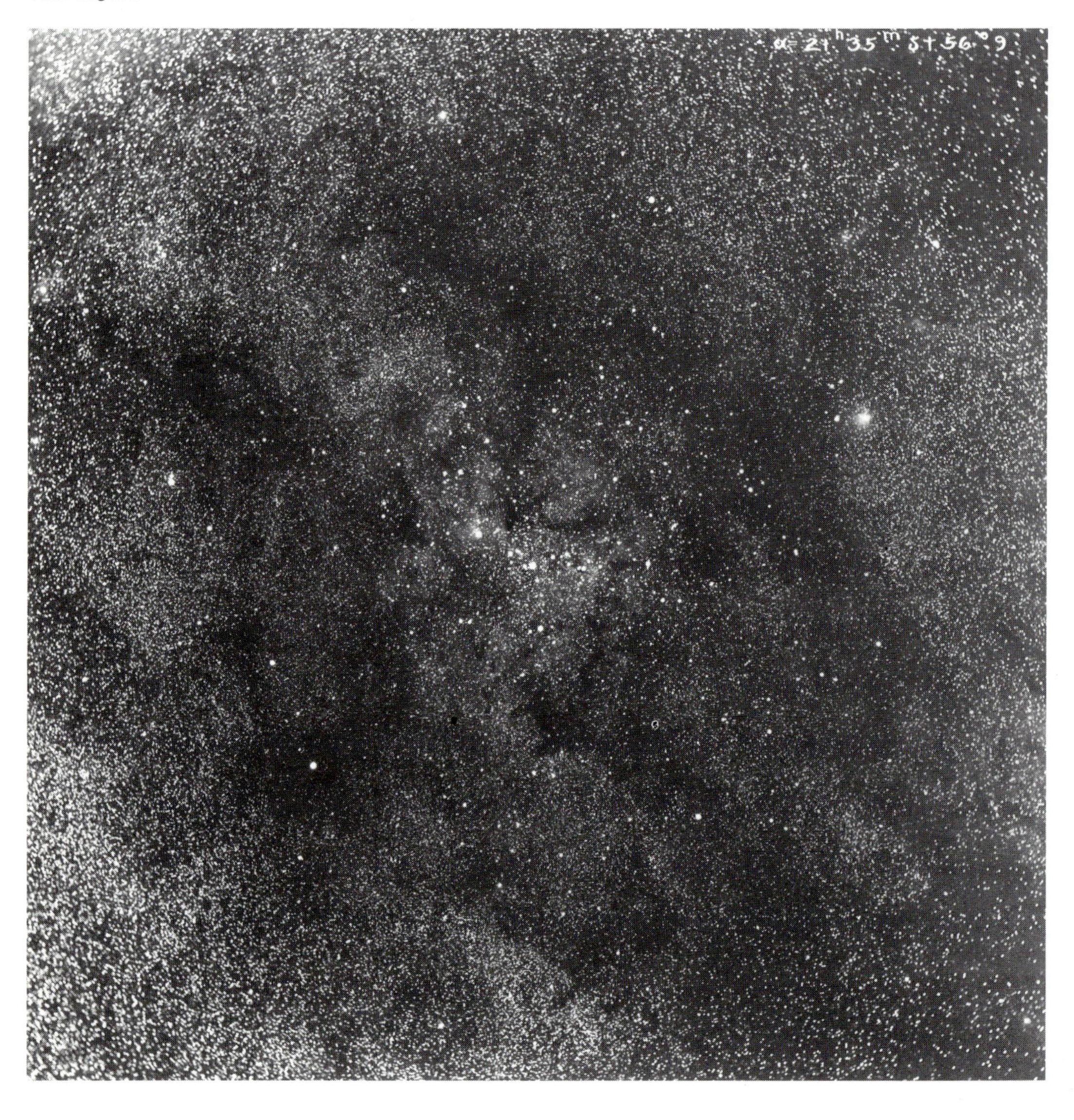

Fig. 1.2. This is a reproduction of Plate 49 from the Barnard *Atlas of Selected Regions of the Milky Way*. The region shown is centered at RA 21 h 34 min, DEC +57 deg 0 min. A complex web of dark and bright nebula is revealed in this 4 h 47 min exposure that was taken at Mt Wilson in September 1905 using the Bruce Telescope. Photograph courtesy of the Carnegie Institution of Washington.

1849 (known as the Willard Lens). By 1887, Barnard was busily surveying much of the sky with this lens using exposures of 2–7 h. He regularly published his newly discovered nebulae in the *Astrophysical Journal*. The 1913 volume of the Lick Observatory Publications contains a sizeable sample of Barnard's early work on the Milky Way and comets. Barnard's last series of photographs were taken using the 10 inch aperture f/5 Bruce refractor of the Yerkes Observatory; they were finally published in 1927, four years after his death. This final publication, *A Photographic Atlas of Selected Regions of the Milky Way*, had been largely completed by Barnard before his death, including preparation of the master negatives, drafting of charts to accompany the 50 plates, and personal supervision of the preparation of the 35 000 prints that were required for the 700 copy edition (Figs 1.2 & 1.3).

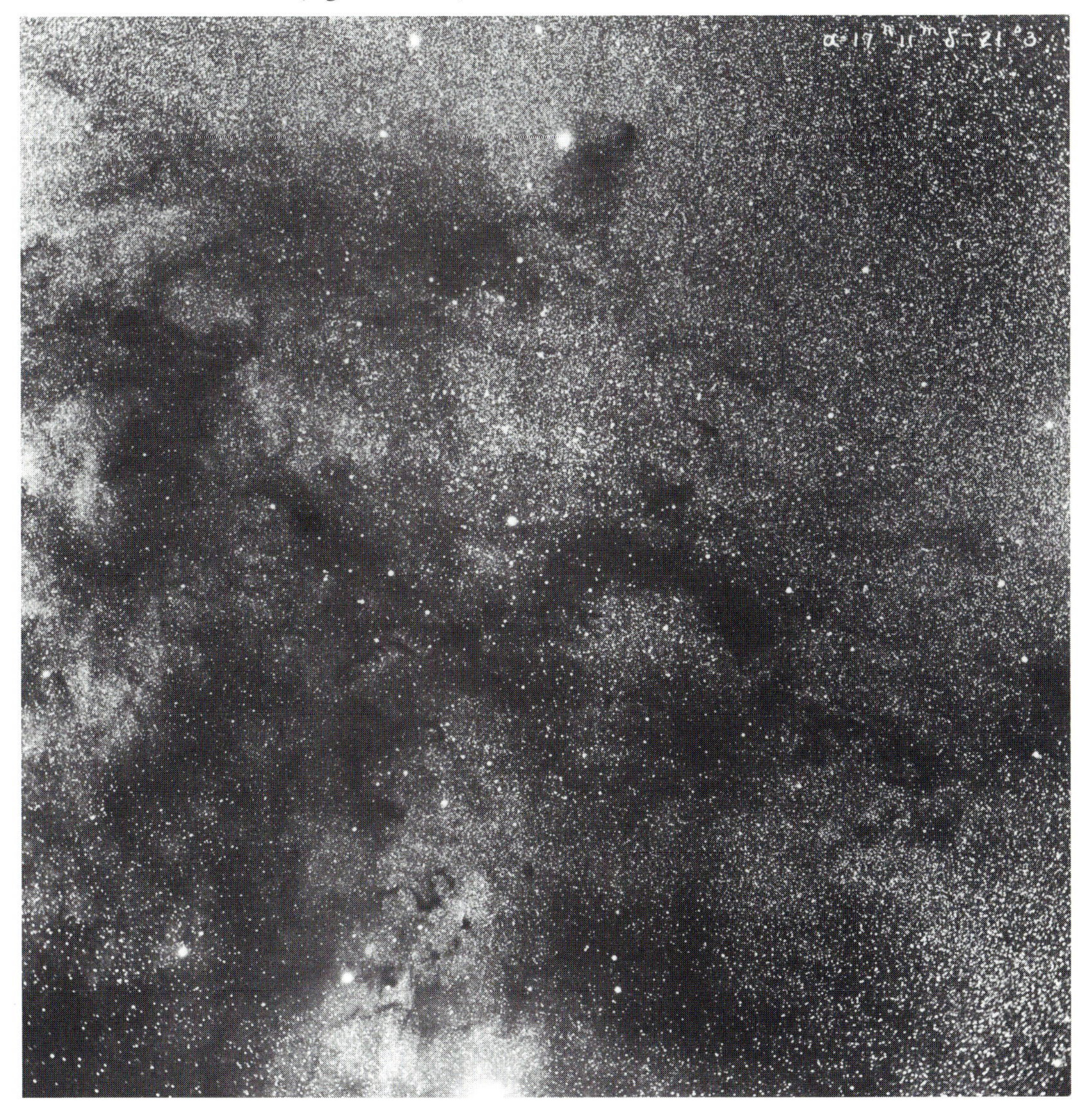

Fig. 1.3. This is a reproduction of Plate 19 from the Barnard *Atlas of Selected Regions of the Milky Way*. The region shown is just north of Theta Ophiuchii and numerous clouds of dark nebulae are shown including the prominent 'S' nebula catalogued as Barnard 72. The 3 h 30 min exposure was taken in May 1905 at Mt Wilson using the Bruce Telescope. Photograph courtesy of the Carnegie Institution of Washington.

In 1885 A.A. Common sold his 36 inch reflector to Edward Crossley whose interest in astronomy waned gradually and, by 1893, he was considering the sale of Common's telescope. At the same time, the Lick Observatory had discovered that their newly installed 36 inch refractor was an excellent visual instrument but did not work well photographically (the Lick trustees had been informed of the great potential of the reflector and of the probable role of reflectors in the future of astronomy in 1876). They quickly purchased the Common telescope from Crossley, with the provision that it be known as the Crossley Reflector, and the reflector and its dome were bound for California. By 1896 the telescope was operational at its new site at the Lick Observatory (Fig. 1.4) but Common's telescope quickly became an object of controversy. The telescope's mounting had been declared totally unsound by one of the prominent staff members (although Common had been able to overcome these difficulties) and Lick's director,

Fig. 1.4. Ainslee Common's 36 inch telescope is seen here shortly after its installation at the Lick Observatory in December of 1895. Once installed at the Lick Observatory, this instrument became known as the Crossley Reflector. Photograph courtesy of the Lick Observatory.

Edward S. Holden, was apparently embroiled in a struggle with the trustees over the purchase of an instrument that was supposedly non-functional. At any rate, Holden resigned in 1897 and was replaced by James E. Keeler who took on the project of making Common's telescope functional once again as a serious undertaking.

Keeler rapidly discovered that the telescope did have some serious problems, but set about to correct them as quickly as possible. When the telescope had been moved from England to Mt Hamilton, a wedge had been made to compensate for the difference in latitude. On Keeler's first evening with the instrument, he found that the polar axis was misaligned by 21′30″! During the next year, Keeler worked on improving the clock drive and the plate holder. Some successful results were obtained in 1898 and, by mid-1899, exposures of up to 4 h were possible with the refurbished instrument. Reading through some of the descriptions of the problems involved in making this instrument functional should be of interest to many, for the problems are classic: flexure, mirror slippage, poor clock drives, polar alignment, instability, etc. During the next year, Keeler began a photographic survey of many of the 'nebulae' and rapidly realized that the number of nebulae was at least an order of magnitude greater than any had ever imagined and also discovered that a large percentage of these nebulae showed a spiral structure. While the extragalactic nature of the 'spiral nebulae' was still unknown, Keeler had recognized that these objects made up a significant portion of the photographic universe. Keeler died at the age of 42 in 1900 before completing the work with the Crossley Reflector, but the work was carried on by C.D. Perrine and published in the 1908 volume of the Lick Observatory Publications. The photographs taken during this period with Common's old telescope are quite spectacular; the mirror (made by Grubb) was of excellent quality and the seeing was apparently excellent during many of the exposures. The quality of many of the photographs is superb even by today's standards (Figs 1.5 & 1.6).

From 1883 until around 1930, the application of photography to astronomy grew steadily and made a number of significant contributions to the advancement of astronomical knowledge and theory. Major photographic milestones of this period include:

1 the first photographic nebula (the Pleiades);
2 the discovery of sunspot cycles;
3 confirmation of spiral structure previously observed and the discovery of spiral structure in many fainter 'nebulae';
4 the discovery of the spiral structure in the Andromeda 'nebula' in 1888 (the dark lanes had been observed since 1847 and had been termed 'canals');
5 the first photographic discovery of a comet (by E.E. Barnard);
6 the first photographic discovery of asteroids;
7 proof that prominences were a solar phenomena and not relics of a lunar atmosphere;

Fig. 1.5. This photograph of the Pleiades nebulosity was taken by James Edward Keeler using the Crossley Reflector. The photograph was taken in December 1899 and was a 4 h exposure. Photograph courtesy of the Lick Observatory.

Fig. 1.6. This photograph of NGC 1977 was taken by James Edward Keeler using the Crossley Reflector. The 2 h 50 min exposure was taken in January 1900. Photograph courtesy of the Lick Observatory.

8 the use of the photographic emulsion for astrometric and photometric measurements;
9 the discovery of the existence of spectroscopic binaries.

Then, in 1928, photography added a major discovery to its growing list of coups: stars were resolved in the Andromeda Galaxy, and it was recognized that the numerous 'spiral nebulae' were actually galaxies containing billions of stars, and it was also apparent that these galaxies were being viewed from an almost unimaginable distance. Photography had suddenly expanded the size of the known universe by an enormous factor. Since that time, photography has continued to expand the size of the known universe and the domain of the astronomer's knowledge. Its list of discoveries has continued to expand and includes: detection and identification of dark nebulae, discovery of Pluto, the discovery of quasars and the identification of highly red-shifted emission lines in the spectra of these anomalous objects.

The installation of the 200 inch telescope at Mt Palomar in 1948 represented the last really significant step in the telescope-size syndrome. From that time on, the only increases in the limiting magnitudes that could be detected using direct photographic methods have come about through research into the theory of the photographic process. Those advances are tremendous. Special emulsions have been created to deal with the problems presented by long-exposure photography. Special developers have been formulated to increase the information yielded by a given exposure. Special instruments and equipment have been developed for handling and measuring astronomical photographs. Special film treatments have been discovered which are used on astronomical films before, during, and after exposure to allow one to extract every minute shred of available data from that precious photographic plate. The research on the problems posed by astronomical photography continues today and advances march on steadily. The photomultiplier tubes and the image orthicon tubes and the charge-coupled devices that are now entering this field are a direct result of this quest for fainter and more distant objects.

One of the major problems that were faced by the photographic researchers was the problem known as reciprocity failure. This is a term that is familiar to even a novice astrophotography enthusiast. However, its true meaning is often not clearly understood. The term 'reciprocity' was coined in 1862 by R.W. Bunsen and H.E. Roscoe who showed by experimental evidence that the time necessary for any exposure was directly proportional to the intensity of the light and that no other factors entered into the problem. This seemingly straightforward proposal meant that a 1 s exposure at f/8 was identical to a 4 s exposure at f/16. That was the Law of Reciprocity. However, like many so-called laws in a newly established field, this Law's existence and acceptance was short lived.

In 1881 P.J.C. Janssen, an astronomer, noticed that the Bunsen–Roscoe Law of Reciprocity did not hold true in most astronomical applications. This discovery of what is now known as 'low-intensity reciprocity failure' (LIRF) was confirmed by Sir Wm de Wiveleslie Abney in 1894 and later was better quantified by Karl Schwarzschild in 1900. In his studies of the sensitivity of photographic plates, Abney also discovered that the Reciprocity Law failed when the photographic emulsion was exposed to high-intensity light during very brief exposures. This effect is now termed 'high intensity reciprocity failure' (HIRF). So, by 1900, it was obvious that the Reciprocity Law is only valid over a narrow range of exposures and that outside of this range the speed of any photographic emulsion is no longer constant. Much of the research on photographic emulsions for astronomical photography today is directed at diminishing or eliminating reciprocity failure during long exposures.

This research has uncovered numerous techniques for lessening the effect of reciprocity failure. One interesting treatment was made quite popular to amateurs in the mid-1960s by two series of articles appearing in *Sky and Telescope* by A.A. Hoag and E. Kreimer. This technique was, of course, cooled-emulsion photography. In this procedure, the film is chilled to very low temperatures by using either dry ice or liquid nitrogen, and a vacuum chamber or glass plug is placed in front of the emulsion to prevent the formation of frost on the emulsion. The results obtained by Hoag and Kreimer were striking and made this device popular enough to entice several manufacturers into the cold camera marketplace. Followers of Hoag and Kreimer also managed to obtain some very striking photographs and the technique is still in use by amateurs today.

It is interesting to note that the principle of cooled-emulsion photography was not a discovery of the mid-1960s; but can be traced back to 1912 when Edward Skinner King of Harvard Observatory noticed that film was noticably faster when exposed on cold, dry nights. King performed a series of investigations aimed at measuring this phenomena and concluded, correctly, that films used for long-exposure astronomical photography were definitely more sensitive when exposed on cold nights and that the sensitivity increased as the ambient temperature dropped. The contributions made by E.S. King to astronomical photography could fill volumes. He was truly one of the major contributors to the growth of this field.

The needs of the professional astronomical community came to the attention of the researchers at Eastman Kodak Co. and the problem of reducing LIRF was taken up. The result of Kodak's research effort was the introduction of a series of astronomical emulsions including those now designated with the '103a-' prefix. The films in this line possess sensitivities which peak in different parts of the spectrum to allow the astronomer the versatility that was needed for astronomical research. The introduction of these 103a- films represented a major leap in the astronomer's plight against low-intensity reciprocity failure. Eastman Kodak has worked hand in hand with the professional astronomer over the decades and has developed numerous films designed specifically for his needs. In the mid-1970s another leap forward was introduced to the professional astronomical community when the first IIIa-emulsion was produced. One major difference between 'normal' films and films designed specifically for astronomical use is the position of the film's maximum speed and the slope of the reciprocity curve in the region of long exposures. The region of maximum speed is shifted towards longer exposures and the slope of the reciprocity curve in the region of long exposures is significantly less than that of a 'normal' film.

During the mid-1950s, Wm C. (Bill) Miller of Hale Observatories produced the first reasonably accurate color photographs of some spectacular celestial objects. The results were quite stunning to lay-public audiences and even professionals were enthused by these beautiful photographs. One astronomer supposedly remarked to the effect that a single color photograph of these objects told him and his students more than a volume of color indices and black-and-white plates. Bill Miller's contributions to virtually every area related to astronomical photography continued until his death in 1981 and the field is greatly indebted to him for the knowledge and guidance that he exerted during that brief period. Bill helped to establish the Working Group for Photographic Materials which established the American Astronomical Society's *AAS Photo-Bulletin* as a voice for papers relating to the advance of astronomical photography.

As astronomical films have been developed and refined, a major concern has been to increase the detective quantum efficiency (DQE) of these films. The DQE is a means of measuring the film's ability to capture and hold photons and to translate this captured energy into film density. Investigations into the film properties that would cause an increase of the DQE yielded the following properties: fine grain and high contrast. This poses an interesting problem as the property of fine grain is not associated with fast emulsions. The IIIa-emulsions that are now being produced for astronomical applications possess both attributes of a film with a high DQE, but they also possess the associated property of being slow compared with the older 103a-emulsions. The limiting magnitude of a IIIa-plate is greater than the limiting magnitude of a 103a-plate exposed under identical conditions but the exposure required to reach the limiting magnitude using the IIIa-plate is considerably longer unless the IIIa-plate has been hypersensitized.

As astronomers have sought fainter and fainter objects, numerous techniques have been explored for increasing the sensitivity of the film. A partial listing of these techniques would include processes like: water hypersensitizing (hypering), ammonia hypering, mercury hypering, dry nitrogen baking, exposure in a vacuum environment, cooled-emulsion techniques, and, finally, hydrogen hypering. By applying one or more of these techniques to the new IIIa-emulsions, the effective speed can be increased until it surpasses the unhypered speed of the 103a-emulsions. The most promising of these techniques appears to be the process of hydrogen hypering. This process was discovered in 1974 by Kodak researchers Babcock and James and was introduced to the amateur astronomer community in 1977 by the present authors. The property of this technique which so intrigued us was the ability of hydrogen to permeate semiporous materials. This property suggested that this hypering process would be applicable to color films where each of the three emulsion layers need to interact with the hypering agent. When 'forming gas' (8 % H_2, 92 % N_2) was later substituted for hydrogen by Alex Smith, this hypothesis was shown to be true. Thus, hydrogen hypering is the first hypering technique that can be applied successfully to color emulsions (the cooled-emulsion process is not a true hypering technique).

Now, optical astronomers have come upon another limiting barrier that they are seeking to break through: even with high DQE emulsions and applied hypering techniques, it appears almost impossible to reach farther out into the universe using conventional photography. In this case, the limiting barrier appears to be the inherent brightness of the Earth's night sky. Even in the remotest areas of the Earth, there is some light present in the night sky owing to the faint emissions of the gasses that compose our atmosphere and the scattering properties of the atmosphere. This barrier does not seem penetrable using conventional photographic techniques. In order to overcome this barrier, astronomers are looking to the cutting edge of our advancing technology: electronic light-capturing devices controlled and interfaced with sophisticated computer software, unmanned space probes that venture out into the solar system to capture images that no man has seen before, and space telescopes that will operate outside of the Earth's atmosphere and allow astronomers to probe the universe as it existed near the time of its birth. The future discoveries promised by this advanced technology are very exciting.

However, for the time being, the photographic plate still holds several significant advantages over its most advanced solid-state competitor, the charge-coupled device (CCD). The CCD does have a better quantum efficiency and thus requires shorter exposures to reach fainter objects, however as CCDs are used more, a number of disadvantages are being uncovered. Computer image processing is required for even the 'simple' readout of a CCD image. The largest CCDs in existence at this time are about 1 inch square making them suitable only for examining extremely small regions of the sky. (Imagine trying to put together the equivalent of a single 14 × 14 inch Schmidt plate using at least 196 CCD images!) Besides the limited size of CCDs, which limits their use in many professional applications, there are still more complex issues to be resolved before the photographic plate can be ruled as obsolete.

Most electronic photodevices use an 8-bit digitization of the recorded light. This forces all data into 256 discreet levels covering the range of grey from black to white. This is far less tonal discrimination than a photographic plate can provide. As computers and chips evolve, it is becoming possible to expand into a 16-bit grey tone digitization of photographic data which will allow electronic devices to approach the range of tonal discrimination available in a photographic plate. This limitation is at least one which technological advances will solve in the next few years. There is, however, a final area in which the photographic plate remains supreme and will quite probably not be surpassed for several decades.

A single 4 × 5 inch plate would contain roughly 26 megabytes of information at present CCD resolution levels (if a 4 × 5 inch CCD could be built). However, at photographic resolution levels, a 4 × 5 inch coarse-grain photographic plate contains over 2.6 gigabytes (2 600 000 000 000) of information while a fine-grained plate may contain over 2600 gigabytes of data. The photographic plate is the most efficient means of data storage in existence at this time (not counting holographs). By contrast, a 5 inch laser disk is capable of storing only 0.4 gigabytes of data. A black-and-white photographic plate that has been processed correctly, and stored and handled properly, has a lifetime of at least a century, the same level of longevity cannot be obtained using any other data storage method.

Some areas of astronomy are already turning to the new electronic technology simply because the photographic plate cannot meet the astronomer's needs. Much will be learned from the use of these devices and electronic technology will undoubtedly advance as a result of these exploratory uses. Other areas of astronomy will continue to use photographic plates and films for years to come. The ranks of professional astronomy will be split over the use of photographic tools while amateur astronomers will continue to use conventional photography almost exclusively in the forseeable future. In any case, it is doubtful that astronomy will ever be free of the debt that it owes to photography.

2

Instrumentation

I speak truth not so much as I would, but as much as I dare, and I dare a little more as I grow older.
Michel Eyquem de Montaigne

Optical systems

In planning a section on photographic optical systems, some limits and directions had to be decided upon in order to reduce the potential scope of such a discussion. We have limited this section to systems that are commonly used by amateurs and have avoided many of the exotic systems that are available to those capable of fabricating elaborate precision optical systems (see Figs 2.1 & 2.2). We are attempting to achieve three goals in this section: to explore the merits and faults of the various systems; to reveal some possible alternative and improvements to some systems; to give some knowledge of what is important to look for in evaluating an optical system in order to help avoid the exaggerated and sometimes false claims of some telescope manufacturers.

While our bibliography is filled with references on optical studies, the works by Willey, Simmons, and Rutten and van Venrooij are highly recommended for their comparative studies. The works by Willey and Simmons are of particular use because apertures, focal lengths, and f/ratios are kept constant for comparative purposes. Comparative studies are of far greater value to readers than a description of a single optical system; these two studies are essentially landmarks in the amateur literature. We would hope that publishers encourage such comparative works whenever contemplating articles describing new optical configurations.

The Newtonian

Isaac Newton was the first to employ a secondary mirror at a 45 deg angle to send the focused beam of light outside the optical support tube, and provide the world with the telescope that now bears his name. This simple addition provided an elegant solution to the problem of using some form of reflecting telescope without standing in the way of the incoming light or requiring the generation of complex optical surfaces to enable one to tilt the primary mirror to allow observing. The single-lens telescope used by Galileo was invented around 1607 while the first discussions of the use of a concave primary mirror were presented just shortly before Isaac Newton provided the solution to the problem of using such a mirror in 1668.

The Newtonian (with its parabolic primary) performs extremely well on-axis but the image quality deteriorates fairly rapidly as one moves off-axis. The main cause of this image deterioration close to the optical axis is coma, while astigmatism dominates farther from the optical axis. The linear size of the 'coma-free field' is totally dependent upon the f/ratio of the system. Thus a 6 inch f/6 and a 100 inch f/6 both have the same 1 inch diameter coma-free photographic field. As one moves away from the optical axis, the effect of coma increases linearly as the angle increases while the effect of astigmatism increases as the square of the off-axis angle.

The comatic image is actually made up of two major components:

Fig. 2.1. Shown are a number of the optical systems discussed in this chapter. All are shown in the f/5.6 configuration with relative tube lengths indicated by the shaded areas. The size of the secondary mirrors and optical curvatures are symbolic and are not meant to show actual sizes or figures.

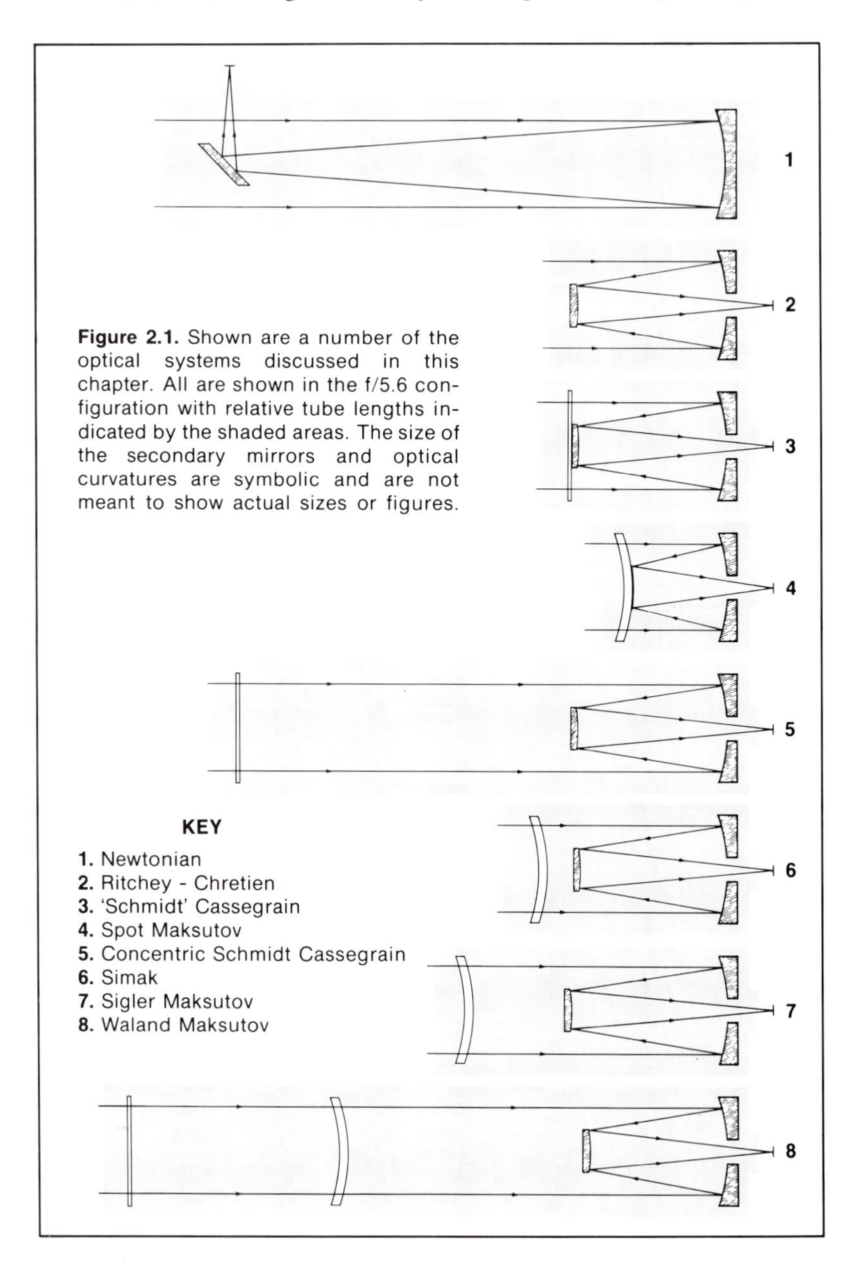

Figure 2.1. Shown are a number of the optical systems discussed in this chapter. All are shown in the f/5.6 configuration with relative tube lengths indicated by the shaded areas. The size of the secondary mirrors and optical curvatures are symbolic and are not meant to show actual sizes or figures.

sagittal coma, formed by the rays from the sagittal plane of the mirror; and tangential coma, formed by the rays from the meridional plane. The comatic image appears to have a bright arrow-shaped head with a broad faint fan-shaped tail. Very little of this faint fantail will show in photographic images of any but the brightest stars. The length of the sagittal comatic image is given by $C_s = y/(16f^2)$, where y is the linear distance off-axis, while the tangential coma image is three times the length of C_s.

Astigmatism is created because the rays from the sagittal plane and those from the tangential plane focus at different distances from the mirror, with the sagittal rays focusing at the greater distance. The result is that the image at the point of best focus is not a point, but a small circular blur. The surface of best focus is a sphere of radius F which is concave to the main mirror (matching the radius of curvature of the focal plane). In the case of most common Newtonians, using no special aperture stops, the smallest astigmatic blur will have a diameter given by: $AB = y^2/(2Ff)$.

It is instructive to note that ray-trace diagrams can be most deceptive when it comes to looking at the effects of various system aberrations upon the actual photographic image. Most ray-trace programs deal with a few rays spaced in a regular array to show the gross characteristics of the system. To look at something that gives an idea of the photographic appearance of an image, it is necessary to look at the density distribution of rays throughout the image using a large number of rays preferably distributed randomly across the aperture of the optical system. By judiciously eliminating the less dense regions of the resulting ray-trace diagram, one can get a better idea of the actual photographic appearance of the resulting image. In the case of comatic images in the Newtonian, the actual photographic images are nearly one-third the length and width of the images calculated, and they are of significantly different shape.

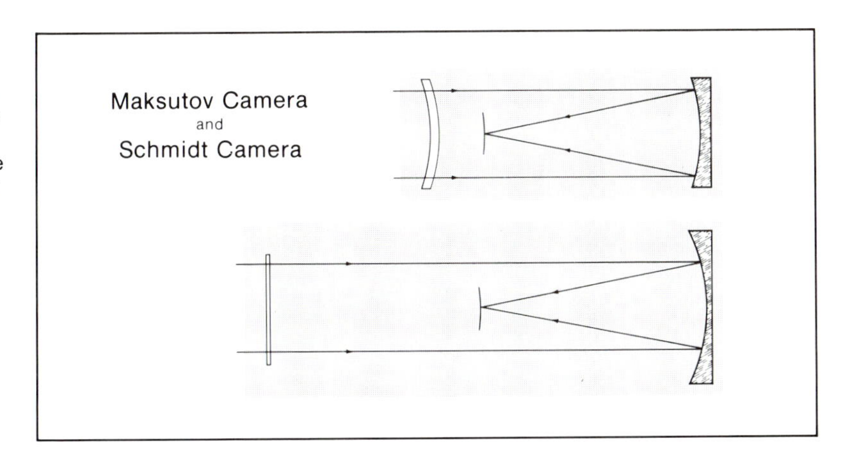

Fig. 2.2. The Maksutov and Schmidt Camera systems discussed in this chapter. Both systems are shown in an f/2.0 configuration with relative tube lengths indicated by the shaded areas. The curvature of optical surfaces is symbolic and not meant to indicate actual optical figures. In both cameras, the focal 'plane' is curved and normally located within the tube structure.

An additional aberration present in the Newtonian (and in most of the other optical systems that we will discuss) is field curvature. The radius of curvature of the focal plane* being approximately equal to the focal length of the primary, and the departure from a flat focal plane being given by

$$dF = F[1 - \cos\theta] \quad (2.1)$$

Thus, in order to optimize the focus over the entire focal plane, it is necessary actually to focus on stars which are displaced from the center of the film by about $\frac{1}{2}$–$\frac{2}{3}$ the large axis of the film format in use. This is especially important when using instruments of shorter focal length. The other solution to this is to create a curved film plane and to press the film against the curved surface during the exposure. This can be accomplished by using a thin 'zero-power' lens or by the use of a vacuum to deform the film.

Newtonian correctors. The Newtonian has been the mainstay of astronomical instruments since the reflector replaced the refractor. The primary reasons behind the success of this instrument are that the main mirror can be large and supported very well to minimize optical distortions, and that the central photographic field is quite excellent at semi-fast f/ratios. However, as the quest for larger instruments continued, the main goal was to have a larger mirror and a secondary goal was to build a fast-imaging system. The focal lengths created by a 200 inch f/6 made the task of mounting such an instrument virtually impossible (not to mention the size of the dome!). The search was on to examine methods of improving the images produced by the Newtonian at the proposed focal ratio of f/3.33.

An image corrector had been designed and built by F.E. Ross for the 100 inch telescope at Mt Wilson earlier, and the results were astounding. The size of the useful photographic field was extended from 1 to 10 inches in diameter and the images were proven to be better for astrometric and photometric measurements! The corrector was quite simple in design; it had virtually no magnifying or reducing power, being very close to a zero-power system, and consisted of two thin lenses with spherical surfaces that were placed close to the focal plane. The corrector eliminated coma, astigmatism, and field curvature but it reintroduced a small amount of spherical aberration (Figs 2.3 & 2.4). With this design, it was difficult to reduce the spherical aberration below the resolving power of the system, but the problem was insignificant when compared with the uncorrected images from the edges of a 10 inch plate. Ross' studies for the 200 inch telescope showed that the two-element corrector would perform very well on the Hale telescope but pointed out that more complex systems (having more elements, aspheric surfaces, and not restricted to zero power) would achieve an even higher degree of correction.

* We will be referring to a 'curved focal plane' throughout this volume as it is a commonly used term. A more technically correct phrase would be something like 'curved focal surface'.

Over the years since Ross' 1935 study, a number of different correction systems have been designed for Newtonians and other optical systems, the latest being the correctors designed by Wynne and Rosin. In these, 3–5 elements are used, 1 or more of the elements may be aspheric, and the power of the corrector may be as high as 1.7. Wynne also examined the possibilities of using an aspheric secondary in conjunction with a two-element corrector and developed several interesting configurations. Other reflector-corrector systems have been investigated by Baker and Wright which depart radically from the Newtonian design. All of these systems which use a corrector at a position other than close to the focal plane suffer from a flaw which diminishes their attractiveness significantly: a large corrector (i.e. the size of the primary) is expensive and difficult to fabricate (especially if it is an aspheric) and the size of such correctors is limited by the flexibility of the glass from which they are made. The cost of such large correctors is a factor which has made them virtually non-existent in the amateur community.

For small focal length systems, the difficulty of designing a field

Fig. 2.3. The optical configuration for a Newtonian with a two-element Ross Corrector is shown along with two examples of Ross–Newtonian optical systems. Both elements of the corrector are made from BK-7 glass in these examples.

ROSS - NEWTONIAN

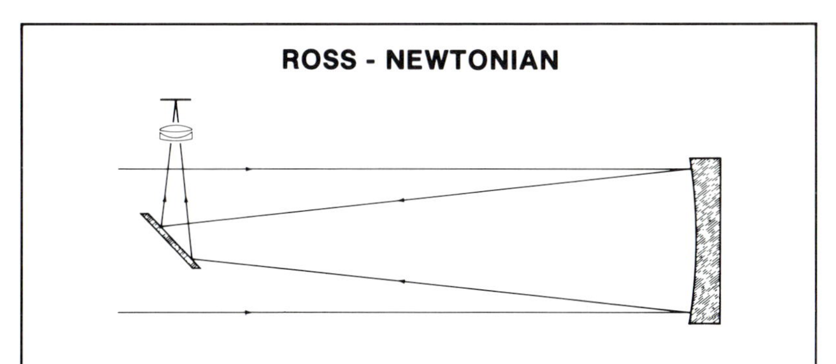

16 INCH APERTURE F/6.0, 1.5 DEGREE FIELD

SURFACE	RADIUS OF CURVATURE	ECCENTRICITY	REFRACTIVE INDEX	THICKNESS
1	-192.00	1.0000	-1.00000	-92.25
2	-6.40	0.0000	-1.51626	-0.12
3	-2.98	0.0000	-1.00000	-0.35
4	-9.90	0.0000	-1.51626	-0.35
5	9.90	0.0000	-1.00000	-----

12.5 INCH APERTURE F/4.20, 2.0 DEGREE FIELD

SURFACE	RADIUS OF CURVATURE	ECCENTRICITY	REFRACTIVE INDEX	THICKNESS
1	-100.00	1.0000	-1.00000	-47.10
2	-4.60	0.0000	-1.51626	-0.12
3	-2.29	0.0000	-1.00000	-0.80
4	-8.80	0.0000	-1.51626	-0.35
5	8.80	0.0000	-1.00000	-----

RADIUS OF CURVATURE and THICKNESS are in inches.

corrector is significantly increased because of the large angular size of the field and the increased effects of coma, spherical aberration, field curvature and astigmatism. In 1983 we began to work on a 16 inch f/6 instrument and decided that we would like to be able to use a larger field than the 35 mm 'coma-free field' and began to look into the possibilities of building a correcting lens system for this instrument. A two-element Ross Corrector was quickly designed but would only correct a 2 inch diameter field without introducing excessive aberrations of its own creation (Fig. 2.3). In designing a two-element Ross Corrector it is necessary to monitor the amount of spherical aberration that is being introduced and to observe image structure over the entire field as the corrector is actually creating a set of aberrations that are intended nearly to cancel the aberrations of the original optical system. This study showed that it was considerably easier to correct a hyperbola than it was to correct the images formed by a parabolic system. If the parabola is only moderately overcorrected ($1.03 < e < 1.8$) a two-element corrector can produce a very well-corrected field in relatively fast optical systems (the line of reflectors introduced recently by Takahashi employs such a system). This fits nicely with the fact that the design of correctors for Ritchey–Chrétien systems is a relatively simple problem and many such systems are in use at observatories around the world. The most notable of all such corrected Ritchey–Chrétien systems is the recently completed 100 inch DuPont telescope which takes 22 × 22 inch plates and produces images of incredible quality at the corners.

In searching for a corrector for the 16 inch f/6 system, we were interested in finding a design which avoided the use of aspheric surfaces and that would allow us to cover a 4–6 inch diameter circle with star images in the 40 μm range at the edges. The literature provided many fascinating possibilities but an article by C.G. Wynne was found that discussed the unpublished three-element Ross Correctors that were designed for a number of large telescopes (i.e. one of Ross' designs was used in the 200 inch f/3.33 system). While Wynne's article would not provide the data needed for our Newtonian system, it did provide conceptual information that was useful to guide our efforts in designing such a corrector. After a few weeks of

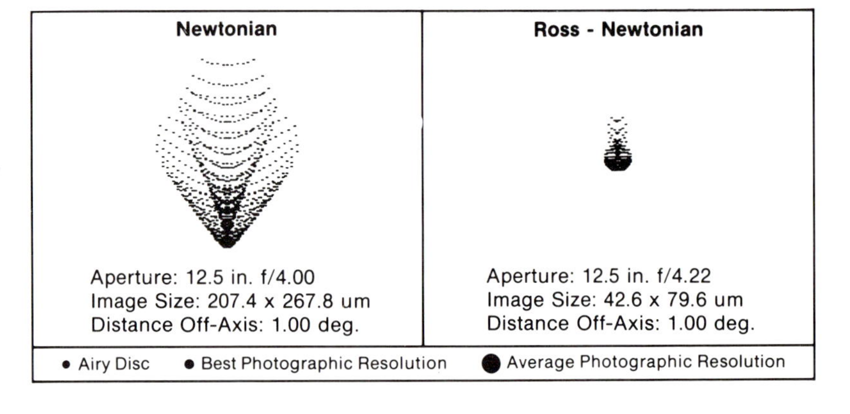

Fig. 2.4. The superior off-axis imaging performance of a Newtonian with a simple two-element Ross Corrector with compared to an uncorrected Newtonian optical system.

work on the computer, a design emerged that would suffice our needs. The front element is a weak concave meniscus lens, the second element is a stronger negative lens and the third element is a double convex lens (Fig. 2.5). The first two elements work together to cancel out spherical aberration, astigmatism and coma and the third element is used as a field flattener. Our bibliography contains many references to correcting systems and to a variety of optical systems that have not appeared in the amateur literature and interested amateurs are urged to look into these sources for a variety of fresh ideas.

The Newtonian is a simple and an elegant system. For the photographic applications of most amateurs, it can be used at f/ratios of f/4 and greater on 35 mm format film and produce excellent results. Only when one needs f/ratios faster than f/4 and/or larger film formats is it necessary to consider alternate designs; the choices that are available to fill either of these requirements are not without significant drawbacks. Of the numerous alternate designs proposed over the decades, only a few warrant any

Fig. 2.5. The optical configuration of a three-element Ross Corrector is shown for a 20-inch f/4.70 Ross–Newtonian system. The optical configuration is given as is an evaluation of the system's performance across its 4 inch diameter useable field. A spot diagram is also shown indicating the system's performance 1 degree off-axis. The 37 μm 'head' of this image is about 1/8 the size of an image created by a similar uncorrected Newtonian this distance off-axis. The coma-like fantail on this spot diagram is less than 1/300 as bright as the 'head' of the image and thus will not be recorded on fainter star images.

THREE ELEMENT ROSS CORRECTOR

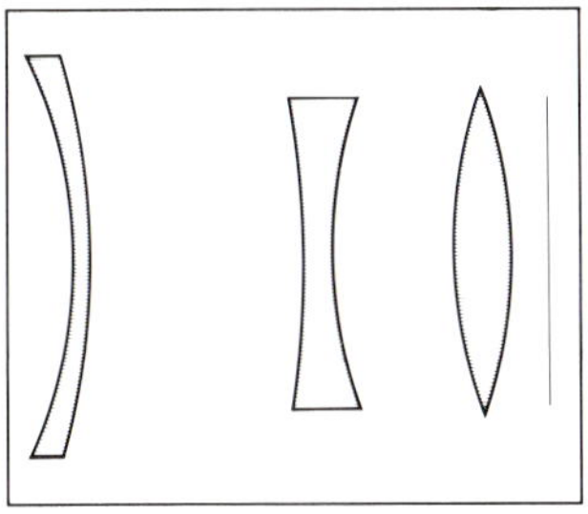

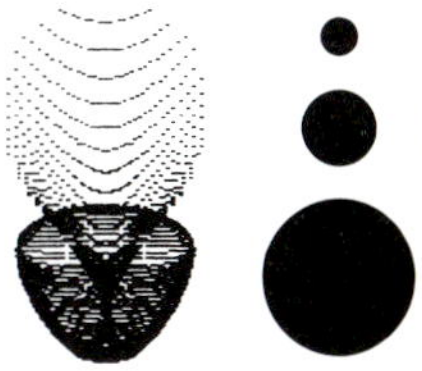

PERFORMANCE OF THREE ELEMENT CORRECTOR

DEGREES OFF-AXIS	0	0.50	1.00	1.22
INCHES OFF-AXIS	0	0.82	1.64	2.00
RAYTRACE IMAGE SIZE	15 um	29 x 72 um	37 x 68 um	48 x 77 um
ESTIMATED FAINT STAR SIZE	18 um	25 um	37 um	48 um

20 INCH APERTURE F/4.70, 2.44 DEGREE FIELD

SURFACE	RADIUS OF CURVATURE	ECCENTRICITY	REFRACTIVE INDEX	THICKNESS
1	-240.00	1.0000	-1.0000	-113.50
2	-8.40	0.0000	-1.5097	-0.20
3	-10.00	0.0000	-1.0000	-2.75
4	-9.50	0.0000	-1.5097	-0.30
5	22.00	0.0000	-1.0000	-1.75
6	12.00	0.0000	-1.5250	-0.50
7	-14.00	0.0000	-1.0000	-----

RADIUS OF CURVATURE and THICKNESS are in inches.

consideration. Many of the 'new' designs seen in amateur publications are merely minor variations on a theme, while others are optical curiosities that merely occupy space in a magazine. We will present some of the designs which offer solutions to some of the problems existing in the basic Newtonian design and we will present the pros and cons of each of these systems. In much of the text which follows this introduction to photographic systems, our basic commentary will be slanted towards the use of the Newtonian system as we feel that the great majority of amateurs interested in pursuing astrophotography in a serious manner will realize the many advantages of this elegant survivor of three centuries of telescope-design evolution.

Cassegrain systems

The Cassegrain design, and its variants, employs a primary mirror and a conic secondary to fold the focus back to a point behind the primary (we will discuss later the Schmidt–Cassegrain variant which also employs a mild corrector to improve the final image quality). The secondary also serves to magnify the image produced by the primary. Thus, with a Cassegrain system, it is possible to fit an f/10 system into an f/2 tube assembly. Compactness is the primary advantage of the Cassegrain. It is uncommon to see Cassegrain systems with focal ratios faster than about f/10 and thus they are generally unsuitable for deep-sky photography (although some continue to make long exposures through such systems). However, the long focal length available in a compact, sturdy tube makes these systems ideal for lunar, planetary, and solar photography or even for guidescopes where an f/10–f/20 focal ratio poses no problem. One difficulty that affects any instrument with a Cassegrain-like design is the fact that the system must be very carefully baffled to protect against stray light or its image contrast will be severely degraded. The instruments that fall into the Cassegrain class include: Cassegrain, Dall–Kirkham, Schmidt–Cassegrain (and its variants), Ritchey–Chrétien and a host of less common similar optical configurations (note that we will be considering the Maksutov variations separately).

Cassegrains. The classic Cassegrain design consists of a parabolic primary and a hyperbolic secondary. The system is free of spherical aberration, but possesses coma equal to that of parabola of the same aperture and effective focal length. The worst aberrations present are astigmatism and field curvature which are equal to that of a paraboloid of equivalent focal length *multiplied by the magnification factor of the secondary.* Since secondaries typically are 3–5 ×, these aberrations are 3–5 times worse than those of an equivalent paraboloid. The added aberrations in the classic Cassegrain design, combined with the problems added by the need for a large secondary obstruction generally limit Cassegrains to speeds slower than about f/10.

The Cassegrain makes an excellent system for lunar, solar and planetary photography provided that it is properly baffled and well constructed. If one is willing to sit through very long exposures, the system can be used at f/10 for deep-sky work although off-axis performance will be considerably worse than an equivalent focus Newtonian.

The *Dall–Kirkham* variation uses an elliptical primary and a spherical secondary but produces off-axis images that suffer from severe comatic distortion making this configuration a very poor choice for anything other than planetary photography or use as a guidescope.

Ritchey–Chrétien. The Ritchey–Chrétien design is essentially the opposite, conceptually, of the Dall–Kirkham optical system. It uses a hyperboloidal primary and a more hyperboloidal secondary than the true Cassegrain. The result is a system in which coma has been virtually eliminated. The remaining astigmatism and spherical aberration produces a curved focal plane with images being slightly better than an equivalent Newtonian but markedly inferior if a flat focal plane is used. The Ritchey–Chrétien is often built as fast as f/8, and f/5 versions are sometimes encountered.

The Ritchey–Chrétien has become very popular as an instrument for large observatories because of some of its optical properties and amateurs could well benefit by investigating the route that professionals have taken. Professionals have managed to take advantage of the fact that spherical aberration and astigmatism are more easily corrected in an imaging system than is coma (see our discussion above of a Newtonian corrector lens for some comments on this phenomenon). The optical designs of Wynne and others have perhaps culminated in the ultimate Ritchey–Chrétien: the 100 inch DuPont telescope in Chile.

Schmidt–Cassegrain. The Schmidt–Cassegrains include a broad class of Cassegrain-type instruments that also employ an aspheric corrector like that of the Schmidt camera. Of the broad range of possible configurations, some are well suited for slow systems in the range of f/10–f/15, some are better suited for fast systems in the f/4–f/7 range, and some provide extremely compact optical systems with tube lengths less than twice the diameter of the primary. After a 1962 article by Willey describing the optical qualities of an excellent f/15 configuration, a compact variety of the Schmidt–Cassegrain found its way into the telescope marketplace and its popularity has spread ever since despite the fact that the various models available commercially have sacrificed much in image quality in order to gain increased compactness. The optical configuration of the common commercial Schmidt–Cassegrains results in a strongly curved focal plane (roughly like the surface of a basketball with a radius of curvature of about 6 inches) that at f/10 will not cover a 35 mm frame with high definition because of the effects of field curvature and coma.

The 'pure' form of the Schmidt–Cassegrain is better described by the name of 'Concentric Schmidt–Cassegrain' (Fig. 2.6). In this form, the primary and secondary are both spherical and share a common center of curvature located at the aspheric correcting plate. This design utilizes Schmidt's principle in its purest form by keeping the corrector at the center of curvature of the primary mirror. The resulting system produces extremely small star images at fast f/ratios over a very large field. The focal plane is mildly curved (approximately equal to the effective focal length of the system) and star images on this surface are well below the size that can be resolved photographically. If a flat focal plane is used, image quality will be sacrificed but will still be considerably better than that produced by an equivalent Newtonian. A simple field-flattening lens can be used to create a flat field without noticeably sacrificing image quality. This form of the

Fig. 2.6. The optical configuration for a concentric Schmidt–Cassegrain is shown along with two examples of the optical system.

CONCENTRIC SCHMIDT CASSEGRAIN

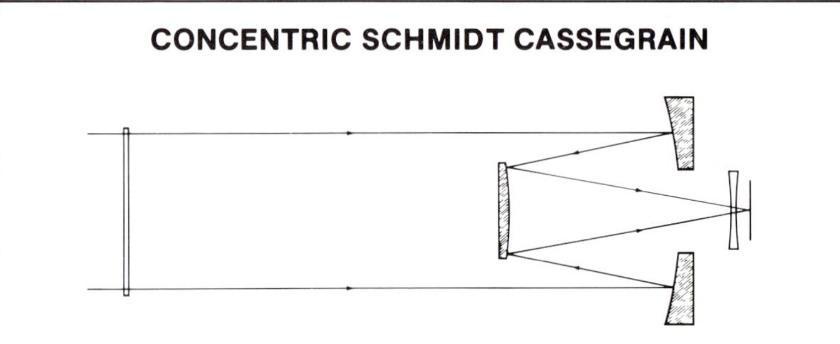

11 INCH APERTURE F/4.0 (14 INCH PRIMARY), 5 DEGREE FIELD

SURFACE	RADIUS OF CURVATURE	ECCENTRICITY	REFRACTIVE INDEX	THICKNESS
1	flat	0.000	1.51626[c]	0.10
2	-2500.00	0.000	1.00000	39.00[d]
3	-39.00	0.000	-1.00000	-11.90
4	-27.10	0.000	1.00000	16.80
5[a]	-30.00	0.000	1.51626[c]	0.15
6[a]	30.00	0.000	1.00000	-----

11 INCH APERTURE F/5.55 (14 INCH PRIMARY), 4-5 DEGREE FIELD

SURFACE	RADIUS OF CURVATURE	ECCENTRICITY	REFRACTIVE INDEX	THICKNESS
1	flat	0.000	1.51626[c]	0.20
2	-2500.00	0.000	1.00000	53.00[e]
3	-53.00	0.000	-1.00000	-16.00
4	-37.00	0.000	1.00000	23.50
5[b]	-41.00	0.000	1.51626[c]	0.15
6[b]	41.00	0.000	1.00000	-----

[a] Field flattener needed for film larger than 6x6 cm.
[b] Field flattener needed for 4x5 in. film.
[c] BK-7
[d] B=4.59 E-06
[e] B=1.78 E-06

RADIUS OF CURVATURE and THICKNESS are in inches.

Schmidt–Cassegrain has appeared only rarely over the years. The reason for this is not obvious as the image-quality advantages of the system overwhelm any of the alternatives. Its main drawbacks are the figuring of the complex Schmidt corrector surface (which certainly does not disturb opticians) and the fact that the tube length is basically the same as that of a Newtonian of the same focal length. This is an excellent system for anyone seeking a high-resolution instrument capable of handling film sizes larger than 35 mm, and we seriously hope that more amateurs and manufacturers will investigate this system in the future.

Numerous variations on the Schmidt–Cassegrain system have appeared over the years but one form is of particular interest as it has a perfectly flat focal plane and does not sacrifice any image quality in order to obtain that desirable feature. The 1982 article by Rutten and van Venrooij describes an optical system which they call the Schmidt–Cassegrain 'R' which they found in a 1944 article by Slevogt on Schmidt–Cassegrain variations. The 'R' system uses spherical primary and secondary mirrors which have nearly the same radius of curvature to obtain the flat focal plane. The only parameter that 'suffers' to attain excellent imaging capabilities over a very wide flat field (an 11 inch f/5.5 model can easily accommodate 6×7 cm film) is the tube length which must be extended to accommodate the corrector plate which is shifted outward from the center of curvature of the primary mirror. Again, we hope that amateurs and manufacturers will take some future interest in this excellent optical configuration.

Maksutovs

Another attempt at an innovative optical system was introduced in 1944 when D.D. Maksutov published an article describing a family of instruments that utilize a spherical primary and a steep curved meniscus corrector lens near the focus to correct spherical aberration. Over the years, several versions of the Cassegrain form of the Maksutov have appeared, but only the Gregory (or spot) Maksutov has gained any popularity. This form uses an aluminized spot on the inner surface of the corrector as its secondary mirror. At about f/23, all of the surfaces can be spherical, but as one approaches f/15 some additional figuring must be applied to the corrector. The system is capable of providing excellent images on axis and the field is about half as curved as an equivalent true Cassegrain, but coma is extreme and severely limits the useful field size. Since these systems are essentially limited to speeds slower than f/15, they are generally useless for deep-sky work but can make excellent instruments for planetary photography. The size and weight of the corrector tends to limit the Maksutov concept to smaller instruments but a few 11 inch models have been built by amateurs and Maksutovs apparently exist with apertures in the 24–30 inch range. The only advantages we can state for this instrument are the compactness offered for an f/15 + system and the fact that an all-closed tube without the diffraction problems introduced by a secondary spider support (loss of contrast for planetary work being a prime concern) provides an attractive option for planetary photography.

In the 1970s, the availability of high-speed computers and ray-trace programs resulted in two refinements of the Gregory–Maksutov system. In 1975, R. Sigler introduced, the Sigler–Maksutov in an f/8 version which produced excellent images over a much wider field than the older spot Maksutov design. Then, in 1980, another variant on the Maksutov theme was published by M. Simmons that could be designed as fast as f/5.0 which he called the Simak. These two designs both utilized the well-known fact that separating the secondary from the meniscus corrector provided a significant improvement in image quality. The off-axis optical performance of the f/8 Sigler–Mak is better than that of an f/5.6 Simak over a flat 35 mm frame, and, when ray-traces of an f/5.6 Sigler–Mak are examined, it still holds a slight edge when images are compared with the Simak. If a curved focal plane is used (or a field flattener), the performance of the two is more equal but the Sigler–Mak appears to have a slight edge as off-axis aberrations are more apparent in the Simak design. However, in both cases, the errors present are near the limits of photographic resolution* when a curved focal plane or field flattener is used, both of these designs represent an improvement over the older spot Maksutov.

The Sigler and Simak designs both represent attempts to improve upon the original Maksutov and will serve well to illustrate a point that we have been trying to communicate for over a decade to all amateurs interested in learning more about various astronomy-related problems that they may face: the professional literature can be extremely useful and should not be ignored! In this case an article describing a general class of fast, flat-fielded Maksutovs appeared in 1961 which not only pre-dates these two designs but actually describes systems that are superior to either of the later designs.

In this 1961 article, R.L. Waland points out that a single-element lens is not capable of correcting the aberrations present in the all-spherical Maksutov design and explores options to that problem. He also discusses the various aberrations of the systems and the compromises that can be made to minimize those aberrations. The 1982 article by Rutten and van Venrooij appears to be the only article in the amateur literature that discusses any systems similar to those described by Waland. The two 'Companar' systems (not to be confused with the Componar lenses made by Schneider for at least two decades) described in this 1982 article possess imaging capabilities that surpass both the Sigler and Simak system by wide margins. The 'Companar'

* The on-axis images of an all spherical Simak are larger than the photographic resolution limit (qualifiers: 'white light', in excellent seeing, with focal lengths less than about 100 inches, etc.) if the f/ratio is faster than about f/6.7. If the system is slower than f/6.7, the on-axis image size drops below 20 μm and it is no longer necessary to aspherize one of the corrector surfaces.

(or Waland–Mak as we call it) also provides additional advantages as it can be built using all spherical optics at f/ratios of f/3.0 or greater. Aspheric models can exceed the f/3 limit and still provide images that are limited only by the photographic resolution of the system (influenced by aperture, film resolution, and/or seeing).

Finally, we come to the prime-focus Maksutov which is a purely photographic instrument. This is basically the design that was suggested in 1944 but has apparently been virtually ignored because of its inability to form an image that can easily be viewed by the eye. It consists of the usual spherical primary and meniscus correcting lens, but no secondary is used. The focal plane is a curved surface between the primary and the corrector like that of the Schmidt camera. In this form, the Maksutov is capable of covering a relatively large field with virtually diffraction-limited images. In addition, the system can be constructed to be as fast as f/1.5, and all of the optical surfaces are spherical. We have never quite understood why this system has been so ignored by amateurs as it is the only alternative to the Schmidt at very fast f/ratios.

Schmidt cameras

The Schmidt camera was developed in 1930 by Bernard Schmidt and, if one excludes exotics like the Baker–Nunn system, it remains the undisputed king of fast, medium-field camera systems. The system utilizes a spherical primary and an aspherical correcting plate to remove spherical aberration. The correcting lens is thin enough so as to not introduce color error into the final image. The one disadvantage of the system is the fact that the focal plane falls on a spherical surface and thus cannot be used visually. The curved focal plane causes few problems for photographic purposes (except for professionals using glass plates) and the relatively wide-field is virtually abberation-free and unvignetted in the classic design that Schmidt created and that has been utilized at observatories around the world. Much literature has appeared on the Schmidt camera over the years, thus we do not intend to spend much time on this classic design. The photographic speeds available and its extremely small images make the classic Schmidt design highly desirable for medium-field astrophotography.

Over the past two decades several variations of the Schmidt have appeared in the literature and one of these variants became widely available in the commercial telescope market during the late 1970s. The commercially popular Schmidt systems utilized a full aperture-corrector variant of the original Schmidt design. This form has a more vignetted field than the classic Schmidt, but is quite acceptable. The primary drawbacks of these commercial models were their extreme speed (f/1.7) which severely limited their usefulness and the poor design and construction of the tube assembly which made it almost impossible to keep these systems properly adjusted and focused. One of the authors worked as a consultant at a college observatory with a 14 inch f/1.7 system for two years and found it impossible to obtain photographs consistently that could not be easily beaten with a 4.25 inch Newtonian of equal focal length. Within a few days of focusing and squaring-on, flexure and mechanical shifts had made the instrument completely useless for photographic work. If one could pick up one of these units at a reasonable price and then rebuild the entire tube assembly, the result would be an excellent photographic system.

A second, less conventional, variation of the Schmidt system appeared in the amateur literature in the late 1970s. This interesting variation was the 'lensless Schmidt'. It utilized an aperture stop in the position of the traditional corrector plate and used no corrector in conjunction with the spherical primary. The theory behind this system relied on the fact that the aperture stop theoretically limited the effects of off-axis rays and the formation of the effects of spherical aberration. The theory is simple, a spherical mirror produces small circular images on-axis. However, the model that led to this design was extremely simplistic. In reality, the aperture stop did decrease the size of the off-axis images but the images were far from ideal and were considerably larger than the tolerances permitted for high-quality astronomical photographs. The images produced by this system would be suitable for 2–3 × enlargements and for survey work or patrol work monitoring galaxies for supernovae, but would be unsuitable for quality work where sharp details are required.

Other alternatives

This is the catch-all category of miscellaneous imaging systems and we will discuss briefly those that amateurs sometimes attempt to use in astrocameras of various sorts. In some cases these configurations are used when one cannot afford a more expensive system while, at other times, these alternatives are tried in search of different views of the skies. A common situation is where a system is being sought that will allow wide-angle coverage instead of the small 1–3 degree fields that are normally encountered when moderate-focal-length telescopes are used for astronomical photography.

The most common situation occurring today involves the use of lenses designed for use on single-lens reflex (SLR) cameras for astronomical photography. Most SLR lenses of 28–60 mm focal length are quite suitable for wide-angle photography, although many must be stopped down in order to produce good-quality images across the entire 35 mm frame. The problem faced by any lens when imaging a field of stars is the most severe test possible of lens quality. The imaging requirements for normal photography are nowhere near as stringent and few lenses are built to such strict requirements for obvious economic reasons. Because of this, many normal lenses will be found to be unsatisfactory for astronomical photography. As one tries to use lenses of longer focal length, the percentage of useful lenses drops dramatically; few 200 mm lenses will be found to pass the astronomical test. The problems of testing and using normal lenses are discussed in more detail elsewhere (pp. 232–7), but we will stress the fact that any and all lenses purchased for use in astrophotography should be thoroughly tested before finalizing a purchase.

Another situation that commonly occurs is the use of any of a wide variety of surplus aerial camera lenses in astrocameras. These lenses were largely designed and manufactured in the 1940s and early 1950s for aerial survey work. Most of the commonly seen aerial lenses do not produce high quality images when used in astrophotography. Some can be used successfully when stopped down and/or when used in conjunction with a yellow or red filter to reduce the effects of inherent aberrations (Fig. 2.7) (the 7 and 12 inch f.l. Aero Ektar lenses fall into this category). Some can be used simply because one is using 4 × 5 film to take photos that will be turned into 8 × 10 prints (i.e. only a 2 × enlargement). The amateur interested in using any such lenses must be very careful as the likelihood of finding a useable lens is small. Any such lens must be tested on a star field before finalizing a purchase, but since such lenses do not simply snap onto a camera body, the problems of quick testing are considerable. For lenses of 6–15 inches focal length, one possible solution may be found in the use of a sturdy press camera body. This allows one to whip up a camera system in a day or so and test the lens without making a major investment in constructing a camera body for a lens that may not be of any practical use for astrophotography.

The final alternative that we will discuss here is essentially the system that is being sought by amateurs trying to get wide-angle astrophotographs using either of the alternatives above. It is possible to find lenses that will cover a wide field without aberrations and which will produce excellent-quality results. However, such optics are expensive and are not found in surplus stores. Some specially corrected lenses do occasionally appear on the surplus market, but the amateur is again taking a chance of purchasing a worthless hunk of optical glass for $1000–$3000. The alternative is to purchase a photographic lens known as an apochromat. These special lenses are designed for copy-camera work and must meet standards that insure their usefulness as astronomical-imaging systems: color-corrected for three visible wavelengths, a large flat field, minimal aberrations at the corners of a large field, and virtually no distortion. The one drawback that an astronomy enthusiast must be aware of is the fact that some commercial apochromats for copy work are not designed to focus at infinity! This capability must be verified before purchasing any process camera lenses. These lenses are not cheap and can run from $500 to $2000, depending on the focal length and the manufacturer (Rodenstock makes one of the best apochromats available at this time). These lenses also have the disadvantage of being very slow by astrophotography standards: currently available commercial apochromat process lenses are not made with f/ratios faster than f/8.0. The cost is high and the exposures long, but the results are outstanding (see Fig. 2.8). A less

Fig. 2.7. The central region of Orion is shown in this photograph by Alan McClure. The 1 h exposure was made on a 103a-E plate using a No. 25 filter with a 12.5 inch f.l Aero-Xenar lens stopped down to f/4.5. This is an example of what can be expected from a good quality aerial lens using a red filter.

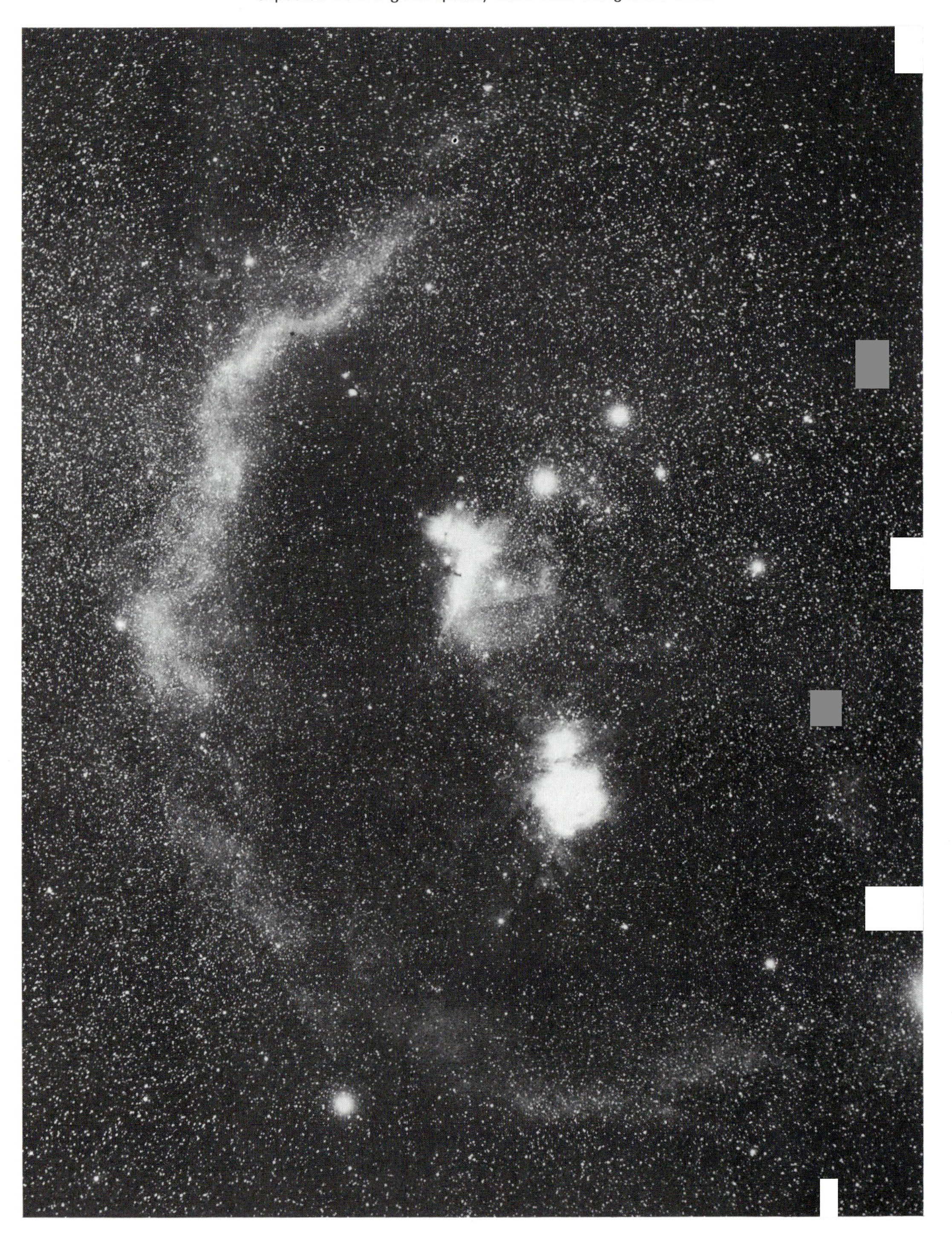

Fig. 2.8. The region around the Scutum Star Cloud is shown in this photograph by Robert Provin and Brad Wallis. The 4 h exposure was made on hypersensitized 2415 Technical Pan using a 300 mm f.l. Apo Rodenstock lens wide open at f/9. This is an example of the image quality that can be obtained from flat field apochromat lenses.

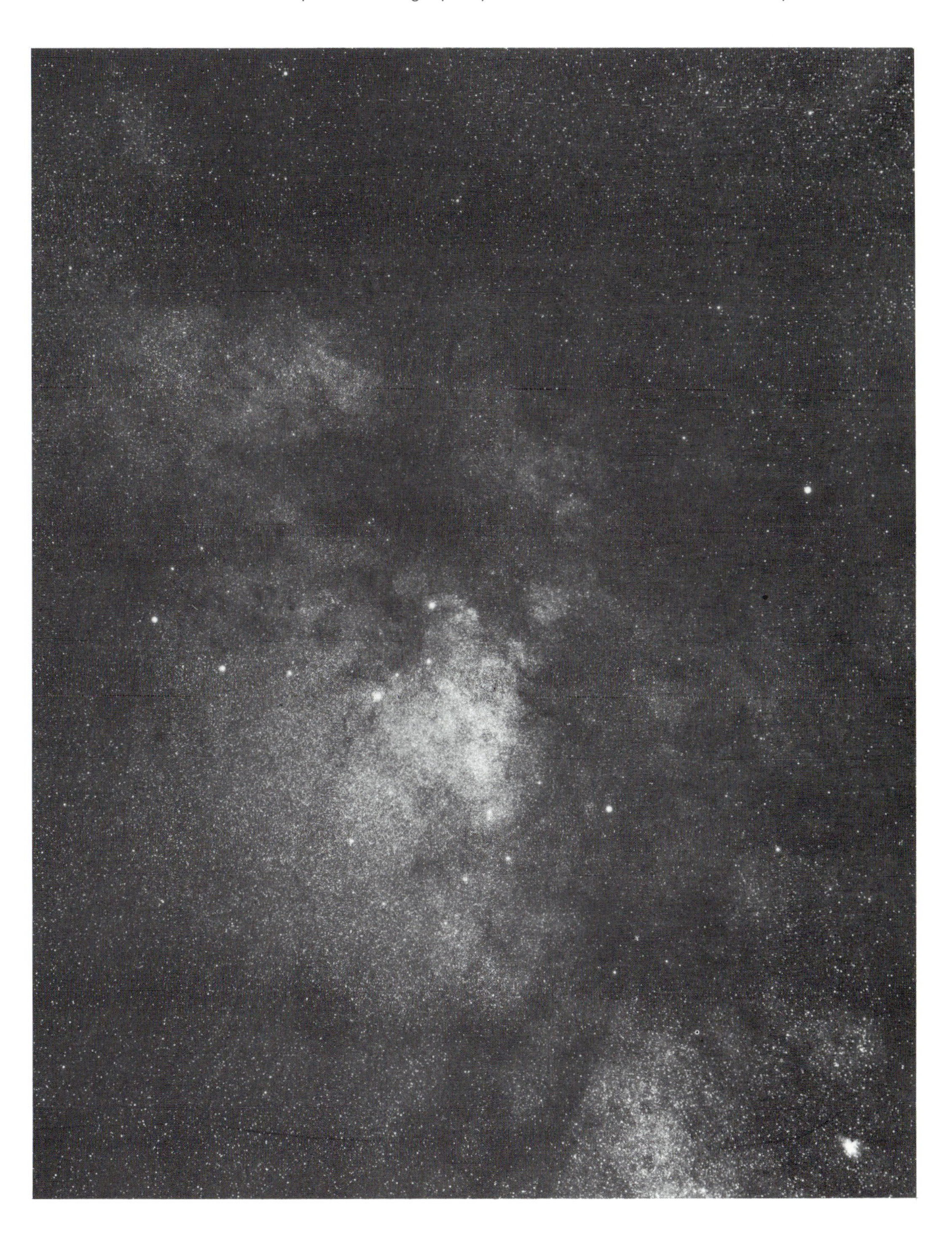

expensive alternative that is less risky than purchasing surplus lenses is to purchase a high-quality view-camera lens for \$300–\$1000+. Many of the results that we have seen indicate that a number of these lenses can be used successfully at f/ratios of f/5.6 and f/6.3; pre-purchase testing is, however, required.

We feel obliged, before concluding this section, to say something about the numerous systems that have come into being over the years seeking an unobstructed, all-reflecting alternative to the refractor (the Scheifspiegler being the primary contender in this field). The systems are all of large f/ratios – f/10 being about the fastest available – and their off-axis imaging capabilities are far from ideal. They can be used for planetary work and are suited for visual work as well but are not suited for any type of deep-sky photography. The objective behind creating these instruments is to eliminate the diffraction effects (and thus the degraded contrast) that result from the presence of a secondary mirror and its required support structure. Much of this contrast degradation can be eliminated by using a small secondary in a planetary instrument; further contrast enhancement can be obtained by eliminating the diffraction effects of a spider support and replacing it with an optical window, which has the added advantage of closing the tube and eliminating tube currents. Thus, the alternative design starts to look like a Maksutov or a concentric Schmidt–Cassegrain and both can and have produced excellent-quality planetary photographs. Our contention is that poor seeing is more to blame for bad planetary photos than the effects of secondary diffraction and that amateurs need to be more conscious of the need for excellent seeing when seeking better high-resolution images.

Mountings and drives

Equatorial mountings

The most important attribute that can be assigned to a photographic equatorial mount is **stability**. The finest optical systems available are rendered all but useless when coupled to a shaky mounting. Although equatorial mountings can take many different forms, the requirements for stability define a number of basic characteristics that are common to each. In addition to basic design and structural characteristics, it is important that the optical system is matched to a properly proportioned mounting with respect to both size and mass.

Although there are probably over a half-dozen different kinds of equatorial mountings, and a number of variations, the majority of mountings in use by both professionals and amateurs are either the German/Fraunhoffer type or the fork mount. For both of these mount types the design principles that insure rigidity and stability are the same and include adequate bearing size (at the appropriate locations), sufficient mass, careful distribution of mass, and adherence to basic principles of structural geometry.

The German equatorial is most commonly employed to support professional refractors, amateur refractors and amateur reflectors. In the basic design, the polar axis is supported by a pier or tripod with the declination axis attached to the poleward end of the polar axis. The telescope is attached to one end of the declination axis while the other end supports an adjustable counterweight to offset the weight of the tube assembly as the declination assembly is rotated about the polar axis. Examining the German equatorial, we find two particularly critical load points that, if constructed carefully, contribute substantially to the stability of the mounting. These critical points include the point where the tube-cradle support joins the declination axis and the junction point of the declination and polar axis. It is critical that the structures comprising these two junctures be as strong as possible which basically translates into thrust bearings with large surface areas at each of these points. A good rule of thumb regarding this matter for instruments with small-to-medium apertures is that the diameter of these bearings should be at least one-half the aperture of the telescope that they support. Since the size of the bearings at the opposite ends of each axis (the drive end of the polar axis and the counterweight end of the declination axis) do not require great strength, the most efficient designs for both the polar and declination axis will make use of a conic structure.

The great advantage of the cone is that it allows large surface bearings where they are needed while conserving mass at the opposite end of each axis where it is not nearly as critical. In addition, the conic shape is very resistant to flexure which further enhances its desirability. However, since equatorials employing cones are difficult and expensive to fabricate, they are rarely encountered. The simple alternative is to employ a cylinder with a diameter equal to the large end of the cone that is needed for support. Although this configuration is nearly as stable, it is more prone to flex and its mass is considerably greater making the mount carry unnecessary weight.

Stability in the German equatorial can also be improved by reducing the length of the declination axis to a minimal functional value. This can be accomplished by keeping the distance between the tube/declination axis junction and the declination/polar axis junction as small as possible. Of course, the minimum distance is limited by the clearance required for declination slow motions and the worm gear and its housing. Further reduction of the declination axis length can be achieved by increasing the mass of the counterweight. It is far better that the counterweight be heavy and mounted on a short declination spindle than to have a lighter counterweight hanging out at the end of a long lever arm. Reduction of the length of the declination axis serves to reduce the amount of bending and to decrease the amplitude and duration of any vibrations imparted to the telescope tube by wind or by inadvertent contact during photographic work. In general, a more compact distribution of movable mass on the mounting (optical assembly and counterweights) will result in fewer unwanted movements and vibrations.

On a classic fork mounting, the declination axle of the German equatorial is replaced by a horseshoe-shaped configuration that rotates about the pole along an axis that, generally, is parallel to its arms. Telescope rotation about an axis connecting the arms and perpendicular to the polar axis defines motion in declination. The polar axis of a fork mounting, for all practical purposes, can be an exact copy of the German polar axis. Again, the surface area of the poleward thrust bearing (supporting the fork and the tube assembly) should be as large as possible for maximum stability. As with the German equatorial, a cone (large end pointing poleward) serves as the most efficient polar axis. A cylinder that is equal in diameter to the large end of the cone makes a useable alternate choice.

It is an unfortunate fact that most commercial and homebuilt fork mountings are terribly underbuilt in the construction of the fork arms. Typically, such mountings possess long spindly arms that are extremely prone to flex under the weight of the optical assembly that must be suspended at the upper end of the fork. The situation is made even more acute when these mountings are used at low latitudes and the arms must support the tube weight at an angle closer to the horizontal. Several of the extremely popular commercial fork mounts made in the USA through most of the 1970s up to the present are so badly designed that they vibrate like a tuning fork in a slight breeze. Properly designed arms should have large cross sections relative to their length. The best and most efficient configuration employs triangular shaped arms with the base of the triangle (which should be wide relative to its height) mounted solidly on the poleward end of the polar axis and the apex connected to the declination trunnions of the tube assembly. It is important that the triangular profile of the arms manifests itself in both dimensions perpendicular to the arm's length (i.e. the arms should take on the aspect of an elongated pyramid). This configuration will reduce flexure when the declination axis is in either a horizontal or a vertical orientation and, at the same time, keep vibrations to a minimum.

A final aspect of the fork mount design concerns the declination trunnions that join the tube assembly to the arms of the fork. It is essential that the tube-mounted trunnions be supported by more than just the strength of the tube alone unless it is of exceptionally heavy construction. A superior arrangement will find the trunnions mounted on a cradle that encompasses at least half (and preferably all) of the tube circumference. The added strength afforded by the cradle will help insure that the two trunnions (mounted on opposite sides of the tube) will maintain alignment relative to each other and thus reduce unwanted tube movements to a minimum. The trunnions themselves should be as large in diameter as design and materials will allow in order to maximize stability. Again, the rule calling for bearing diameters equal to roughly one-half the aperture of the telescope that they support will serve well in this application.

In outlining some specific elements that can contribute to the stability (or instability) of the German and fork designs, a common unifying idea developed that deserves further discussion: time and again references are made to cones, triangles, the concept of compactness, and, to some extent, the conservation of mass. All this leads to some general rules of structural design which, when followed, will result in enhanced stability. The basic concept distills down to the fact that certain geometric shapes and mass distributions are more resistant to bending and twisting forces than are others.

Triangles, pyramids, and cones (a cone being no more than a plane triangle rotated about one axis to form a solid) are among the most stable shapes that can be incorporated into a telescope design and they should be used wherever possible and appropriate. The stiffness of triangular structures make them particularly resistant to bending and can be incorporated at a number of locations in the mounting to suppress flexure. The stiffening qualities of the triangle become greater as its base is enlarged relative to its height. The application of this principle becomes extremely important in the design and construction of fork arms and tripod legs.

Another important principle of structural geometry is concerned with the distribution of mass and the compactness of a given structure. One very effective way to reduce flexure and vibrational amplitudes in a mounting is to reduce the length of the members (and thereby make the structure more compact) that support massive objects such as the tube assembly and/or the counterweight. Flexure and vibration increase as the stiffness of a beam (i.e. a fork arm or a declination axle) decreases. Since, all other variables remaining constant, the stiffness of a beam varies inversely with the cube of its length, it becomes readily apparent that substantial gains can be realized by simply making reductions in the length of supporting structures. By way of example, consider that reducing the length of a declination axle by a factor of two results in an eight-fold increase in stiffness. In the case of a fork mounting, it is readily apparent that considerable gains in stability can be obtained by making the fork arms as short as the required amount of tube clearance will allow.

A discussion that is frequently heard concerns the matter of sheer simple mass in the design of a telescope mounting. There are two schools of thought on this issue; one school says that there simply is no substitute for mass when it comes to producing stability, while the other contends that it is all but unnecessary if the mounting is constructed from the right materials and designed with sound engineering principles. Our position on this is that both schools deserve some consideration but the structuralist argument is considerably stronger. For example it is well known that certain hollow structures, when just slightly scaled up, can have the same strength and stabilizing characteristics as much more massive solids of the same shape. Thus, many mountings employ hollow declination and/or polar 'shafts' with considerable savings in weight and no sacrifice in strength. However, mass does have its redeeming qualities, the most important one being inertia. While some will argue that massive mountings can develop high-amplitude low-frequency vibrations that damp out quite slowly, they ignore the fact that massive objects have a large inertia. This translates into the fact that it requires a proportionally larger force to impart vibration to a more massive object and thus they are less likely to occur. A well-designed mounting needs to incorporate sound engineering principles along with sufficient mass to produce a sturdy foundation and a structure that is able to resist vibration. Neither component by itself can produce maximum stability but the judicious application of basic design principles and moderate mass will yield the best results.

For the amateur without the necessary machinery to construct an ultrastable mounting, the alternative is to select a commercially made mounting that will serve adequately. This poses a problem as most commercial mounts are designed for simplicity and for marketability which generally means that they are made to be sold at a relatively low price and that they are not designed to the standards of the connoisseur. In the 1960s and 1970s, a line of well-made German equatorials was produced in the USA by a company known as The Optical Craftsman and used mountings can still be found now and then for reasonable prices. In the early 1980s, Edward Byers produced an exceptional mounting called the Byers 812 that utilized conic bearings and a sector drive (Fig. 2.9). This mount was very sturdy, yet quite portable and the price, when introduced, was comparable to commercial mountings of considerably less stability and quality. Unfortunately, the Byers 812 was not well received in the amateur market and it is only available as a custom mount. Today, the array of commercial mountings that are readily available to amateurs in the USA presents no outstanding options for the serious amateur. A few small companies (like Ed Byers and Thomas Mathis) do make some excellent quality custom German equatorials and very stable fork mounts. If one must purchase one of the common mounts available, our advice is threefold.

1 To avoid the purchase of the flimsy fork mounts that several US companies are producing for their Schmidt-Cassegrain instruments. Perhaps some day these will be redesigned and gain some stability but, at this time, they should be avoided by all needing the stability that astrophotograpy demands.
2 When purchasing any mass-produced German equatorial: find the mounting that the manufacturer normally sells with the telescope size that you plan to use and then purchase the next size mounting that is made. If the manufacturer makes a deluxe line, purchase a mount from that line.

3 If you can avoid buying a mass-produced mount and can afford to purchase a mounting from a firm that specializes in building custom telescope mountings, do so. The mounting will be better designed and will generally possess a drive that is at least 10 times better than anything that will ever appear on a mass-produced mount.

The Poncet mounting

The Poncet mounting has somewhat limited photographic applications. However, its unique design and simplicity deserve some comment here. At first appearance, one would hardly guess that Poncet's simple

Fig. 2.9. The Byers 812 mount is shown here as an example of a very stable and well-designed equatorial. The conic polar axis of the 812 allows it to support telescopes of 12 inches aperture while still remaining lightweight. (Photograph by Ernest D. Wallis.)

platform could track like an equatorial. However, as Poncet points out, the design is based on a simple mechanical theorem: 'a rigid body utilizing three points, one of which acts as a fixed pivot and the other two being constrainted to move in a fixed plane, can only rotate around an imaginary axis perpendicular to the plane and passing through the pivot'. Of course, this 'imaginary axis' is, when properly pointed, the platform's polar axis. The mounting is inherently stable since the tracking platform rests on three points and sits very close to the ground. However, the real beauty of this mounting lies in the fact that any altazimuth-mounted telescope, or camera, can simply be set on the platform and instantly be capable of tracking celestial objects just like any equatorially mounted telescope.

The most common form of the Poncet (there are a number of variants) can typically track from 60 to 90 min before the mount must be reset for tracking to continue. Tracking time is primarily limited by the separation between the mount's base and the tracking platform and by the amount of tilt away from the horizontal that can be tolerated before stresses undermine the mount's stability. Although most Poncet mounts employ tangent-arm drives, a gear assembly at the pivot point would provide a more stable drive rate, greater tracking accuracy, and would also eliminate the necessity for periodic resetting of the drive.

The primary limitation of the Poncet is that it is only capable of motion about one axis (the polar axis) and is thus unable to make true motions in declination whenever one uses the azimuthal-pointing capabilities of the altazimuth mounting. For photographic applications, this poses a considerable problem since declination corrections required to compensate for misalignment on the pole and atmospheric refraction are only possible in a very limited region of the sky. It does not require much thought to see that while the altazimuth motions of a telescope resting on a Poncet platform can be used to correct declination drift, these movements will only constitute true motion in declination when the telescope is pointed at the meridian and when the telescope is oriented in such a way that the elevation motion corresponds to a north–south direction. Elevation motions in other azimuthal orientations will not be truly north–south and declination corrections will require movement in both elevation and azimuth. Not only is this a more complex motion for making corrections, but this type of 'declination correction' will always result in field rotation at the film plane. Thus, with long focal lengths, successful deep-sky photography on a Poncet will require rather stringent constraints, i.e. nearly perfect polar alignment and relatively short exposure times with all exposures being limited to areas within an hour or so of the meridian.*

* The problems can be largely overcome by using a guiding head near the position of the film plane. A mechanism that can impart x,y motions to the film plane, when oriented properly, is capable of true motion in both RA and DEC at any part of the sky. This solution not only can be applied to telescopes mounted on poncet platforms but can also be used with telescopes mounted on more conventional equatorials as well.

Difficulties aside, the Poncet will serve well as a tracking platform for cameras of short focal length. With good polar alignment and an accurate drive system, wide-angle photography with anything up to a moderate telephoto lens can be accomplished with relative ease. Since the platform can carry any number of cameras depending upon its size, full-sky coverage is feasible for meteor photography, or for nova patrol work, using a single platform. In fact, the Poncet seems to be ideally suited to those applications where tracking accuracy is not paramount and, given adequate polar alignment, declination corrections can be eliminated.

Clock drives (or . . . as the worm turns)

A standard piece of information seen in numerous books and articles on telescope making is that the drive must make one turn each 1436 min if the telescope is to track accurately. Is this really a mark of the quality of a drive, and does this ability insure good tracking for astrophotography? To answer part of this question, one must realize that the driving rate of virtually all of today's commercial clock drives is determined by the speed of a synchronous motor which is regulated by the frequency of the incoming electrical current. Another alternative which is coming into existence, and which may well grow in popularity as the use of small computers spreads, is the use of stepping motors. This option would allow altazimuth mountings to be used for observation purposes (in order to photograph with such a setup, the camera must also be driven, as the field of view rotates). In the USA, this input frequency is a very regular 60 cycles/s (60 Hz). If you are just interested in observing an object for a short period, and you wish to plug into an electrical outlet somewhere, a drive with a 1436 rate will probably suffice to keep the object in a medium power field of view.

However, if you are interested in keeping a star confined to a very small area of the field (for photography or photometry), you will not be able to just plug into any wall socket. Telescope drives typically do not track at a constant rate. Virtually all drives turn the telescope at a variable pace. If this is a smooth, long-period, cyclical variation, it is termed a periodic error. If this variability is random, and/or of short duration, it is termed erratic. Usually, one sees an erratic error superimposed upon a periodic error. It is the magnitude of these combined errors which can make the task of guiding a photograph miserable or pleasant. Because of these driving errors, it is necessary to vary the speed of the drive in order to track any object with great precision. The easiest way is to obtain an 'inverter' which converts a 12 VDC (volts DC) and/or 120 VAC (volts AC) into 120 V with a frequency that can be changed slightly to compensate for minor variations of the driving rate and that can be altered radically to move objects around in the field of view. Such devices typically have a tracking range of 55–65 Hz which can be varied continuously and is sufficient to accommodate driving rates from 1321 to 1550 min/rev. This is not to suggest that one should purposely construct drives with rates other than 1436, but it should be clear that this 'sacred number' is not a mark of the quality of a clock drive. The ability to vary radically the speed of the drive is normally provided by a pair of pushbuttons for 'slow' and 'fast'. These control rates can be as small as ±10 Hz for fine adjustments or as crude as ±60 Hz (i.e. double speed and stop) for gross changes.

Refraction provides another reason for not expecting to be able to use a constant drive rate throughout a long exposure: as the elevation and azimuth of any star changes, its rate of motion varies and its actual rate of movement is a function of its declination and hour angle. E.S. King studied this phenomenon extensively and published the complex results of that study in 1901 and again in 1931. His goal was to take astronomical photographs without the need for guiding, using only equations to solve the problem of the motion of a celestial body across the sky. King published photographs showing exactly how close he came, but concluded that the quality of the clock drives of the day was inadequate for the purpose. With today's computers, King's goal might well be achieved, but the flexure of the instrument would need to be analyzed in great detail in order to compensate for this complex and apparently (but not truly) erratic motion. In King's time, the lack of such sophisticated equipment made the use of his equations impractical unless the exposures were made within a few hours of the meridian; at that time, exposure times of 24 h and more were often required to reveal details that can be captured with an hour-long exposure today. In any case, refraction causes a star's rate of movement to vary from about 1435 to over 1440, which is easily compensated for by using the frequency range that is available on virtually all variable-frequency inverters available today.

The accuracy of any clock drive can be evaluated by placing an object of known angular size (i.e. a planet or a double star) on the crosshair of the eyepiece. First, adjust the driving rate of the telescope so that the position of the object is relatively stable. Then watch the motion and estimate the magnitude of the variation in the object's right ascension (RA) position. If the motion is on the order of 5–15 arc sec you can consider the drive to be of excellent quality. However, most of the drive units that are supplied on virtually all commercial telescopes are not of such quality, and much greater errors will undoubtedly be discovered. It has been our experience that **errors on the order of 1–3 arc min are to be expected from common commercial drive units.** Such units are designed to be inexpensive, and quality is sacrificed for low cost and profits. In fact, it is difficult to see how the price of such units is justified when the cost of the materials is accounted for. There are several manufacturers of custom telescope drive units in the USA, and the drives produced by these craftsmen will probably out-track a common, mass-produced drive by a factor of about 50! Of these custom-drive builders, the works of Edward R. Byers are generally of the best quality, although others (like Mathis) produce gears of extremely good quality. If you own an instrument with an error of 1 or 2 arc min, and a large erratic error component, it would be wise to consider replacing the drive with a custom-made unit. These units are not cheap, but the money spent on high-quality gears saves a lot of wear and tear on the eyes and nerves when making long exposures on cold winter nights!

If a drive test is performed photographically, a much better idea of the errors present in the drive can be obtained and the magnitude of the errors can be accurately measured. In addition, the nature of the tracks produced can be used to isolate the source of the error in the drive unit to a particular gear. To perform such a test, the polar axis should be offset from the pole by about 5 deg in azimuth. Then, find a field of faint stars near the equator and set the drive rate just a little slow and begin an exposure. The exposure should be long enough to allow the worm to make 2 or 3 revolutions (20–30 min is usually an adequate exposure time) and you should not attempt to guide the exposure. It will also be necessary to record the exact duration of the exposure in order to perform calculations later).

When the film is processed, you will find a series of parallel wiggly lines like those shown in Fig. 2.10. The regular wavy pattern is the periodic error while the tiny bumps superimposed upon those wavy lines are the erratic errors. If the total exposure time was E and the lengths of P and a are measured, then the period of the periodic error is given by:

$$P' = E(P/a) \quad \mathbf{(2.2)}$$

The period of this error will almost always be found to equal the period of the worm. The magnitude of the error can be found by measuring the height of the sinuous trail x (in millimeters) and then multiplying this value by the image scale of the instrument to yield:

$$\mathrm{d}R\ (\text{min}) = [135/F\ (\text{inches})]x \quad \mathbf{(2.3)}$$

If the error is mostly periodic and larger than about 30 arc sec, there is a method of removing much of that error by making the worm regulate its own rate of revolution. However, it is important to realize that it is not really the periodic error which creates the greatest difficulties when guiding photographs. At any rate, the method is relatively simple and a detailed

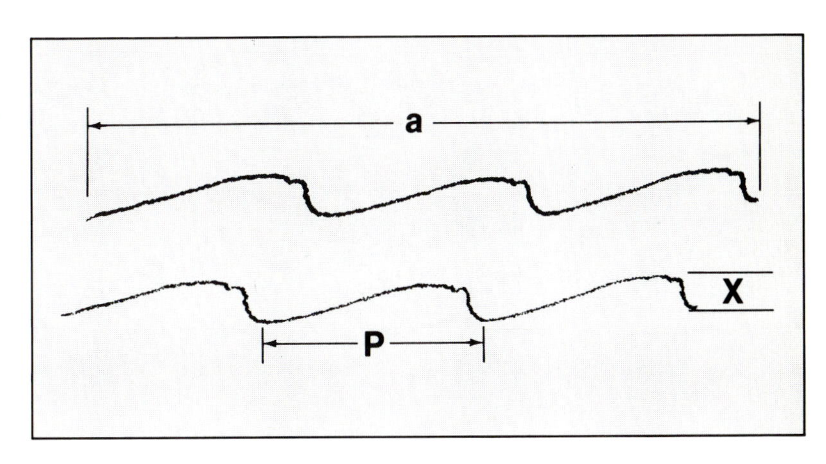

Fig. 2.10. Sample star trails are shown which were obtained from a drive test. The total length of the trail (a) and the length of one complete trace of the periodic error (P) are labelled as is amplitude of the drive error (x).

description can be found in an article in *Sky and Telescope* (May 1978). In this method, the speed of the synchronous motor is regulated by varying the frequency of the input current by linking a potentiometer to the turning worm. By a careful selection of the potentiometer, and by carefully adjusting the linkage between the worm and the pot, the frequency can be varied so that, as the telescope would normally be speeding up, the frequency is being varied so that the motor slows down. Again, it is important to keep in mind that the periodic error is not responsible for those sudden speed changes that often catch one off guard and have the potential of ruining an exposure that has been going well for many minutes. If your drive has a lot of erratic errors, this solution may not make your guiding any easier. It is necessary to consider carefully the nature of a drive's tracking problem before deciding on the approach required to remedy that problem. It is also important to realize that most of the 'bad guiding' experienced by novice and experienced astrophotographers is really due to flexure and must be addressed separately from the problem of the quality of the clock drive.

Drive correctors and slow motions

As mentioned earlier, some means of controlling (and varying) the speed of the clock drive motor is an absolute necessity in order to compensate for drive irregularities, refraction, differential flexure and sometimes even the motion of the body being photographed. The method which is currently popular (and very effective) is to vary the frequency of the current driving the synchronous motor by a small amount. These variable-frequency drive controls work extremely well, are small and compact, and are generally trouble free because of their solid-state construction.

When selecting a drive control to purchase or to construct, there are several points to consider which may bear on the successful use of this unit for astrophotography. The primary concern is that the frequency range on the unit (fast/slow and fine-tuning controls) be well suited to the guiding task. For most situations, if one is using a good-quality clock drive, a fine-tuning frequency range need not exceed ± 5 Hz; this small fine-tuning range will make it easier to set the unit at the desired frequency while ranges of ± 30 Hz will generally make it more difficult to tune to a particular frequency.

With the fast/slow pushbutton controls, the situation is slightly different. If the range is too small, it will take forever to make corrections at high declinations and to 'slew' to position a star on the crosshairs. If the range is too great, the slewing problem is solved, but it will be difficult to control the actual corrections owing to the extreme rate of motion of the guidestar. A range of ± 5 Hz is a very fine guiding control for most applications, while a range of ± 30 Hz is better suited for the positioning of a guidestar on the crosshairs. The best solution is to use a drive control which is capable of being switched to select a fast/slow range suited to the particular guiding task. Such capabilities are currently not available except in custom units, but would be most welcome for this task. The 'ideal unit' would possess a fine-tuning control of ± 5–10 Hz and fast/slow ranges of ± 5, ± 10, ± 15, and ± 30 Hz.

If a unit is not available with a selectable fast/slow range, and the existing fast/slow range is $\sim \pm 5$ Hz, it is easily possible that a situation may arise where the range of motions will be inadequate for the purpose of guiding an exposure. This situation arises in the case of comets which have close approaches to the Earth and is aggravated even more when these bodies reach declinations greater than $\sim$45–50 deg. In these situations, the motion of the comet may require fast/slow motions of $\pm$20–30 Hz to properly guide the telescope. Fortunately, this motion will only persist for a night or two, but this may well be the most spectacular time for photography. For these very special occasions it may be desirable to construct a second control paddle which has a coarser range of fast/slow speeds and to keep this paddle ready for the rare comets that grace our skies from time to time.

The drive used for making declination corrections also deserves some thought as it is just as necessary for astrophotography as the clock drive. The easiest means of assembling a declination drive is to construct a tangent arm with a screw drive. This is also the standard assembly provided on many commercial mounts. Quite often, the screw drive that is used on many mounts will be found to be too coarse. A common arrangement is to use a 16 TPI (threads per inch) screw with a 6 or 7 inch tangent arm. With such an arrangement, a single turn of the screw drive amounts to as much as 35.8 arc min! By using a 12 inch tangent arm with a 40 TPI screw drive, a single turn of the screw amounts to only 7.16 arc min. The added control provided by this latter arrangement allows for finer adjustment of the declination drive and makes the act of correcting declination drift less subject to overcorrection.

This arrangement of using a tangent arm with a screw drive works quite well even though it lacks in precision. One of the biggest problems with this arrangement is the fact that the screw drive suffers from a great deal of backlash. If some remedy is not provided, the backlash can cause serious problems during an evening's work. Fortunately, the solution for this drive is quite simple and very effective: by attaching a strong spring to the tangent arm and then anchoring it to the telescope saddle, the backlash can be completely eliminated (Fig. 2.11).

Mechanical and design considerations

Optical-support system

The optical-support mechanisms are extremely important elements in the design of a photographic telescope, whether it is a Newtonian or some other system. Improper design in this area will result in image shifts during long exposures which will ruin many hours' work at the telescope. For the Newtonian, and most other mirror systems, we are concerned with the primary and secondary mirror mount as well as the secondary-mount-support structure.

A solid optical-support system is important not only from the viewpoint of preventing flexure, it is also a basic requirement for keeping the optical system correctly collimated. Poor collimation is more than just another annoyance, it can be responsible for ruining many photographs since it has the effect of tilting the focal plane with respect to the film plane. Many just accept the fact that their Newtonian or Cassegrain goes out of collimation from time to time as an inevitable event: it does not have to be! If the spider support and secondary mirror mounting structure are designed and built well any telescope (even those with fiberglass tubes) can stay collimated for years without any need for adjustment.

Most commercial optical supports do not meet the stringent requirements of a photographic telescope. While slight shifts, and even some large shifts, of the optical components can go unnoticed in visual instruments, even minute shifts or tilts of either mirror during a long-exposure photograph can have ruinous results. For example, a 6 inch mirror of 30 inches focal length can shift no more than $\pm 0.000\,55$ (14 μm) inches during an exposure if the resultant trails of the star images are to remain restricted to a circle less than 50 % greater than the 'resolving power' of that instrument. For this same system, the allowable tilt of the primary mirror is only 1/10 of the allowable shift (the ratio of R/F)*, or $\pm 0.000\,055$ (1.4 μm) inches! Given a reasonably 'normal' secondary/focuser arrangement, the

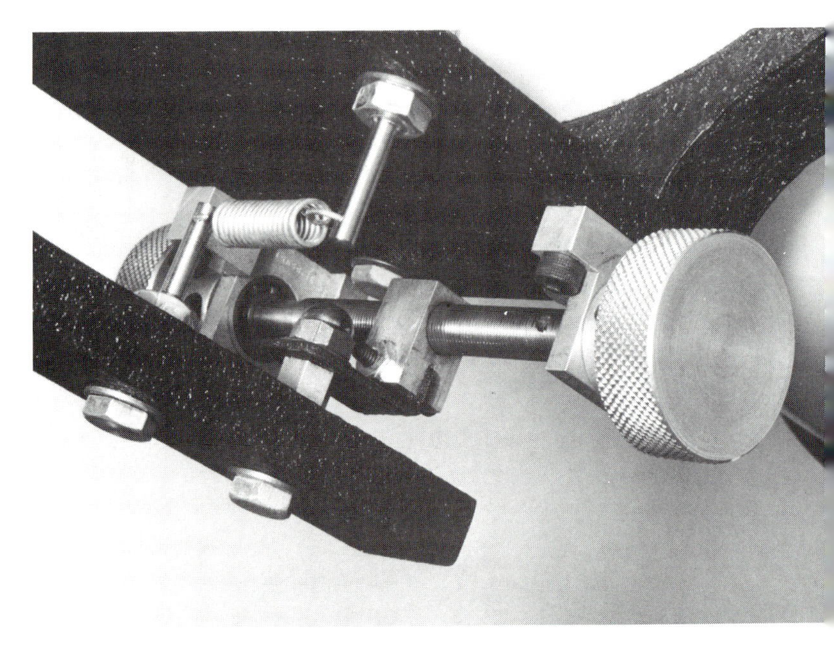

Fig. 2.11. A declination tangent drive assembly is shown. The fine threads on this unit allow very small corrections to be made. The high-tension spring serves to eliminate backlash from the drive system.

* R, radius of mirror; F, focal length.

allowable secondary tilt is around ±0.0001 inches (2.54 μm)! Given the mass of the primary mirror, a reasonably sturdy primary mirror mount and the typically frail secondary support system, the area in a typical instrument which is most likely to produce image shifts is generally the secondary support system.

In addition to the image-shift problem, one must also be aware that focus shifts can occur as a result of movement of any optical component. Usually, such focus shifts do not occur without some accompanying image motion, but a singular effect can occur. The 6 inch f/5 instrument referred to previously can tolerate a focus shift of no more than ±0.0020 (50 μm) inches before star images enlarge beyond the allowable limits. These examples should suffice to point out the necessity for well-designed and rigid optical supports.

Most commercial and amateur-made primary mirror mounts are basically sound in concept and design, but usually lack sufficient strength in critical areas to maintain proper alignment during long exposures. The most common deficiency lies in the small size of the collimation and locking screws on the mirror mount (many mirror-mount designs do not even employ locking screws, but rely upon springs to maintain the 'final' mirror position!) Since these screws must **rigidly** hold the full weight of the primary mirror, they should be as large as the mirror mount will permit. The upper limit for the size of these screws is primarily a function of the design of the rear component of the mirror mount since these holes should not be so large as to weaken the strength or structural integrity of the mirror mount.

Another problem area with many mirror mounts results from the movement allowed by spring-loaded collimation screws. While this arrangement makes collimation quick and convenient, it is a particularly bad arrangement for holding optics with the stability required for photography. A better system involves a push–pull arrangement whereby three screws are used to collimate (120 deg apart) and three others work in opposition (offset by 60 deg) to lock the adjustments in place. All six screws are further secured by the addition of locking nuts. Although this system is slightly more difficult to collimate than a spring-loaded system, it is much less likely to flex or move during an exposure and virtually never requires recollimation once final adjustments are completed.

A final problem encountered with many primary mirror mounts is lateral movement of the mirror within an oversized cell. Many commercial cells are up to $\frac{1}{4}$ inch larger than the mirror diameter that they are designed to accommodate. Besides this, the mirror clips cannot (or should not) be pressed against the mirror with much force without causing some mirror deformation. With all this, it can be readily appreciated that lateral slip is a real possibility. A solution to this problem involves the placement of three nylon or teflon screws placed equally between the mirror clips around the cell and about halfway up the side of the mirror. When these screws are tightened moderately against the edge of the mirror they will serve to eliminate any chance of mirror shift. To be certain that the mirror clips and/or the nylon screws are not pressing too tightly on the main mirror, a star image should be examined both inside and outside of focus with a high-power eyepiece. Any excess pressure from overtightening the screws or the clips will typically result in oval or triangular star images that will rotate 90 deg as the eyepiece is racked inside and outside of focus. These imaging defects should be eliminated at all costs as they seriously affect photographic resolution. Several of the many articles by R.E. Cox address the design of adequate mirror cells and there are several alternate designs which would be more substantial than the mirror cells which are currently available commercially.

Up to this point our discussion has been concerned with stationary mirror supports. However, many instruments (most notably a number of Maksutovs and Schmidt–Cassegrains on the market) employ a moving primary as a means of focusing and this is worthy of some commentary. Many of these instruments produce images that are trailed and/or out of focus when used for long-exposure photography because of the mirror 'flop' induced by this non-rigid focusing system. This flop occurs owing to the shifting weight of the mirror as the orientation of the instrument changes during long exposures. The use of an off-axis guider is mandatory to avoid trailed star images; even this will not cure the problem of focus shifts. Some manufacturers provide an optional set of set-screws that can be used to lock the primary mirror in position prior to beginning an exposure. This system is easy to use, but it can result in bending the mirror which will introduce some unwanted aberrations; another possible draw back of this system is the actual chipping of the primary mirror due to the pressure of these set-screws. An equally effective solution without either of these drawbacks entails the insertion of a weak spring over the baffle tube behind the main mirror. This spring provides just enough tension to prevent mirror flop during long exposures without encumbering the workings of the focus mechanism.

While most commercial and amateur-made primary mirror mounts are at least serviceable, the situation with respect to secondary mirror mounts and spiders is all but dismal. After tube flexure and poor guiding, sagging or movement of the secondary mirror is probably the most common cause of trailed star images on long exposure astrophotos. The problem can be traced directly to flimsy spiders and poorly designed mirror mounts. One remedy for the spider problem is to construct your own solid support. The traditional four-vane type with heavier-than-normal construction should be adequate, provided that the vanes are made from stainless steel or some other high-tensile strength material and can be tightened to very high tension (more on this later). A better solution is to build a spider with rigid vanes that can be up to $D/30$ in thickness without causing undue diffraction effects. A spider made from rigid vanes welded to a central piece designed to hold the secondary mirror mount and then bolted to an aluminium tube would come close to approximating a solid object, which is the ideal situation.

Another classic departure from sound optical mounting design is the all-too-common practice of mounting a secondary mirror in a metal or fibreglass cylinder of one kind or another and then holding the mirror in place by stuffing the remaining void with cotton or even with tissue paper! This may suffice for visual work but the effects of temperature and humidity change the amount of pressure that this stuffing applies to the back of the mirror. The likely result of all this will be unwanted movement of the secondary mirror itself. A more effective mounting method is shown in Fig. 2.12. The secondary is mounted on a truncated rod of plexiglass using a layer of felt or cork to separate the two surfaces, and allow for independent expansion and contraction, and the secondary is held in place by fitting a truncated cylinder, which has a narrow lip around the slanted end, over the whole assembly and securing it at several points. This truncated fitting can be made from either plexiglass or aluminium and should be just slightly larger than the truncated plexiglass rod and the minor axis of the secondary.

The above comments on the design of a secondary support system for a Newtonian are equally applied to Cassegrain-type systems which require a spider support. In these configurations, the secondary functions as a type of reflecting barlow to increase the effective focal length of the system. This configuration is even more demanding of a solid support system for the secondary mirror since the secondary mirror will also magnify the effects of any movement. In optical systems where a corrector is present, the corrector itself can be used to support the secondary mirror which makes a very solid support for the secondary mirror.

Flexure

The amateur astrophotographer who states: 'Don't blame the telescope, you are the principal reason for success or failure' is displaying an extreme ignorance of this field. However, using a very broad interpretation, this arrogant ignorance can be turned into almost constructive criticism. If you are getting photographs with long or multiple star images (or any other persistent problems), then you know that you have a problem and that you should solve that problem. If you do not pursue the solution, or give up after only a few attempts, or just decide that trailed images do not look so bad, the problem that exists in the instrument is there because you do not wish to solve it and you are the one to blame.

If you use a separate guidescope when photographing and continually obtain elongated star images despite the fact that you are keeping the guidestar on, or very near, the crosshair, it is probable that many of your 'badly guided' photos are really the result of 'differential flexure' between the guidescope and one or more of the many components which comprise the main telescope. If your star images occasionally show a 'U' or a 'J' shape,

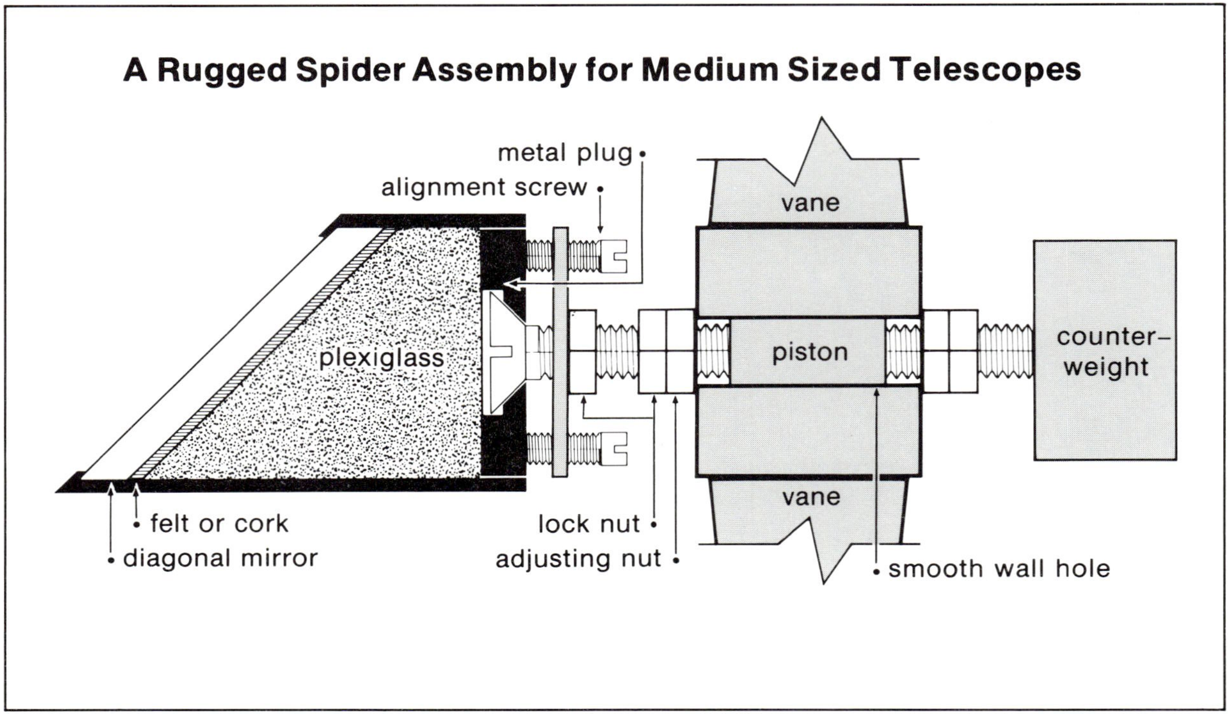

Fig. 2.12. Shows details of a Newtonian Secondary Support System suitable for use with small to medium size diagonals (up to approximately 3–5 inch minor axis). The plexiglass support is cut so as to nearly fill the space behind the secondary and then, after the secondary is fixed to this plug using a layer of felt or cork, the 'cowl' that fits over the secondary and the base of the mirror holder is used to hold the mirror snugly in place.

appear like a knotted line, or even exhibit multiple images, you are seeing the classic tell-tale signs of flexure.

If you are uncertain about just how well your photographs are guided, there is another simple means of diagnosing flexure. The only equipment required for this test is two guiding eyepieces with crosshairs. Before conducting this test, you should calculate the resolution of your system using the equations presented in Chapter 4 and then examine the size, separation, and general appearance of a double star having a separation equal to one-half of your expected photographic resolution. This is basically to get acquainted with the visual appearance of an object or space which approximates the movement that can be tolerated during a photograph without registering on the film.

To carry out this flexure test, you must place one of the guiding eyepieces on the guidescope and the other on the main instrument. The magnification on both instruments should be similar for this test. First, place a star on the crosshair of the main instrument and then line up the crosshair of the guidescope on the same star (line up the two scopes on this star exactly, spending some time on this task). Next, move to another star that is about one hour away in RA and center the star in the eyepiece of the guidescope and then look into the eyepiece of the main scope. If the star is still on the crosshair, you can continue the test; move off to another star that is another hour away and recenter the star in the guidescope and inspect the image position in the main scope. You should also allow the scope to sit for a few minutes in each new position and even give a few light taps to the instrument to see if anything is under some stress or tension that will show only after some time has passed.

There is a pretty good chance that your 'guidestar' will not be on the crosshair of the main scope after the first step of the test. With most commercial scopes, or scopes built by amateurs for normal observing, this will probably be the case. If the star is displaced, you can decide if the movement would record on film by comparing the image shift to the mental image you have of that double star that you examined before starting this test (a crosshair with graduations would be helpful to evaluate the actual size of the shift). If your image shift due to flexure is greater than half the photographic resolution of your system, you now know that flexure, and not guiding, is responsible for some of the 'badly guided' photographs that have been causing your ulcers to develop. To discover the full extent of the flexure problem, this test should be repeated at a number of different declinations, and the motions of the 'guidestar' should be mapped as the test progresses. You may discover, as we did at one point, that flexure is the main cause of ruined photographs over the whole sky. You may also find that the occurrence and magnitude of flexure in any given area of the sky is not very predictable, and you should not be surprised to see image shifts of 20 arc sec or more if you are using an instrument built to the common standard accepted for amateur telescopes.

If you use an off-axis guider, you may think that flexure cannot affect you. However, if you still obtain badly guided photos now and then, you should probably run this same test. If your guiding head has any loose parts in critical places, this test will reveal them. It is also difficult to obtain accurate polar alignment with a scope which flexes badly, it is a very good idea to eliminate as much flexure as is possible from instruments which use off-axis guiders.

The results of this test provide good news and bad news. The good news is that you are not totally responsible for all those 'badly guided' photographs and that you aren't really incompetent at all this. The bad news is that the test cannot tell you what is flexing! If you have just discovered that your off-axis guider has a problem, it is at least isolated to just a few parts, you will soon learn why an off-axis guider is such a pleasure to use for most aspects of astrophotography. However, it is important to eliminate as much flexure from any telescopic system, even if an off-axis guider is to be used. An off axis-guider alone is not the answer to all problems, and it is essential to have a solid photographic instrument.

You should now start asking some very basic questions about your scope and checking on the answers. Questions you should ask include: is the focuser attached solidly to the main tube? Is the main mirror securely in its cell? Are the diagonal vanes tight? Is the guidescope objective mounted securely? Is there any way for the guiding eyepiece or its reticle to shift? Is the diagonal mirror held securely in its cell? Can the various parts of the main mirror cell possibly be shifting? What parts of the scope can move when light pressure is applied (it only takes a shift of about 7 μm (0.007 mm) on the film plane to ruin a photograph)? You should check out everything that touches or supports either of the optical assemblies. You should be able to lock up all movable parts on either system prior to taking a photograph. You should be suspicious of every single part of the instrument and not overlook anything just because 'that couldn't possibly move'. After checking, repairing, and double checking the entire system, repeat the flexure test and take some photographs to see if your efforts have had any effects. You will probably see a substantial improvement, but the problem may not yet have been reduced below the limit of photographic resolution. If it is still large enough to be resolved, it is time for some more drastic measures.

In many of the scopes that we have examined, the guidescope rings are a very weak link in the system. Some rings are constructed of thin metal, some are not securely attached to the main telescope, others are not even securely assembled. The guidescope mounting must be a sturdy inflexible unit and it must also be securely attached to the main instrument. If the guidescope rings are mounted on the mounting rings that hold the main instrument, another possibly weak link is introduced into an already complex system.

One way of increasing the stability of the guidescope rings is to combine them into a single unit. The two rings can be mounted on U-channel and the rings can then be linked together at the top by aluminium or steel rods. Once this unit has been constructed, it must be securely attached to the main tube. This can be done using bolts with very large washers on the inside of the tube or, even sturdier, a rod or bar can be placed inside the tube which runs the entire length of the guidescope ring unit. If additional rigidity is desired, guy wires can be run from this unit to the main telescope tube to insure that the guidescope cannot shift from side to side during long exposures.

Finally, we have come to the one area of the scope that no one wants to suspect of flexing: the fiberglass tube. This link in the system is always taken for granted, yet, as we found, it may well be one of the weakest links and may be the cause of most of your woes. Fiberglass is a material that varies in composition from one manufacturer to another, and it is therefore impossible to cite any hard figures on its strength. However, a tube made from aluminum can be anywhere from 4 to 20 times stronger than a fiberglass tube of the same thickness. After replacing one fiberglass tube with an aluminum tube, with a thinner wall, we found that over 80 % of our flexure problems were eliminated. The cost of an aluminum tube (1981 prices), for scopes from 6 to 10 inches aperture, is quite comparable to the cost of a fiberglass tube. The weight is also in the same range, but will usually be less than the weight of an available fiberglass tube since the wall thickness for smaller scopes can be reduced. For larger apertures, thin-walled tubing is not available, and sheet metal must be rolled to form a tube. Even if you are using an off-axis guider on your large scope, it is important to eliminate tube flexure as it is difficult to polar-align with a flexible telescope tube.

A common issue which arises when discussing aluminum telescope tubes is the question of focus shifts due to temperature changes. After looking at some calculations, it was clear that a 10 deg F shift would be required to create a measurable focus shift using a 48 inch length of aluminium tubing. After starting to photograph using this tube, we found that the calculations were validated and, in fact, found that they were on the conservative side. So, when we begin an evening's work, we note the temperature at the start of a session and then watch for a 10–12 deg F shift. Only after such a temperature shift is noted, will we attempt to readjust the focus.

Newtonian diagonal: size, placement and illumination

For simple observing, it is quite easy to calculate the minimum size of the minor axis of the Newtonian secondary that is required to fully illuminate just the central point of the focal plane. If observing is the only application that is planned for a Newtonian, it is not unusual just to add a fraction of an inch to this minimum value and then to select the next larger size diagonal that is available. For photographic applications, this method is not satisfactory and a more detailed approach is required to find an acceptable solution.

The main item of concern in choosing a secondary size for a photographic Newtonian is the size of the fully illuminated field. In addition, some thought should be given to the amount of light loss which occurs at the corners of the photographic frame. The basic minimum requirement that we have used as our standard for many years is to fully illuminate the minor axis of the film format that is being used. In the case of most amateur instruments, the film format is 35 mm (24 × 36 mm) and it is necessary to fully illuminate a circle 1 inch in diameter. It is, of course, unnecessary to reiterate that 2 inch internal diameter (ID) low-profile focusers are necessary to obtain a fully illuminated 1 inch field and to keep the secondary size to a minimum. With apertures under ~12–14 inches, the size of the secondary is disproportionately large because of the large ratio of focuser height to the mirror diameter. For larger instruments with f/ratios greater than about f/7, or when using a Newtonian equipped with a field-correction system, it is possible to use 120 or 220 size films (and larger) and one of the following film formats: (4.5 × 6 cm), (6 × 6 cm), or (6 × 7 cm). The formulas derived below can be used to calculate the diagonal size that is required for any system and show the illumination across the entire film plane.

Looking at Fig. 2.13 below, we will define the following parameters: D is the diameter of the primary mirror; F is the focal length of the system; I is the diameter of the fully illuminated field; d is the minor axis of the desired secondary mirror; S is the separation between the central axis of the primary and the film plane. S is the sum of three other quantities: the radius of the outer surface of the telescope tube, the focuser height, and the depth of the camera body with its focuser adapter (or T-mount) attached (or, the distance from the focuser to the film plane). It is easy to derive the relationship that gives the size of the required secondary. The formula so derived is:

$$d = I + S[(D - I)/F]. \tag{2.4}$$

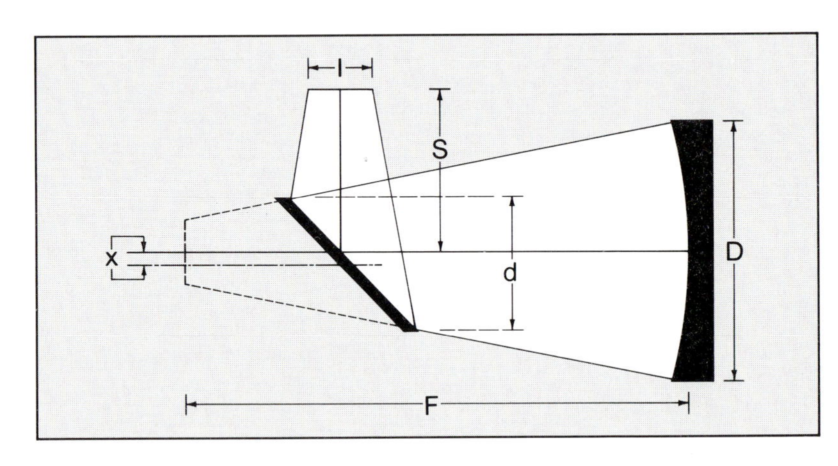

Fig. 2.13. Illustrates the optical layout of a Newtonian. Various basic parameters are labelled including: D, the primary diameter; F, the focal length; and d, the size of the minor axis of the secondary. Also labelled are parameters relating to: I, the calculation of the size of the fully illuminated field and x, the required offset for the secondary in such a system.

It is also possible, and instructive, to derive the 'inverse' of the above formula. By this, we are referring to the following problem: given a Newtonian system that is already constructed, what is the size of the fully illuminated field? Solving Eq (**2.4**) for I yields the following expression:

$$I = (Fd - SD)/(F - S). \qquad \textbf{(2.5)}$$

Figs 2.14 and 2.15 show nomograms based on Eq (**2.4**) and serve to illustrate the effects of varying the focuser height and the size of the fully illuminated field. These nomograms can be used to obtain an approximate value of the minor axis of the diagonal required and to see what sizes are available commercially. Each of these nomograms is based on the assumption that the telescope tube in use is 2 inches larger than the main

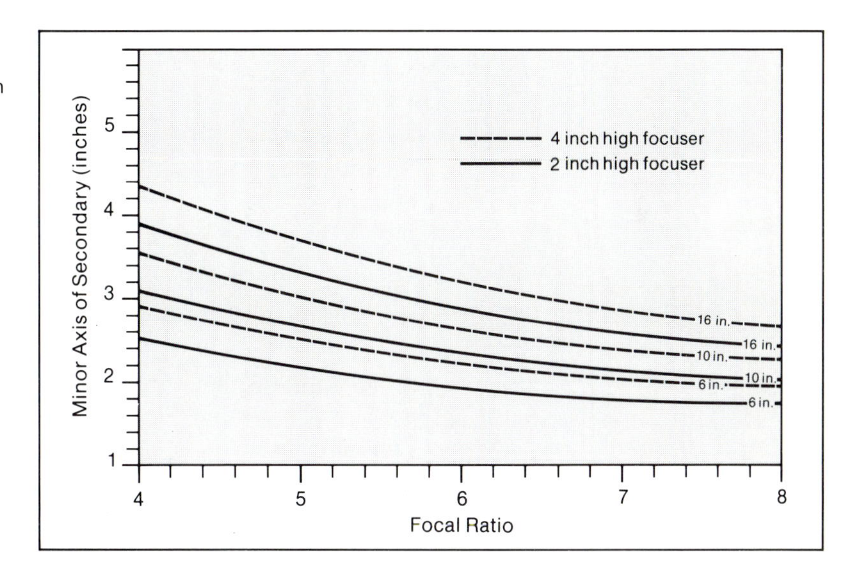

Fig. 2.14. Indicates the secondary size necessary to fully illuminate a 1 inch diameter field as a function of f/ratio for selected apertures. Two sets of curves are shown to illustrate the effect of focuser height on diagonal size. An SLR body is assumed to be used with the focuser.

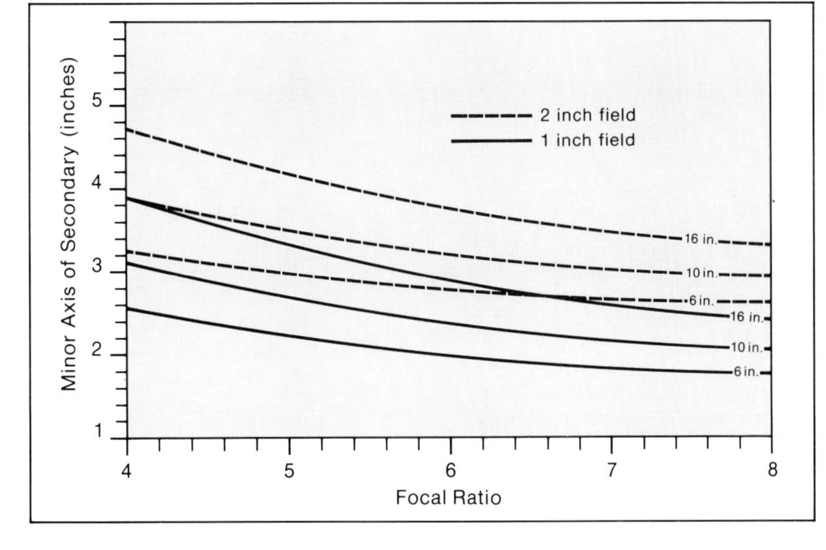

Fig. 2.15. Illustrates the increase in secondary size required to fully illuminate a 2 inch diameter field as opposed to a 1 inch diameter field for selected apertures. A 2 inch high focuser has been assumed to prepare these curves. An SLR body is assumed to be used with the focuser.

mirror diameter. For telescopes using larger or smaller tubes and/or focusers, Eq (**2.4**) should be used.

Fig. 2.13 shows another factor that can affect the illumination at the focal plane. Because the diagonal mirror is intersecting a cone of light that is gradually truncating, it can be seen that the secondary needs to be offset slightly from the center of the converging cone. The size of the required offset is largely dependent upon the inverse of the f/ratio (f) of the telescope; for slower instruments this offset can be ignored. The amount of the required offset is given exactly by the formula:

$$x = d(D - I)/(4F). \quad \textbf{(2.6)}$$

This can be reasonably approximated by the equation:

$$x = Dd/4F. \quad \textbf{(2.7)}$$

This value is usually less than 0.25 inch for virtually all systems in use by amateurs and most of the secondary supports on the market allow for enough adjustment to accommodate this offset.

These formulas (**2.4** and **2.5**) can be applied to a number of problems involving illumination at the focal plane. They can be used to calculate minimum sizes for baffle tubes in refractors or catadioptric systems. They can be used to calculate the minimum size secondary required in a catadioptric system, or they can be used to see the effects of different size focusing tubes and T-mounts on the size of the fully illuminated field. The vignetting caused by T-mounts and focuser tubes is quite severe owing to the nearness of these limiting apertures to the focal plane, as illustrated in Fig. 2.16, in almost all studies concerning the illumination across the focal plane, these important limiting factors have been ignored. In folded optical

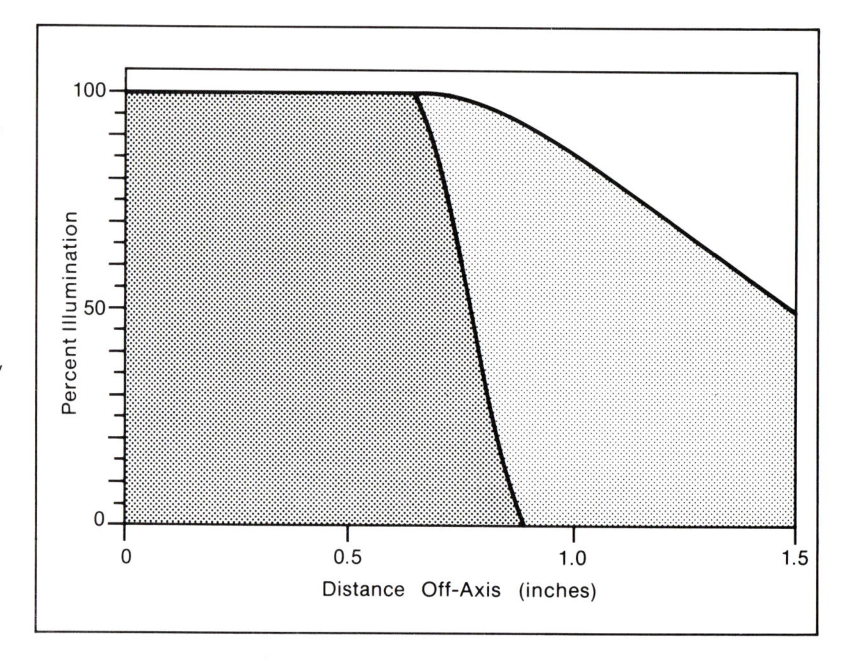

Fig. 2.16. Shows the vignetting effect of a common 'T-mount' on an otherwise well-designed 8-inch f/6.0 photographic system. The T-mount in question has a 1.50 inch diameter aperture and is situated 1.40 inches from the film plane. Heavily shaded area shows the effects of the *T*-mount while the lighter region shows the vignetting at the film plane without any *T*-mount.

systems, like the Cassegrain and its variations, it is important to consider the size of the secondary and the length and aperture of the central baffle tube from the standpoint of the size of the fully illuminated field and the elimination of stray light from the entire film plane.

The use of aperture stops

When many first choose an instrument for astrophotography, photographic speed is their prime concern. This, plus the thought of having a richest field telescope (RFT) that will yield splendid views of the night sky, often lures amateurs into purchasing Newtonians with f/ratios as fast as f/4. As the novelty of such a fast system wears off, the desire for less coma begins to take hold and a general dissatisfaction with the instrument (in terms of its photographic use) sets in. An instrument of f/5 produces a pretty reasonable field on 35 mm film, but the aberrations of faster paraboloids produce a very small usable photographic field.

At one point, we were using an f/4 instrument for photographing small galaxies and just using the center of the field. However, we were interested in using an instrument of that focal length to photograph a nebula that would nicely fill the field but that required a coma-free field over the entire 35 mm frame. The obvious solution was to use an aperture stop placed right on top of the mirror but somehow it seemed too easy. Later, with ray-trace programs, we could see that this was the right solution but (at that time) we had to use the trial-and-error method to see the effects. We stopped the 6 inch f/4 instrument down to a 4 inch f/6 and found that we had an instrument that could be used for several different functions by just adding or removing the aperture stop (Fig. 2.17).

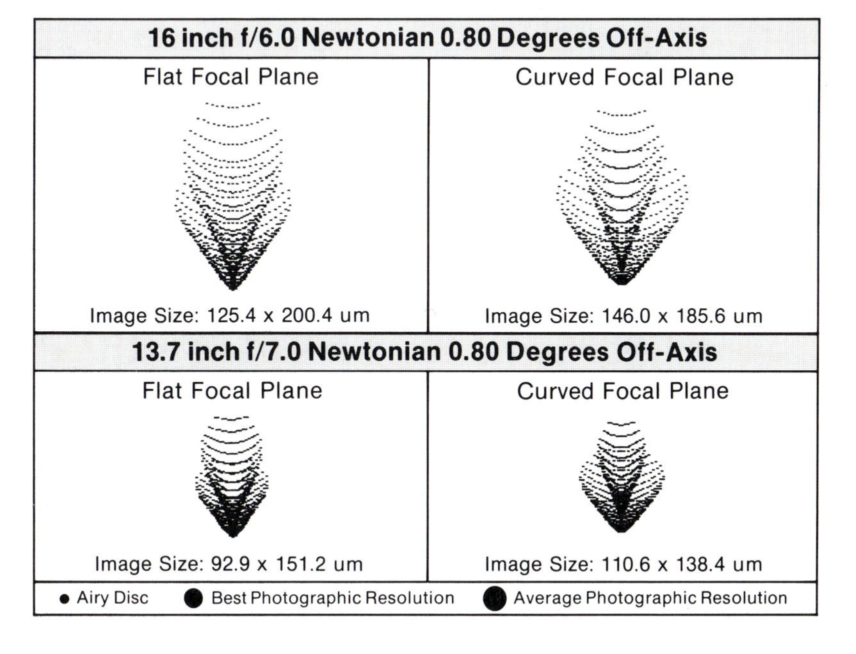

Fig. 2.17. Illustrates the effect on image size and structure 0.80 deg off-axis when a 16 inch f/6.0 Newtonian is stopped down to a 13.7 inch f/7.0 system. The gains in image size and structure would have been even more dramatic if an f/4 system had been stopped down to an f/5. Images are shown for both a flat focal plane and a curved focal plane.

As we used the stopped-down RFT for photographic purposes, we started getting the impression that more than just image quality had been improved by adding the aperture stop. When we compared negatives taken at f/4 and at f/6 it seemed fairly obvious that the field of the f/6 was more evenly illuminated. As we thought about the geometric effects of using an aperture stop, it became obvious that this was indeed an added benefit. We had decreased the diameter of the cone of light that struck the secondary and then passed through the aperture of the focuser and thereby decreased the amount of vignetting that occurred at each of these points. The effects of such an aperture stop are shown in Fig. 2.18.

The number of optical systems, other than the paraboloid, that can be improved by the use of an aperture stop (and where such a stop should go) is currently unknown to the authors. This would be an excellent study for someone with a reasonable ray-trace program. However, another obvious use for such a stop is with the extremely fast commercial Schmidt cameras that are on the market. These f/1.7 systems are simply too fast to be used for anything but photographs of hydrogen nebulae, using deep red filters. Even in dark skies with slow-color emulsions, the limiting exposure is on the order of 20 min and the contrast is very low. By stopping the system down to a reasonable f/2.5 or f/3 (some experimenting is required to find the optimum) one can gain image contrast and perhaps improve image quality and field illumination. Additional benefits of stopping down an f/1.7 system are that vignetting is improved and depth of field is increased, thus it should be easier to keep these units in focus.

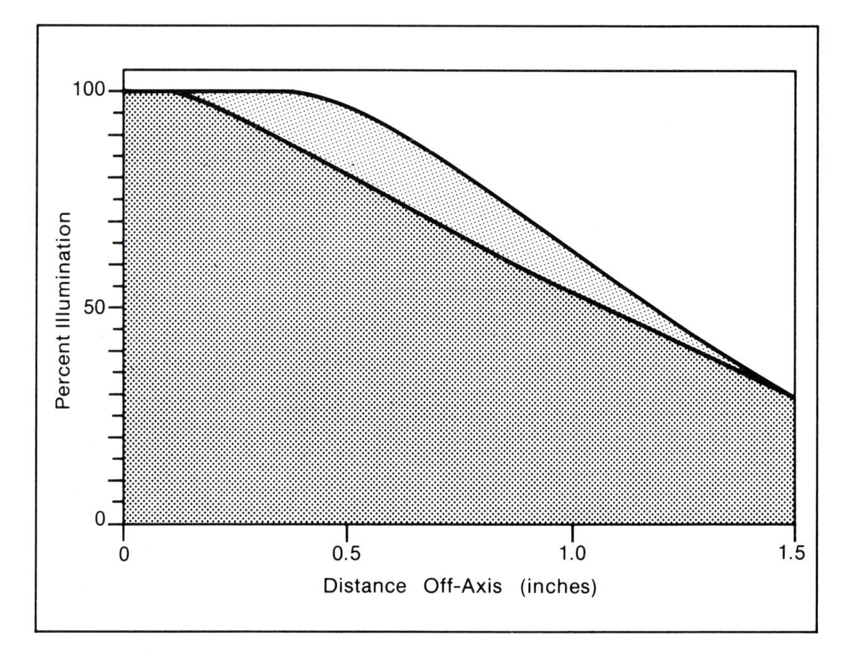

Fig. 2.18. Shows the effect on field illumination of stopping a minimally designed 6 inch f/5.0 system down to a 4.5-inch f/6.66 system. The upper curve (lightly shaded area) represents the gain that is obtained by stopping this system down. Notice that the 0.20 inch diameter fully illuminated field of the original system grows to a 0.80 inch diameter fully illuminated field when the system is stopped down.

Tube supports

Although a sufficiently thick-walled aluminium tube will provide adequate rigidity to maintain optical alignment during long exposures, added strength can be achieved with a well-designed tube cradle. The cradle should be constructed so as to span as much of the distance between the focusing mount and the main mirror mounting as is possible. By extending the cradle beyond what is normally seen in most commercial mounts, the probability of tube flexure is greatly diminished. If the cradle is one which is constructed so as to form rings at either end when the tube is secured in position, added support can be gained by adding four metal struts which span the distance between the two rings. Such struts serve to lock the tube rings into a more solid structure which then adds support to the telescope tube. On larger instruments, it may become necessary to add a third set of rings in the center of the cradle to prevent the tube from sagging in the middle.

Another important aspect of tube-cradle support applies to those designs without provisions for the rotation of the telescope tube. With these designs, the castings (or wooden structures, etc.) that link the tube to the cradle plate are usually made far too small. For minimal support in photographic applications, the linking structures should wrap one-third to halfway around the circumference of the tube and should also be securely bolted to the tube at three equidistant points. When these structures span less than about one-third of the tube circumference, the cradle cannot provide full support for the tube when placed in any of a number of extreme positions. As with rotating-tube assemblies, metal struts will give added strength to cradles of this type. The struts should be placed near the top of each end of the cradle structure to add the most support.

Tube strength and cradle supports are particularly important when using a separate guiding telescope. Because there is no way of knowing if the main instrument is flexing when guiding through a separate telescope, both tube assemblies and supports cannot be made too strong. In addition to rigid aluminium tubes (for both the photographic instrument and the guidescope), it is recommended that the cradle be constructed so that the tube is firmly secured to the saddle so that it cannot be rotated. With off-axis guiders, the situation is more relaxed since flexure can be monitored and compensated for during the exposure. In fact, the ability to rotate the tube is virtually a necessity when off-axis guiders are being used since eyepiece accessibility is critical to comfortable guiding.

Telescope accessories

Guidescopes

When purchasing, or constructing, a guidescope, a number of factors should enter into the selection of an objective. The basic facts should be obvious: the larger the aperture, the more guidestars are available and any given guidestar will appear brighter using a larger aperture instrument.

However, some obvious drawbacks also come to mind: larger instruments weigh more and thus create more stress when mounted and they are more cumbersome to handle and adjust. Practical guidescope apertures run from as small as 2.4 inches for small photographic instruments on up to 6 inches aperture for a photographic instrument of 16–24 inches aperture. An aperture of 3 inches is quite useful for most amateur applications using photographic instruments from 6 to 12 inches aperture. As long as the aperture of the main instrument is less than about 10 inches, such guidescopes are most satisfactory but, for telescopes with larger apertures, an off-axis guiding system is a more satisfactory approach to the guiding problem.

When considering a guidescope, there is one fact which must be kept foremost in the mind: the guidescope must be a rigid unit, all of its parts must be securely combined into a single solid structure. It is absolutely imperative that none of the optical parts of the telescope shift during exposures. Most of the popular commercial catadioptric units that are available today are not suitable for use as a guidescope because of the poor-quality focusing mechanism which causes the main mirror to shift continuously during long exposures. However, most could be made quite rigid by glueing the primary mirror in position with epoxy and attaching an independent focusing mount at the rear of the instrument.

The focal length of the guiding telescope should also be given some consideration although there are no rigid requirements on this point. The general rule of thumb states that the focal length of the guidescope should be at least as long as that of the main instrument. There is no real problem with this guideline, but it is very demanding and it can be relaxed. We have found no problems with systems using a guidescope having a focal length of 60–75 % of that of the main instrument, and a lower limit of 50 % is probably not out of line. However, the focal length of the guidescope is tied to the guiding magnification; the longer the focal length of the objective, the longer the focal length of the guiding eyepiece which can be used to obtain a useful guiding magnification (see discussion of this issue in Chapter 3).

Off-axis guiding units

For instruments larger than about 10 inches aperture, off-axis guiding is preferable to using a separate guidescope because of the extreme difficulties of constructing a flexure-proof system (including the financial aspects of such a project). The size, weight, and tooling necessary to construct a fully operational off-axis guider also make its use more reasonable on such instruments. However, a number of small off-axis guider units have been made available for smaller telescopes which can greatly speed up the time lapse between owning a telescope and obtaining good photographs through that same instrument. Of these, the units made by Lumicon provide very useful features in a small, lightweight guider which adds very little additional distance between the top of the focuser and the focal plane, thus having little effect upon the illumination across the photographic field. Any such units must have at least a 2 inch ID focuser tube in order to be useful for photography, and the telescope should be an f/6 system or slower to provide off-axis star images which are not overly distorted by coma and astigmatism.

An off-axis guider will solve a number of problems instantly, but it will also create some at the same time. Among those solved are the following:

1 the areas where differential flexure can occur are now isolated to the off-axis guider itself (buy or build a sturdy unit);
2 many more guidestars are now available for use since the guiding aperture will normally be greatly increased by the use of an off-axis guiding system;
3 the weight which the mounting has to support should be reduced considerably by the lack of a guidescope.

One problem which is created by the use of off-axis guiders is that the guiding eyepiece can no longer be situated at virtually any height in order to maximize comfort during long exposures. In fact, with Newtonians, one often ends up standing or sitting on a step ladder in order to reach the guiding eyepiece. There is also an increased possibility of ending up in some rather awkward positions. Rotating tube rings become almost a necessity, but it is then critical that the tube be balanced radially and that the balancing process be carried out with great care. It should be obvious that an off-axis guider is not a remedy for a poor-quality clock drive. In fact, if one is to be perched atop a ladder all night, an excellent clock drive sounds like an essential addition.

If flexure is not largely eliminated from an instrument with an off-axis guider, another problem that will not go away is that of obtaining accurate polar alignment. Since extremely accurate polar-alignment techniques require careful observation of the direction and magnitude of a star's drift, it can readily be seen that random star motions due to flexure can seriously interfere with the process of achieving accurate polar alignment. It is our experience that it is virtually impossible to achieve accurate polar alignment when using an off-axis guider in conjunction with a tube assembly known to have serious flexure problems.

You may also have discovered a major disadvantage of relying exclusively upon an off-axis guider: you cannot guide on an object through the off-axis guider and have that same object in the photograph (i.e. a comet). Because of this fact, it is necessary to have a well-built, flexure-free, telescope and have an auxiliary guidescope ready for use in the event of a visit by any of these spectacular voyagers. Thus, the off-axis guider cannot be adopted in place of taking the effort to build a solid instrument for astrophotography (for additional discussions of some of the possible problems associated with this careless approach to this field, see the discussions on flexure and comet photography, pp. 245–8).

Cameras for astrophotography

Although there are a number of specialized cameras designed for specific types of astronomical work (the cold camera is but one example), the great majority of work is done with ordinary 35 mm cameras used either with commercial lenses or coupled to the telescope. Given the great variety of available cameras, it would seem that choosing a camera for astronomical photography would be a confusing task at best. However, the fact is that, once one is familiar with the requirements for such a camera, the number of viable alternatives is considerably narrowed.

First a SLR camera is generally preferred for a number of reasons. A rangefinder camera is usable, but does not allow direct viewing through the photographic optical system. This means that it is more difficult to point and orient the photographic field. SLRs usually also have available a wider selection of lenses and accessories that enhance their potential for astronomical work. However, among the large number of SLR cameras available only a few will meet the requirements for astronomical applications. Some primary considerations include cold-weather performance, focusing systems, viewing prisms, and some minor mechanical considerations.

A major trend that has developed over the past few years in 35 mm camera technology is the growing emphasis on electronics. While everything from exposure measurement, focusing, shutter action and film advance is being driven electronically on more and more cameras, these cameras are becoming all but useless for most astronomical work. The reason for this is that the batteries being used to power all of these camera functions do not perform well in the cold. Once temperatures drop below about 40 °F, most camera functions will become impaired to some degree. If the shutter is driven electronically, cold weather will make it impossible to keep the shutter open for more than a few minutes before the current drain weakens the batteries and the shutter closes because of lack of power. Although some camera manufacturers offer special power supplies for cold-weather applications, these units do tend to be bulky and heavy. Therefore, it is advisable to select an old-fashioned mechanical action camera over one of the computerized models, for this application. However, some of these electronic models do have mechanical overrides which allow for limited functions without any battery power; such models are usable provided the mechanical override allows one to make 'bulb' or 'time' exposures with the camera. Even with a mechanical override, the battery-power issue may still cause problems for lunar and planetary work in cold-weather situations.

If, for whatever reason, an electronically powered camera is all that is available, the astrophotographer is left with only two options: to purchase the manufacturer's power supply or to build his own at considerable saving. Our experience with a very fine 6 cm × 7 cm electronically controlled camera serves as a good illustration of the second alternative. The camera in question, while optically superb and a very good all-round performer, consumed one $12.00 (US) battery for every two hours exposure time! In cold weather (below 40°F) the camera simply did not function (except possibly as a very expensive paper weight!). The solution, of course, was to replace the small and expensive 6 V battery that this camera used with a much larger power source of the same voltage (important: the voltage must be the same). In this case a wooden dowel of the same diameter as the original battery was cut to the appropriate length to fit in the camera's battery compartment. Tacks were put in both ends of the dowel (to serve as electrical contacts) and were wired to a 6 V lattern battery. The greater current flow and storage reserves of the lattern battery allow long exposures at very low temperatures. After two years of use, the battery is still going strong. Although this kind of solution is not possible with all cameras it serves as an example of what can be done, given a little ingenuity.

The standard focusing screens that are provided with nearly all off-the-shelf SLRs usually consist of a matte screen with a central split-image and/or microprism focusing aid. These are excellent for normal photography but are of little use for astronomical purposes. It is impossible to focus using a split screen or a microprism on a star or planet, or even on the moon. These focusing devices were designed for entirely different applications, and do not function well in these unique applications. Better arrangements include the use of either a plain matte screen or a clear screen with a fine matte central circle for focusing at the telescope. These screens are also much easier to use when aligning fields and are essential when trying for critical focus with the use of a magnifier. Although only a few camera models allow for screen interchangeability, several manufacturers offer a selection of different focusing screens and will substitute your preferred screen for the standard one at a local service center.

Interchangeable viewfinders are not an absolute necessity but they can add to a camera's versatility for astronomical work. With those cameras that will accept interchangeable viewfinders, the astrophotographer has the option of using either the standard pentaprism, a waist-level finder, or a special magnifying finder. If knife-edge focusing is used, or a magnifying eyepiece is attached for focusing, then the standard pentaprism will serve for most situations. The standard prism is particularly well suited to telescopic work, especially with a Newtonian, since viewing is similar to using an ordinary eyepiece.

The special magnifying viewfinders available for most 'system cameras' are the best alternative to knife-edge focusing. Most of these provide a 6 × magnification and allow one to see the entire field. They provide a considerable improvement over the use of magnifying attachments that provide only a 3 × enlargement of a small portion of the field. Unlike these snap-on magnifiers, the special magnifying viewfinders function very well with f/10 and slower systems. The disadvantages of these magnifying finders include high cost and the possibility of putting the user in inconvenient and uncomfortable positions while focusing.

Finally, any SLR to be used for astronomical work needs to incorporate a mirror lock-up mechanism and the largest lens-mount aperture that is available. A major source of shutter vibration in SLR cameras is due to mirror flop caused as the reflex mirror flips up out of the optical path before the shutter opens. The impact caused by this rapid action is most destructive for lunar and planetary work, and any work on a slightly instable mount. By locking the reflex mirror up prior to the exposure, the mirror-flop vibration is eliminated and one is left with only the vibration caused by the shutter itself, which is of much smaller intensity.

Since the lens-mounting flange on an SLR body is the primary bottleneck restricting light reaching the film plane, it is most important to select a camera with the greatest possible aperture at this critical point. Even with large-aperture, low-profile focusers, severe vignetting of the corners of the photographic field result from the restricted aperture at the lens mount. Surprisingly, the aperture at this point varies considerably between different camera brands. The advice here is to be aware of the differences and of the importance of this factor. If all other factors seem equal when comparing two or more cameras, lens-flange aperture could easily be the deciding factor.

To summarize, a good basic camera for astronomical work should be a single-lens reflex model and, probably, a mechanical model rather than an electronic one (for better cold-weather performance). The camera should either accept interchangeable screens or should allow the user a choice of screens when buying the camera. A fine matte screen is preferred; split-screen or microprism focusers are to be avoided. The standard pentaprism viewfinder should accept a magnifying eyepiece attachment unless one intends to use a knife edge for focusing. Although interchangeable viewfinders add considerably to the versatility of an astronomical camera, they are not seen as a necessity for excellent work. Finally, an astronomical camera should have the widest lens-mount aperture that is available and the provision for mirror lock-up is highly desirable for several applications.

3

Techniques at the telescope

Polar alignment

The need for accuracy

In most articles on astrophotography, polar alignment is often mentioned as being critical for obtaining good photographs. Many of these same articles avoid any mention of methods for obtaining accurate polar alignment. From this one begins to think that the ability to align an equatorial mounting on the celestial pole is an innate trait of astrophotographers. Various methods are employed by amateurs with portable mountings to obtain approximate alignment: special reticles are used in finders to offset from Polaris; setting circles are used to adjust the polar axis; and, sometimes, one just stands behind the mount and estimates where the pole is.

When polar alignment is achieved using any of the above methods, a star that is placed on the crosshair of a guiding eyepiece will quickly drift off the crosshair, either to the north or to the south, even when low power is used for guiding. Polar alignment which would be more than adequate for observing (even if one uses setting circles for observing) can only be considered as crude when it comes to guiding long exposures. When a telescope is aligned accurately for long-exposure astrophotography a guide star should drift less than 10 arc sec during a 2 h exposure (and preferably less than 5 arc sec).

Field rotation

When many amateurs first dabble in the mystic ways of long-exposure astrophotography, they often receive a number of unpleasant surprises. The first attempts of most are 'piggy-back' shots, using a normal or wide-angle lens. The telescope is aligned on the celestial pole before starting the exposure, the shutter is opened for 20–30 min, the declination drive is used to correct for declination (Dec.) drift and RA-drive controls are used to keep the star quite close to the crosshair. According to everything which they have read, that is all that is necessary. The declination slow motion is all that is needed to correct declination drift. The results of these early attempts are often disappointing and confusing and much of the advice given about curing the problem is usually inaccurate and muddies the subject even more.

If the images on many of these photographs appear like polar star trails, with the star that was used as the guidestar perfectly still and the rest of the field filled with star trails in small arcs, you have just encountered the effect known as 'field rotation'. The effect arises from the fact that the drift rate of stars in various areas of the sky is not uniform and the declination corrections only act to keep one star (the guidestar) stationary during the exposure. This can affect photographs taken with any focal length and it is caused by inadequate polar alignment.* Contrary to what is written in some articles, field rotation can be prevented quite easily by the proper and exact alignment of the polar axis of the telescope. When the number and size of the declination corrections made over the course of the exposure is very small, field rotation will be insignificant and undetectable in nearly all situations normally encountered.

Field rotation can be demonstrated quite easily by using a drawing such as that in Fig. 3.1. First, a transparent overlay (or a tracing made on vellum) is made of the lower portion of Fig. 3.1 (the points P′ and A′ and the ray passing through those points) and this then is set on Fig. 3.1, with the origin P′ set in register over point F and point A′ in register over point A. Let the point P represent the true north celestial pole, then all stars will appear to circle around this point (use your imagination!). Then, let F represent the point in the sky where the telescope's polar axis is pointed. The driven telescope will follow along arcs that are concentric to F while the stars will move along arcs that are concentric to P. By following the instructions below it will be possible to demonstrate the effects of poor polar alignment and produce an example of field rotation:

1 Lay the transparency over Fig. 3.1 with point P′ directly over point F and A′ over A.
2 Using a fine point pen, trace the positions of 6–10 'field stars' onto the overlay and circle them for reference.
3 Now, use a pen point to pin the overlay to 3.1 at point P and rotate the overlay about P for one major division of the arc passing through A.
4 Next, shift the pen point to pin the overlay to 3.1 at point F and rotate the overlay until point A is under the line joining P′ and A′.
5 By this time, A will have slid below the arc corresponding to a line of constant declination and you will need to make a declination correction.
6 This 'declination correction' *must* be done by sliding the overlay up along the line A′P′ so that the 'guide star' (A′) is back on the declination arc and the distance between the line A′P′ and the point F remains constant.

* Differential refraction can cause trailing of stars which are near the horizon. This effect is usually encountered when one tries to photograph comets which are only a few degrees above the horizon. This effect should not be confused with field rotation.

7 Now, carefully retrace the positions of the 'field stars' that are marked on the overlay.
8 Repeat steps 3–7 five or six times.

After going through the above procedure you will see that the 'field stars' have formed short arcs concentric to A′ and are becoming longer as the distance from A′ increases. If you repeat this process for more than 6 hours of RA you will also see that the 'stars' will begin to retrace their arcs, and this process will repeat every 6 hours.

From this approximation and from observing Fig. 3.1 it is easy to see

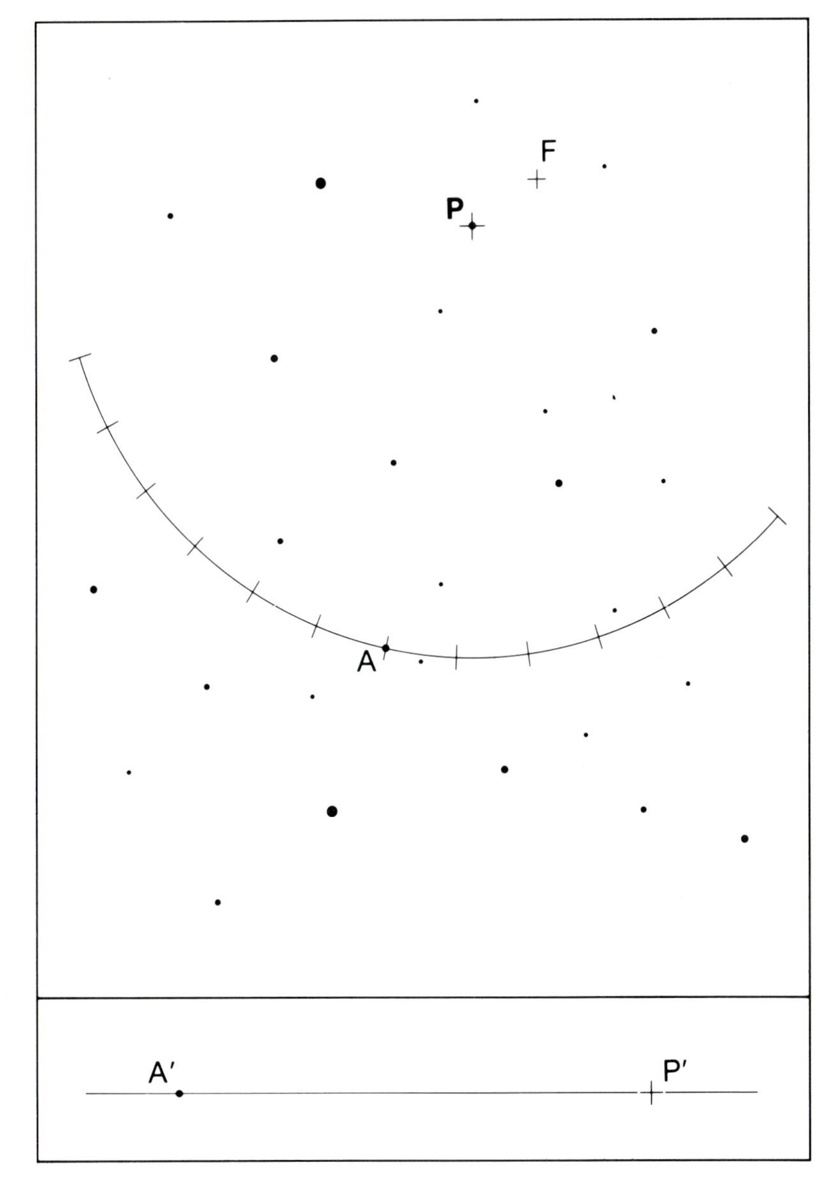

Fig. 3.1. This figure provides the reader with a means of illustrating field rotation when used in conjunction with the instructions that are presented in the text.

that declination drift will vary in different areas of the sky. Along a line joining P and P', the light and dark lines will be parallel and there will be no declination drift. However, along a line perpendicular to PP', these lines will diverge at a maximum rate and declination drift will be at a maximum.

At this point, you probably understand why proper polar alignment is so important and you can also see that declination corrections do not solve the problem created by imperfect polar alignment. Very accurate polar alignment will minimize declination drift (but cannot totally eliminate it because of refraction), and reduce field rotation below the point where it can be recognized in virtually any photograph regardless of the extent of the photographic field, focal length, or the amount of offset from the guidestar (exposures of extreme duration (4 + hours) may still show 'field rotation' effects when wide-field cameras are used). We will now describe a polar-alignment method which can be used to align a portable instrument on the celestial pole with extreme accuracy with less than an hour's work.

Polar alignment for photography

Unlike some of the less precise methods which require accurate setting circles and a star catalogue which gives star positions for the current year, the method which we have adopted requires no special equipment beyond the basic equatorial mounting, with a telescope and a guiding eyepiece. We were introduced to the basics of this method in the early 1960s by R.T. Little and later found references to similar methods in Sidgewick's classic work. Roth's work also contains a reference to an 1897 article by Scheiner which describes the basic method. As we have used the method over the years, we have made refinements which will allow one to achieve the highest degree of accuracy in aligning on the mean refracted celestial pole. Using this method, it is possible to routinely take hour-long exposures using focal lengths up to 200 mm without making a single declination correction. With such accuracy, and with a very accurate drive, one can do piggy-back photography with a 135 mm or 200 mm lens without even guiding!

Before we get to the actual procedure, it will be necessary to define a term that we coined when adding refinements to the method that allows precise alignment on the 'mean refracted pole' for a given latitude. This is the 'mean photographic declination' and it varies with the latitude of the observer. This corresponds to the average declination that one would normally expect to be taking photographs. The normal lower declination limit for photography is about 15 deg above the southern horizon (for workers in the north hemisphere), while the normal upper declination limit is about +80 deg declination (again for the northern hemisphere). For an observer at 35 deg north latitude, the lower declination limit would be around 35 deg south, which would mean that our 'mean photographic declination' is near +22.5 deg declination.

To begin, set up the telescope so that the polar axis is pointed close to north by sighting along the axis. Next, locate a star that is near the meridian and near what we have termed the 'mean photographic declination', and center that star on the crosshairs using the magnification that you would normally use for guiding. As you watch the star, it will drift off the crosshairs towards either the north or the south; this drift will tell you what to do to improve the polar alignment. If the star drifts to the north, the polar axis is too far to the west; if the star drifts to the south, the polar axis is too far to the east. Make an azimuth shift in the appropriate direction, recenter the guide star, and watch again. Each time you make a correction, the drift will slow. After the drift has been slowed to the point where several minutes pass before the star leaves the crosshair, your polar axis azimuth is now approaching that of the celestial pole. Before adjusting the azimuth any further, go on to the elevation-adjustment step since the adjustment of the elevation will undoubtedly disturb the azimuth setting and because the final positioning of either axis cannot be done properly with the other axis improperly positioned.*

Now, locate a star that is near either the eastern horizon or the western horizon, and at least 12–15 deg above the horizon at an hour angle close to 6 h. Center this star on the crosshair and again observe the drift. For a star near the eastern horizon, drift to the north means that the polar axis is too high, while drift to the south means that the polar axis is too low. If the star is on the western horizon, reverse these directions. Adjust the elevation of the polar axis, recenter the guidestar and watch for the drift. Repeat this process a number of times until the drift slows to the point where 5–10 min are required for the star to leave the crosshair. Once the altitude is precisely adjusted, go back and make the final azimuth adjustments to the same level of accuracy.*

Using this technique, the magnification that is used while aligning (and the patience of the observer) will determine the ultimate accuracy of the final alignment. We have found it possible to reduce declination drift over much of the sky to about 1–3 arc sec in 15–20 mins. The technique may sound time consuming, which it will be the first few times, but it can be carried out in 45–60 min if one starts from scratch, and there are a number of possible shortcuts that can help to reduce the required time. The one disadvantage in using these shortcuts is that one needs to return to the same observing site each time.

If you use a site which has a clear view at a horizon that is 1–2 miles away, you can make a drastic cut in polar-alignment time, provided that the mount can be set up in very nearly the same place each time and provided that you can identify distinctive features on that horizon. First, set up the scope for the night's work and align on the pole very carefully. Then, the next day, set the declination axis exactly level and turn the scope in declination until the optical axis of the main telescope intersects with the hozorizon and make a sketch showing the position of this intersection point and a number of identifiable objects in the near vicinity. Finally, place a camera on the telescope and photograph the horizon. After the photograph

* For southern hemisphere use, the instructions presented here must be reversed.

is processed and printed, mark the position where the telescope's optical axis intersected the horizon (which should theoretically be on the north–south line running through your instrument). Next time you go to that site, set up in the same location and realign the optical axis on the same point by shifting the azimuth of the mount with the declination axis perfectly level. You will still need to do some minor adjustments each time you set up, but the time required will be reduced greatly. Each time you set up and align the scope at that site, repeat the process of marking the photograph at the point the optical axis intersects the horizon. After repeating this process a number of times, the points will cluster about one azimuth on the photograph. Using this information, you can set up the telescope in daylight and be able to photograph an early evening object (e.g. a comet) without the need of spending time on polar alignment while the object slips towards the western horizon.

If you repeatedly use the same site but cannot use a distant horizon as a reference, you can still cut the time required for polar alignment. Begin by always reassembling your mounting in exactly the same manner. If your pedestal consists of a metal pier with removable legs, you can mark the legs and the pier with codes to allow you always to put the legs back in the same position. As you have already seen, each leg of the mount must be equipped with altitude-adjustment screws. These should be as large as possible and the bottom of each screw should be filed or machined flat. In addition, each bolt should be equipped with a locking nut so that it will not 'slip' when the mount is moved to and from the site. Now, a way must be found that will allow the equatorial head to be positioned repeatedly on the pier close to the same orientation. The simplest solution to this is to put two index marks on the pier and the equatorial head, about 120 deg apart, to mark the orientation.

The final problem that has to be solved is how to set the mount on the ground in exactly the same position time after time. If the site has a concrete or asphalt area that the scope can be set up on, the problem is solved. If the site is just a flat dirt area, it will be necessary to install some permanent solid support beneath each of the pedestal legs. One solution that we have used at several sites 'off the beaten track' is to set bricks into concrete with the surfaces flush to the ground. Then, after the telescope has been aligned, marks are placed on each of the bricks indicating the position of each of the leveling screws. When these wilderness sites are not in use, we simply cover the bricks with some dirt and grass restoring the site to a natural and undisturbed appearance.

Polar alignment: a photographic method

If the telescope is located at a permanent site and set on a permanent mounting, the polar alignment can be fine tuned for long-exposure photography using the photographic method of E.S. King. This process can be repeated a number of times over weeks or months to refine the alignment to any degree desired. This process is carried out using a star field near the celestial pole. However, it is important to realize that the apparent position of the refracted pole will vary in both altitude and azimuth as one observes stars at varying declinations and hour angles. Because of this, there will still be declination drift in various areas of the sky which can never be eliminated. Thus, 'perfect polar alignment' is an illusion and to pursue it is to chase an apparition.

To use King's method, set the RA at the 00 hour angle with the telescope on the east side of the mounting and select a star field centered on the celestial pole, after carefully adjusting the driving rate of the telescope. Place a camera on the telescope (photographing through the main instrument) and make an exposure of 10–15 mins. Then turn off the drive *with the camera shutter still open* and allow the stars to trail for a few minutes. Any polar-alignment errors will cause the star images to trail even when the driving rate has been perfectly adjusted. These undriven star trails are added for the purpose of indicating the direction in which the driven trails are moving: the driven trails will all be parallel to one another while the diurnal trails will circle around the celestial pole which should be in the center of the photographic field (see Fig. 3.2).

After processing the film, examine it with the emulsion side away from the eye and held as it was when exposed in the instrument (with respect to the horizon); the following information can be learned from the direction of the driven star trails (assuming that the instrument is a refracting system):

1 if the driven trails show an upward drift, the polar axis is directed to the west of the pole;
2 if the driven trails show a downward drift, the polar axis is directed to the east of the pole;
3 if the driven trails show a drift to the right, the polar axis is directed above the pole;
4 if the driven trails show a drift to the left, the polar axis is directed below the pole.

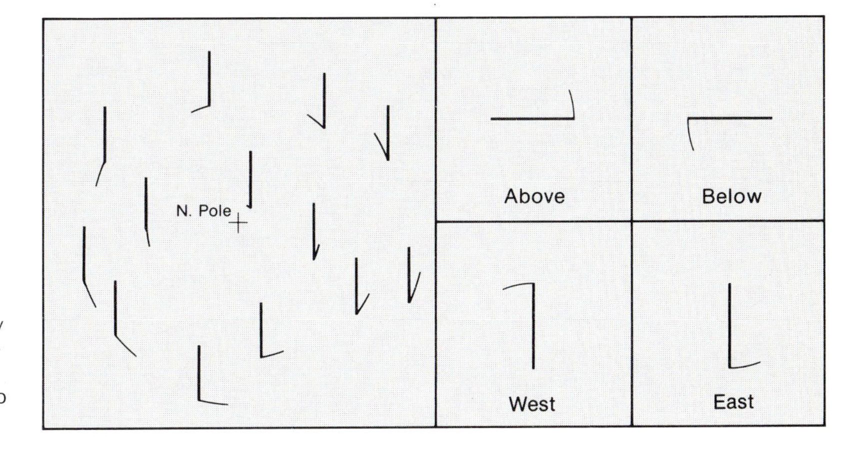

Fig. 3.2. An illustration of the results that one might obtain from a photographic polar-alignment test. The parallel driven star trails provide the information necessary to correct the instrument's polar alignment while the undriven trails that circle the pole indicate the end of the exposure and thus provide information needed to determine the direction of drift. The key at the right shows where the axis of the mounting is pointed with respect to the actual pole.

If the film is viewed as above, but rotated one-quarter of a turn in the clockwise direction, the direction of the guided trails will point to the direction to which the polar axis must be moved to correct the error. In general, the trails will point along some diagonal showing that the axis is off in a combination of directions. If the polar axis can be shifted in an exact manner in both azimuth and elevation, it will be possible to determine just how far the polar axis must be moved to correct the alignment error. To do this, make a correction in the required direction and record the exact amount of the shift (e.g. three turns of a given screw), then make another photographic test using the same combination of exposures as was used in the first exposure. By comparing the two photographs and by measuring the trails on the second photograph, one can derive a close approximation to the amount of correction required.

That which does not kill us makes us stronger.
Nietzsche

Guiding

The possession of an excellent-quality clock drive greatly reduces the strain of guiding long exposures. The fatigue that is induced while struggling to keep a faint star on the crosshair is a major contributor to badly guided photographs. When the observer is tired, the probability of making serious mistakes is greatly increased. One of the most common mistakes induced by fatigue is the error of making corrections in the wrong direction because of forgetting the correct directions. Here we discuss the various aspects of guiding in the hope of reducing this fatigue.

One of the major factors in guiding is the guidescope itself. We have already discussed the primary requirements which need to be considered in selecting a guidescope (pp. 59–61) and the reader is advised to review the information presented in that section.

The guiding magnification is an important area to consider. If the clock drive is of excellent quality, a magnification equal to twice the camera's focal length in inches ($2F$) or three times the focal length ($3F$) will be found to be adequate. If a magnification of $2F$ is used with even the best clock drives, meticulous care must be taken to be certain that the star does not move off the crosshairs during the exposure. For this reason, a magnification of $3F$ will be found to be more relaxing during an evening's work. If the drive imposes a number of erratic errors on the periodic ones, this magnification range will be completely inadequate, and magnifications from $5F$ to $8F$ will be required. The reason for the increasing magnification when dealing with a large erratic error is to allow one to respond to the sudden changes in the driving rate; when a star begins to race away from the crosshairs, there is a response time required before the photographer can react and the increased magnification allows one to more quickly react to a potential disaster.

If your drive is in the excellent class, you will probably be able to employ a $2\times$ or $3\times$ barlow with your guiding eyepiece to obtain long eye relief at the required magnification. However, if you should wish to photograph using a 2 × or 3 × teleconverter in front of the camera, you will again find yourself in need of obtaining magnifications in the $4F$ to $10F$ range. Such magnifications create a number of inconvenient and potentially serious problems: if short focal length eyepieces are used, eye relief is virtually non existent, and even a 3 × barlow provides little help since an 8 mm focal length eyepiece is still required to yield $9F$ (assuming the focal length of the main instrument and the guidescope to be equal). Our experience has shown that 16–25 mm focal length eyepieces are the most suitable for guiding since they offer comfortable eye relief and thus allow for more freedom of movement during the exposure.

In order to obtain these high powers, a different type of guiding eyepiece is needed. One that will allow the use of a 16 mm or a 25 mm focal length eyepiece with long eye relief and still yield high magnification and excellent image quality. Our solution was to replace the simple eyepiece with a microscope (Fig. 3.3). Unfortunately, coming up with this idea was easy compared with the problem of calculating the magnification of the combined telescope–microscope system. The use of very high magnifications for guiding does make it more difficult to guide on faint stars and the only real solution is to use a better-quality drive. The difference between guiding at $10F$ with a poor drive and guiding at $3F$ with an excellent drive is like the difference between day and night. However, assuming that one does have a need for a high-quality eyepiece that has long eye relief and a very short effective focal length, we will proceed with the description of such a system.

In its simplest form, the total magnification of the system is given by:

$$M_s = (1.39FML)/f, \tag{3.1}$$

where

f = focal length of eyepiece in millimeters,
F = focal length of guidescope objective in inches,
L = separation between the microscope objective and the eyepiece in millimeters,
M = magnification of the microscope objective (i.e. 5 ×, 10 ×, etc. is standard labeling).

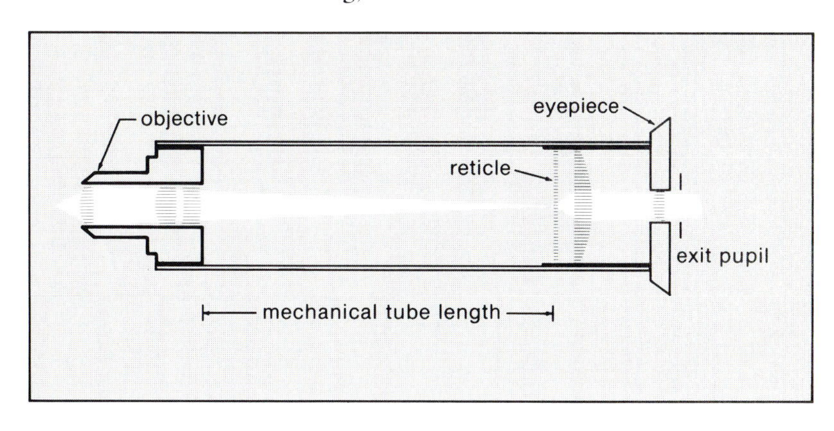

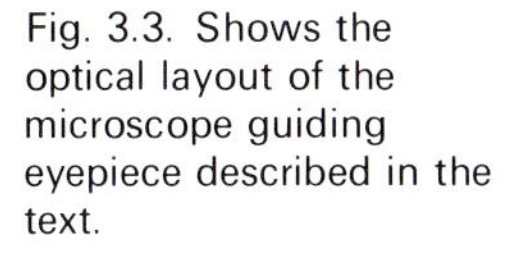
Fig. 3.3. Shows the optical layout of the microscope guiding eyepiece described in the text.

An 'eyepiece' such as this provides an excellent means of achieving very high power while allowing one to retain the luxury of long eye relief. This is also an excellent-quality observing eyepiece when high powers are required. For planetary viewing when the seeing is essentially perfect, or for viewing diffraction disks to make seeing estimates, or even for trying to split those close doubles when the seeing is fantastic, it is hard to find a higher-quality eyepiece than one using a good-quality microscope objective in conjunction with a 20 mm or 25 mm orthoscopic eyepiece.

The choice of a guidestar can also be important to the outcome of a photograph. A double star or one star in a field of stars of nearly equal brightness is a bad choice if you are at all tired because of the probability of forgetting which star is which! With a guidestar that is reasonably bright, it is convenient to align the crosshairs along RA and Dec. so that the guidestar moves parallel to the crosshair lines. However, when using a faint guidestar, this method is unsatisfactory and it is much wiser to guide with the crosshair lines oriented at 45 deg angles to the RA and Dec. lines. Faint guidestars should be avoided as much as possible when tired as they always seem to fade away slowly during the course of the exposure. This phenomenon is actually the result of continuous staring at the crosshairs causing the eye receptors (rods in this case) to fatigue and desensitize. It is a good idea to look up and around fairly often to avoid such eye fatigue. If you know that you will need to use guidestars that are 2–4 deg away from the object being photographed, you should spend some extra time on your polar alignment as field rotation will ruin such photographs if the total declination corrections exceed a few arc sec.

It is a poor practice to guide on out-of-focus star images unless one is guiding a piggy-back photograph using a relatively short f.l. lens. This does make guiding easier (on bright stars), but it sacrifices accuracy for convenience by making errors less perceivable in the guiding eyepiece. This practice also introduces the problem of parallax (the star image appears to move around behind the crosshair as you move your head around) which will add to the guiding errors that normally occur during the course of any exposure.

The ability to guide accurately can also be influenced by the choice of reticle. Fig. 3.4 shows nine different reticle configurations that are commonly found in guiding eyepieces. Of these designs, A, B, and C are quite satisfactory for guiding eyepieces while D–I are not at all satisfactory. The reticles in styles A–C require fine lines and reasonably low magnification as well, so as to not lose the guidestar behind the lines. For comfort and ease of memory recall (e.g. which way is up?) a guiding reticle should divide the field up into quadrants and, most important, it should mark the center of the field in some manner. If we had a choice, we would prefer to use the reticle shown as style B over any of the others. Styles D, E, and G do not mark the center and thus make accurate guiding impossible. Style F does mark the center but the lack of other references in the field makes it unsuitable for use. Styles H and I (and also G to some extent) are just too complex and cluttered for easy use over many hours. Some readers may not agree with us in these evaluations, and some may even successfully use one of the designs that we have just discarded. However, we have tried many of these and believe that most astrophotographers would find the switch to style B a welcome one.

Another area of interest when constructing, buying or modifying a guiding eyepiece is the ease of use and the possible eyestrain that can result from hours of use. If stray light illuminates much of the field and the edges of the field are bright compared with the reticle, eyestrain will develop quickly (always examine guiding eyepieces in darkness and wait for your eyes to fully adapt before rendering any opinions). This problem usually can be solved with little effort by painting the entire rim of the reticle with silver paint except for a small area where the light enters the glass and then covering the silver paint with a layer of black paint and placing a small field stop on top of the reticle. Eyestrain can also be reduced by providing more contrast between the colors of the reticle and the guidestar. We normally use a red light-emitting diode (LED) source to illuminate the crosshair but a green one would work as well (there are very few deep red stars, and no green ones). Another trick to help ease eyestrain is to keep the illumination

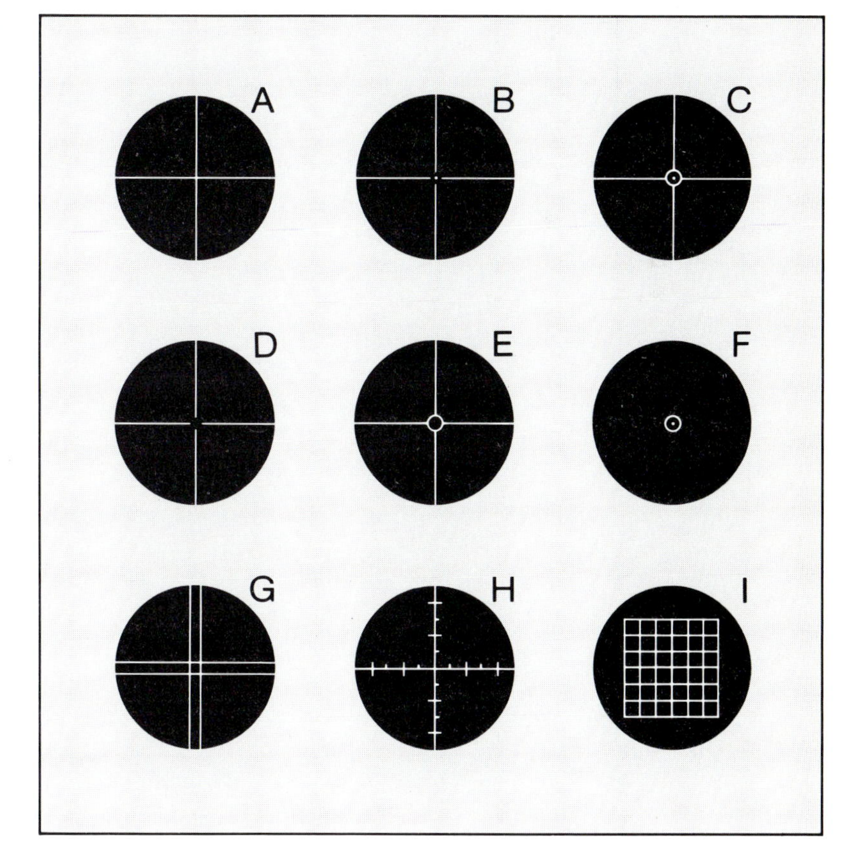

Fig. 3.4. Nine commonly used reticle designs are shown. The merits and problems of each are discussed in the text. Designs A and B have distinct advantages over the remaining styles.

of the reticle down so that it does not overwhelm the guidestar: the reticle should always be fainter than the guidestar, adjust its brightness for maximum comfort. With some units one may not be able to dim the reticle enough to use faint guide stars, it is a simple matter to replace the rheostat with one of greater resistance.

Other techniques that can be of help in avoiding fatigue include looking up and inspecting the sky and the night scenery as much as one can safely afford in order to provide relief from staring at the guidestar and the reticle. You may well be surprised by all the things that you can see on a nice dark night (it is a real surprise to look up and find a coyote or a skunk sitting calmly next to you in the dark of the night . . . don't panic and don't move!). If you cannot afford the time to inspect the night's wonders, move your eyes around the field of the eyepiece to help ward off the fatigue that develops. If your drive is of really superb quality, you can even afford to get up and stretch or walk around briefly if you get too stiff (don't gamble if you aren't sure of your drive!). During a particularly 'long' evening, one of the present authors actually fell asleep for 5–10 min and woke up as his head bumped the guiding eyepiece; the photograph came out just fine!

Comfort and guiding

We have mentioned fatigue several times as a factor which can affect one's ability to guide and discussed a few ways to avoid tiring the eyes. Physical and mental fatigue, caused by the late hours and/or the cold, also drains energy and increases the probability of making errors that may ruin the photograph. It is essential to be comfortable during the course of any exposure, and comfort implies warmth. Before starting an exposure, spend some time evaluating your position at the telescope: will you be able to stay in that position comfortably for an hour and will you still be nice and warm in an hour? An adjustable guiding stool or chair can help insure a reasonably comfortable sitting position. Sitting on a layer of ensulite can help to keep the cold of the guiding seat from penetrating through layers of clothing.

During the past decade, a number of new insulating materials have been developed that can be used in warm clothing. Down and Holofil clothes are quite warm and the new synthetic materials do not require the special care which down clothes need. The new polypropylene undergarments offer additional protection against the cold. Gloves usually pose the toughest problem for working in the cold at the telescope. It is necessary to have warm gloves which will allow the telescope controls to be operated without any incumbrance. Mittens and heavy 'ski gloves' are too clumsy for this purpose, and wool or closely knit acrylic gloves are to be preferred.

Focusing

Focusing is often approached in a carefree manner by many astrophotography enthusiasts. A 35 mm camera body is snapped into position, a semi-bright star is placed on the ground glass and the camera is focused in less than a minute. A careful practitioner will repeat this process several times during an evening's work. If this focusing process were repeated after each photograph and the diameters of the faint star images measured with a microscope, a great deal of variability would undoubtedly be found. With excellent eyesight, it might be possible to pass this exacting test. However we will proceed on the assumption that perfect focus cannot be achieved using the above technique and assume that one is interested in achieving a perfectly focused photograph every time.

How can focus be improved? If you prefer to continue focusing using the ground glass, it should be replaced with the finest texture screen that is available for the SLR in use. If screens cannot be interchanged in the model that you are using, you may need to change to a different camera. Then, when focusing, the star should be set about 6 mm off center to help insure good focus over the entire field (with reflectors it is far better for the center of the field to be inside of the true focus by 0.001–0.003 inches than it is for the center to be outside of the true focus because of the inherent field curvature of the optical system). Many SLR manufacturers make an attachment that snaps on the viewing eyepiece and which magnifies the image 2 × or 3 ×. These attachments can be very helpful and may well be the *easiest* solution to the problem, however the best solution to any focusing problem is to use a knife-edge focuser. A camera that is focused very carefully with a ground glass will usually be fairly well in focus, while a camera that has been focused with a well-made knife-edge focuser is always in perfect focus!

Knife-edge focusing

Focusing with a knife edge is an extremely accurate and sensitive method. The principle is identical to that of the Foucault Test when used with a collimated beam of light: as the knife edge cuts the focused beam of light at the focus point, the illuminated disk of the mirror will 'black out' instantly. When the knife edge is inside of the focus point, the dark shadow will appear to move across the mirror from the same direction as the knife edge. Outside of the focus point, the shadow will appear to move across the mirror from the direction opposite that of the knife edge.

One can always build a knife-edge focuser to use at the back of the camera before putting in the film and starting a night's work. One problem with this is that the focuser must be of a type that can be locked and maintain a position for an entire evening or even for several nights without moving as much as 50 μm (0.002 inches). Another problem is the fact that the telescope tube may expand or contract by more than the allowable focus tolerance. If two or more camera bodies of the same make are available, one can be used for focusing while the other is used for photography. We have found that camera bodies of the same model and manufactured by the same company are virtually identical and can be interchanged in this manner without any problem. One should check the depths of all camera bodies in use however just to be certain that the body depth falls within the required focus tolerance for your system. The best approach is simply to construct a knife-edge focusing device that can be used in place of the camera during a night's work at the telescope.

Such a device is easy to construct and can be made from available materials if a lathe is not available. The basic requirements that must be met by a stand-alone focusing device are:

1 the plane of the knife edge must be parallel to the focal plane of the telescope;
2 the height of the device must be adjustable in order to allow for initial focusing;
3 it must be possible to lock the device into position after initial focusing.

The design of a knife-edge focuser can be quite simple and does not require the use of micrometers to build, as most articles in journals would seem to indicate. In fact, we managed to find standard polyvinylchloride (PVC) piping pieces that can be assembled to make a knife-edge focuser for just a few dollars! After making a knife-edge focusing device, one simply focuses the camera using a knife edge at the film plane and then replaces the camera with the device and adjusts the device until its focal plane matches that of the camera. Once this initial adjustment is completed, the focusing device is locked into a permanent position.

It is quite easy to use a knife edge for focusing once you get used to it. Choose a star of magnitude 1 or 2 which is well above the horizon in order to avoid seeing or scintillation problems that occur near the horizon. If the seeing is good to excellent you should have no problem achieving perfect focus in just a few minutes. As the seeing deteriorates, you will find it more difficult to determine the exact focus; this should tell you something about what a star will look like on the film as the seeing changes. When you find it extremely difficult to tell where the exact focus is with the knife edge you should change your evening plans to accommodate the bad seeing: sit back with a little wine and watch the stars twinkle!

To focus for an evening's work, place the star about 5–6 mm from the center of the field (with a 35 mm camera) in order to pick a focus that will be optimum for most of the 35 mm frame, set the knife edge close to the star and move your eye up close to the knife edge (within ~12 mm). Now, slowly move the knife edge towards the star until it starts to intercept the light from the star. Then, as you **slowly** move the knife edge into the cone of light, watch the shadow cover the mirror. If the shadow moves in the same direction as the knife edge, you are inside the true focus. If the shadow moves in the opposite direction, you are outside of the true focus. The closer the knife edge approaches the actual focus point, the more difficult it will become to detect the direction of shadow motion; it will thus require more care to move the knife edge **slowly** into the beam of light. When the knife edge is at the exact focus point, it will become impossible to detect any direction of motion and the entire mirror will black out instantaneously.

We have used several different knife-edge designs in our work and all have worked well. A knife edge of metal can be used, but this can damage a camera back or a plexiglass cold camera plug. A better design is one that uses a thin sheet of plexiglass and a strip of metallic foil sensing tape like that used on reel-to-reel tapes as end-of-tape markers. Another alternative to a knife edge is to use a ronchi grating in its place. Although some prefer the ronchi grating; our preference is for the knife edge. The following section goes into more detail on the considerations that must be given to the construction of a knife edge that is to be used at the back of a 35 mm camera.

Addendum on focusing

When planning to build and use a knife-edge focuser, considerable thought should be given to the way that film is supported or held in the camera. In the case of the 35 mm camera, paying attention to the camera's design can avert considerable disappointment.

When looking at the back of a camera, with the back open, one will see a number of features relating to the support of the film. The most prominent structure will be the rectangular aperture, which allows light to reach the film, and which will be closed by the focal plane shutter (if your camera uses one). At the top and bottom of this aperture, you will find two raised 'rails' which are approximately 2 mm wide and machined. Upon close examination, or measurement, it will be found that the outer set of rails is several thousandths of an inch higher than the inner set. On the inside of the camera back is a flat plate which, when the back is in place, is pressed against the outer set of rails by springs.

When film is loaded in the camera, the film sits between the two outer rails. The flat plate presses against the outer rails and the film lies flat against this plate. The gap between the back plate and the inner rails is just slightly more than the film thickness (so the film can move freely) but not large enough to allow the film to buckle.

In making a knife-edge device to use at the back of the camera, this film-support mechanism must be considered. If the device is constructed to sit on the inner rails, the film plane will lie just slightly outside the actual focus. If the device is constructed to sit on the outer rails, the knife edge must sit at just the right distance 'below' the rails to match the position where the film surface will be when there is film in the camera. Since most of the roll films used in astronomical photography are close to 0.11–0.16 mm thick, the knife edge must sit this distance below the outer rails.

One of the simplest designs for a knife-edge focuser that is to be used at the back of a camera employs a thin sheet of glass or plexiglas which rests on the outer rails. The knife edge is then attached to this transparent support. The knife edge can simply be a slice of film, or a piece of shim metal stock can be used as a more permanent knife edge. With any knife edge, it is important that the edge is not beveled in a way that would shift the cutting edge closer to the transparent plate. With a shim stock knife edge this can be done by putting a slight bevel on the metal with a knife-sharpening tool and then mounting the shim stock with the bevel facing the glass plate.

Focusing tolerances

With all cameras and imaging systems, there exists a range of distance that the focus plane can be moved without making a visible change in the image that is formed on film. In the telescope, this is due to the fact that the smallest star images on the film form a circle which is significantly larger than the theoretical size of the Airy disk. The finite size of the image on film is dependent upon a number of factors, including seeing and film structure. As long as the diameter of the cone of light falling on the film stays well below the minimum star-image size, virtually no difference will be seen in the final image. Using this information, it is possible to calculate the focus tolerance distance for any given focul ratios. Table 3.1 shows some practical focus tolerances for a range of focal ratios.

Useful information

Record keeping

As one progresses in the field of astrophotography, it is important to be able to learn from mistakes. In order to find out what has gone wrong when faced with a ruined piece of film, it is necessary to know what went on at the telescope when the frame was exposed. This principle also applies to doing something right: with a perfectly exposed frame of a particular nebula, it is good to be able to recall the exposure that was used. In order to have this information, it is necessary to develop the good habit of keeping records. It is important to record exposure information, seeing conditions, unusual

Table 3.1. *Depth of focus as a function of focal ratio*[a]

	103a-films (inches ±)
f/4	0.0020
f/5	0.0025
f/6	0.0030
f/7	0.0035
f/10	0.0050

[a]This table shows the maximum depth of focus for a range of f/ratios and for coarse-grain films. To insure maximum results, these values should be divided by 2 to obtain practical depth-of-focus limits.

events (e.g. drive slips or trips over the mounting during the exposure) and other information pertaining to the actual exposure conditions (e.g. frost forming on the cold-camera window during the exposure). It is also important to record the handling and processing of the film away from the telescope: hypering procedures, processing formulas and times. Later, the information recorded before, during and after the exposure can be combined with information about the quality and usefulness of the final images to form a very complete information log that may be of considerable use in future applications.

From this discussion, you should be able to see that there is more to record keeping than just jotting down the duration of all exposures at the telescope. To make full use of such logs, one must pay careful attention to everything that goes on at the telescope during each exposure and also keep track of what goes on in the darkroom before and after the film is exposed at the telescope. Over the years, we have designed a number of forms to help us to keep better and more useful records. A sample form appears in Fig. 3.5.

Thin clouds and photo quality

Often there are nights when conditions would be perfect except for a thin, almost transparent, layer of cirrus clouds. In some cases, this cloud layer is so thin that there is no certainty that there really is a cloud layer, but the sky 'just doesn't look right'. Sometimes the scattering caused by such cloud layers can be seen around brighter stars using binoculars or the main telescope. Under these conditions, there is often temptation to open up and photograph. The results will look just like the stars did through the telescope: each of the brighter star images will be surrounded by a hazy glow, and the size of each halo will depend upon the brightness of the star. In general, it is better to resist the temptation to photograph on such nights unless there is some pressing need for the photographs. The only exceptions to this rule should be photographs with wide-angle lenses under 60 mm focal length.

Twilight: the limits of an evening

As the sky slowly darkens following sunset, the question of when to begin the first exposure of the evening is considered by everyone wanting to photograph the night's splendors. The exact time is partially dependent upon the f/ratio and the focal length of the lens in use, the film used, and the distance of the photographic field from the western horizon. However, it will usually be found to be 'safe' to photograph regions near the zenith when the sun is between 9 and 15 deg below the horizon. Systems with longer focal lengths and slower f/ratios can be opened up earlier, while faster systems covering more of the sky will need darker sky conditions before exposures can be started.

It is easy to test for this 'safe time' by pointing the camera near the zenith and starting a test exposure of average duration at successively earlier times each evening. When these test exposures start to show unacceptable fog, or pass the established fog limit ($D > \sim 0.4$–0.45), the twilight limit has been established. By keeping good records and calculating the sun's position at the start time of each test exposure, the distance of the sun below the horizon can be determined when it is safe to open up for an evening's work. Once this distance has been determined, the actual time can be calculated on any evening of the year.

There are some circumstances when it is desirable to continue exposures past the 'safety margin' and to continue on to the absolute limit without ruining the photograph. Examples of this situation include the

Fig. 3.5. A sample exposure form. This has been designed to allow entry of all necessary data to fully describe the conditions surrounding an evening's exposures.

ASTROPHOTOGRAPHIC EXPOSURE DATA

LOCATION: ____

DATES: ____

FILM: ____

INSTRUMENT: ____

SEEING: ____

OBJECT	NO.	EXPOSURE	COMMENTS
	1		
	2		
	3		
	4		
	5		
	6		
	7		
	8		
	9		
	10		
	11		
	12		
	13		
	14		
	15		
	16		
	17		
	18		
	19		
	20		
	21		
	22		
	23		
	24		

ADDITIONAL COMMENTS: ____

____

DEVELOPER, TIME AND TEMPERATURE: ____

photography of early-evening and late-morning comets. In each of these cases, the twilight limit is pushed to the absolute limit (and sometimes beyond), the goal behind taking such a risk is to gather as much detail as is possible without ruining the exposure with too much added fog. The only sure way of determining this absolute limit for your own particular system is to run a series of test exposures like those described above. Our own experience with relatively bright objects using f/5 to f/6 Newtonians with 103a-F emulsions has shown that exposures can safely be continued as long as fine-structure detail can be discerned using binoculars to observe the object being photographed.

Humidity and film sensitivity

On evenings when the humidity causes heavy dewing to occur, there is an additional effect which can cause photographic results to suffer, especially if hypersensitized films are being used. Humidity has an adverse effect on film sensitivity, and this effect is greatly magnified with films that have been hypersensitized using one of the methods which relies (partially or wholly) upon removing sensitivity contaminants (like oxygen and moisture) from the emulsion layer. When film which has been hypersensitized in such a manner is exposed to an extremely humid environment, the effects gained by the hypersensitizing process may be entirely lost in a few hours' time. With spectroscopic emulsions that have not been hypered, an exposure on an extremely humid evening may show a sensitivity loss amounting to a magnitude or more when compared with an identical exposure on a dry evening.

If one uses hypered films and encounters humid evenings fairly often, or if one lives in an area where the humidity is always high, it would probably be a reasonable idea to construct a camera attachment that would allow the film to remain in a dry environment while being exposed. By placing an optical window at the end of the tube which attaches the camera to the telescope and then flooding the resulting chamber with a dry inert gas, like nitrogen, the film is immersed in a dry environment which will help to protect its sensitivity. This provides a very simple solution to a problem which can undo an evening's work by desensitizing a hypered film like 2415. If the film needs to be stored for the next evening's work, it should be stored in a nitrogen environment.

If an optical window cannot be placed in front of the film for some reason, another alternative to avoid some of the problems created by the effects of moisture is to advance the film one extra frame between exposures. This solution does not eliminate the desensitization of the frame being exposed, but it will eliminate much of the desensitization of the succeeding frame.

4

The first step to knowledge is to know that we are ignorant.
Lord David Cecil

Theoretical points of interest

In general, the books that have been written on astrophotography for amateurs have ignored topics relating to astronomical photography beyond the mechanical aspects of putting film in the camera and exposing it at the telescope. The magazines that cater to the amateur community are guilty of the same oversight, although articles on seeing, limiting magnitudes, hypersensitizing, color photography, and ray tracing various optical systems have occasionally appeared. In some cases, in-depth theoretical articles have appeared but generally the articles have described specific applications of a technique or a theoretical area and have avoided elaborate discussions. Because of these shortcomings amateur astrophotography advanced very little over a 20 year period, until the late 1970s when some amateurs began to see that their work could be improved significantly by learning more about the field and investigating more techniques.

The exact reasons for this sudden advance (which seems to have slowed in the last few years) are not obvious. Some advanced amateurs began publishing works that seem to have appealed to the competitive nature of others. Some articles offered new ideas that greatly improved on equipment that was available and some manufacturers were quick to copy the devices and put them on the market (the originator usually received no payment for the idea). In the United States, a number of amateur conventions started to emphasize papers on astrophotography and some of the better astrophotographers from around the country began sharing their techniques. In addition, one company (known as Lumicon) was founded that catered exclusively to the needs of the amateur astrophotographer and offered a variety of filters and photographic devices to the amateur community.

However, this advance and the accompanying wave of new technology also created a certain amount of backlash. Some amateurs thought that all the technology was alienating newcomers to astrophotography, and a vigorous campaign started off expounding a simple approach to astrophotography (using cold cameras and hypering techniques with only a cookbook background). Another obstacle that helped to slow this advance came in the form of a flood of low-priced, super portable telescopes that were deceptively marketed for astrophotographic use by a number of manufacturers. Photographs with these instruments are almost always identifiable by the large fuzzy star images over the entire field and a general 'soft' appearance (in both stellar and planetary photographs). The widespread popularity of these instruments, due to their portability and price, resulted in the amateur literature being flooded with inferior-quality photographs which basically offset the increased capabilities that other amateurs were taking advantage of.

Finally, some of these manufacturers have responded to growing criticism by introducing 'deluxe' models for their more sophisticated customers. The use of advanced techniques without proper knowledge, the simplistic approach to astrophotography, and the uncritical use of mass-produced optical systems have slowed the advance of amateur astrophotography. We feel that the only available direction should be forward, and that the desire for more simple ways of doing things is a retreat to the past. With few exceptions, the work of some advanced amateurs of 1985 is equal to or superior to that of many professionals and we feel that amateurs are capable of producing work of even better quality if they are willing to learn more about the media that they are using and to take a little of extra time and care in taking photographs at the telescope.

With all of this in mind, we will begin our discussions of non-instrument-related topics with several explorations of some theoretical aspects of astronomical photography which deserve consideration. Two of these topics (photographic resolution and limiting magnitudes) involve the question of 'how far can I go?'; the third (ray-trace interpretation) presents some information and precautions about interpreting information derived from a technique that is being used with ever-growing frequency in the amateur literature.

A look into photographic resolution

Much of our work with astrophotography has involved using small-aperture Newtonians (6–8 inches aperture, with f/5 to f/6 being typical for our use) and we have always been determined to obtain the absolute maximum from our instruments before we were satisfied with our results. In addition to using 'correct' film/exposure/developer combinations, some sophisticated darkroom techniques, and applying knowledge of photographic materials to our work, we have found that the task of obtaining maximum resolution is particularly critical to the use of small-aperture instruments of moderate focal length.

We have found that several conditions must be met to achieve results that approach the limits of resolution:

1 excellent focus;
2 excellent seeing;
3 very good guiding;
4 good optics (with a minimum amount of aberration that would throw light into the outer diffraction rings).

Excellent seeing and focus are of virtually equal importance. If you can't focus the camera accurately, you should not expect sharp images: when the seeing is bad the images will never be sharp. Guiding photographs allows for some tolerance because of the finite size of a star image on and in the film's emulsion layer. The optical system should be of reasonable quality, but the main concern is to eliminate aberrations which will throw light into the outer diffraction rings and result in enlarged images on film. Irregular optical surfaces with numerous minute zones and ridges seem to be one cause of such problems. The fact that these four conditions are all necessary to achieve high-resolution photographs should really be apparent but there are some who refuse to acknowledge that seeing has an effect on the results of long-exposure deep-sky photography. In addition it is not easy to tell when you have achieved the maximum possible resolution from a given photographic system. Other studies which have addressed this problem have not yielded results which agreed with images taken in ideal-seeing conditions with small-aperture instruments such as the ones we use.

Basic considerations

It seemed obvious at first that the actual resolution of the system must depend on the film's ability to resolve closely spaced lines or points (R_f) as well as the image scale of the photographic system (S). Tying these factors together into an equation yields: $R = R_f S$. Typically, 103a-type films can resolve about 80 lines/mm (0.0005 in or 0.013 mm) while films like Kodak Technical Pan Film 2415 can resolve about 320 lines/mm. Therefore, a photographic system using a 103a-emulsion can resolve no better than: $R = R_f \times 206264.8/F$ arc sec (where R_f and F are in the same units and F is the focal length of the system), or:

$$R = 103/F(\text{in}) = 2575/F(\text{mm}) \text{ arc sec.} \quad \textbf{(4.1)}$$

This formula seems reasonable but we can see that an important qualification must be added immediately. For instance, in a 6 inch aperture f/5.0 system, the photographic resolution would be: $R = 103/30 = 3.43$ arc sec. The upper limit for visual resolution (R_v) for all 6 inch aperture (A) instruments is:

$$R_v = 4.56/A = 4.56/6 = 0.76 \text{ arc sec.} \quad \textbf{(4.2)}$$

It is now obvious that if this formula for photographic resolution (R_p) is applied to a system with an f/ratio greater than f/22.6 that the predicted photographic resolution would exceed the instrument's theoretical resolution. Therefore, we need to qualify the formula to read:

$$R_p = \text{AMAX}(R\,,\ R_v) = \text{AMAX}[(103/F)\,,\ (4.56/A)], \quad \textbf{(4.3)}$$

where A and F are in inches, the film used is a 103a-emulsion, and AMAX is a Fortran statement indicating that the largest value is to be used.

New data

This appears to be a reasonable solution at first glance and it predicts values that can be tested. Late in 1980 we began using Kodak 2415

Technical Pan film and the doubts that we had regarding this formula were substantiated by numerical data. More thinking and calculating was required to find a relationship that would explain the new data that we had available.

The problem which was apparent with the formula that we had derived is the prediction that the resolving capability of any telescope should increase directly as the resolving capacity of the film increases. When 103a-type films are compared with Kodak 2415 Technical Pan film the formula predicts a resolution increase of a factor of four; what we were able to measure was only a 10 % increase (see Figs 4.1 & 4.2). The Kodak 2415 Technical Pan film resulted in extremely fine grain and more continuous tone in areas of nebulosity, but the star sizes only decreased by 10 % to an 18 μm diameter. Therefore, while this film has the obvious advantage over 103a-type emulsions in reducing grain and making prints of nebulosity more aesthetically pleasing, it will not help those who insist on 40 ×

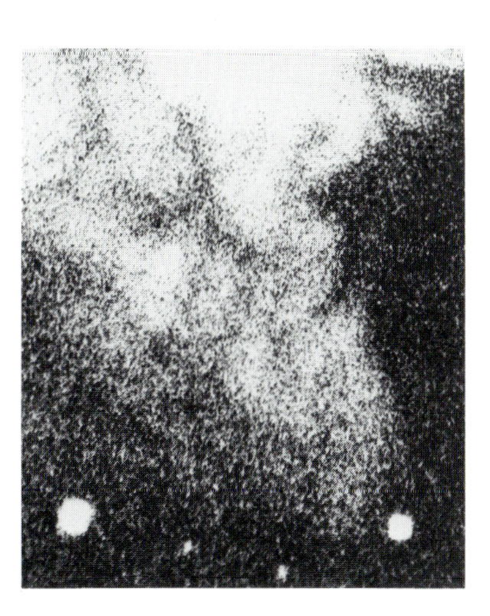
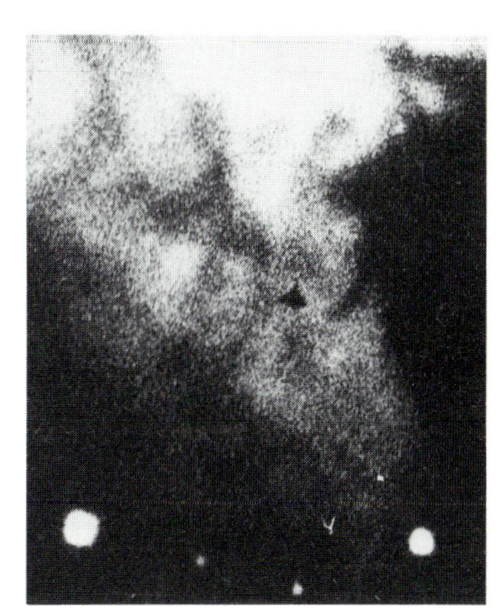
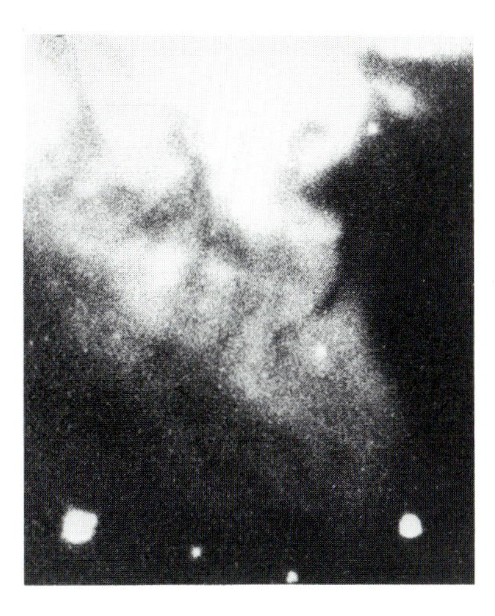
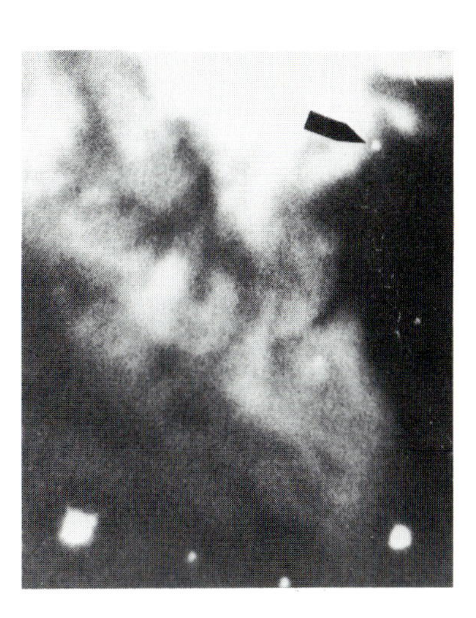

Fig. 4.1. The resolution gains obtained by using high-resolution films for nebular objects is clearly shown in these photographs (and those of Fig. 4.2) by Edgar Everhart. However, while nebulosity gains in clarity, stellar resolution is basically unchanged, indicating a physical limit for the photographic resolution of stellar objects (Photographs courtesy of Edgar Everhart.)

enlargements. (The other advantage that this film has is the expected increase in DQE due to its higher contrast and its extremely fine grain.)

To solve the problem and to find a formula which more closely reflected reality, we had to look at the structure of the actual star image that is being projected onto the film. Fig. 4.3 shows a graphic display of the classic textbook formula for the diffraction pattern of a star image and a function that has been modified to throw more energy into the first diffraction ring. This more closely matches diffraction patterns encountered in the real world. Fig. 4.4 shows this modified Airy function at three wavelengths over the sensitivity range of a panchromatic photographic emulsion (400 nm, 550 nm, 700 nm).

To look at the net effect of white light on a panchromatic film, we took the modified diffraction patterns for 400, 475, 550, 625, and 700 nm and found the mean curve of all wavelengths to obtain a representation of the actual white light diffraction pattern, as seen in Fig. 4.5. Using this graphical representation we found that the first and second minima were located at radii of $6.78/A$ and $12.39/A$ arc sec, which would correspond to photographic resolutions of:

$$R_{p1} = 13.6/A, \text{ and } R_{p2} = 24.8/A \text{ arc sec.} \quad \mathbf{(4.4)}$$

If the photographic resolution of the faintest on-axis star image is dependent on the structure of the diffraction pattern, then one of these formulas should match the empirical data obtained (or, perhaps, each would apply in certain sets of circumstances).

We had some old data from our 6 inch f/5 and f/6 instruments using 103a-emulsions which yielded star images of about 20 μm and which

Fig. 4.2. For caption see Fig. 4.1. (Photographs courtesy of Edgar Everhart.)

matched the values that we expected for a coarse-grained photographic emulsion. We also had some new data from our 8 inch f/6 instruments using Kodak 2415 Technical Pan emulsion . When these new images were measured, the smallest star images were about 18 μm which is 3.1 arc sec, as predicted by the formula for R_{p2}. Each of these measurements were taken from negatives exposed under virtually ideal seeing conditions. Another interesting fact arose during this exercise regarding the actual image size of the smallest stars on film. To convert from micrometers to seconds of arc, we use the formula for image scale,

$$S = 8.1/F \text{ arc sec}/\mu\text{m}. \quad (4.5)$$

When this is combined with the formula for R_{p2} to yield the photographic resolution of any system yields,

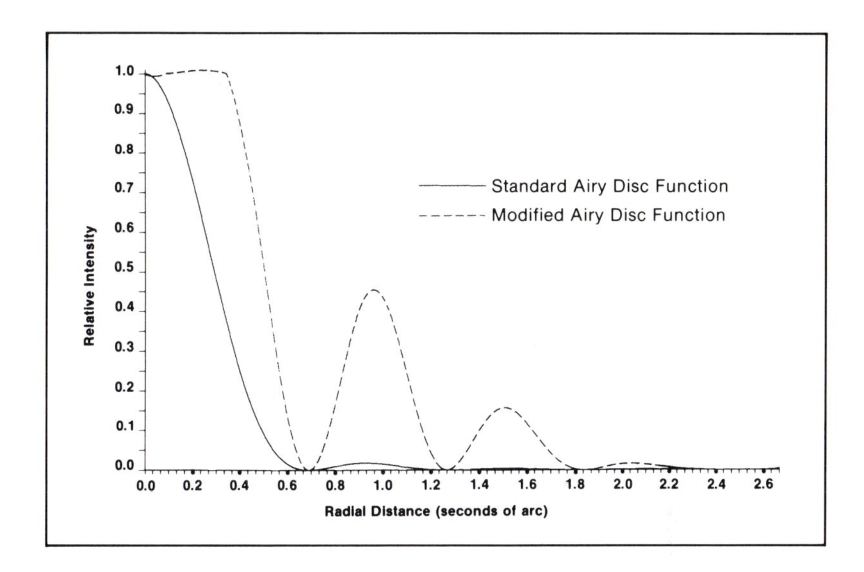

Fig. 4.3. The theoretical Airy Disk function is shown for a wavelength of 400 nm along with a curve that has been modified to more closely reflect the actual visual appearance of a diffraction pattern produced in a 6 inch aperture unobstructed system.

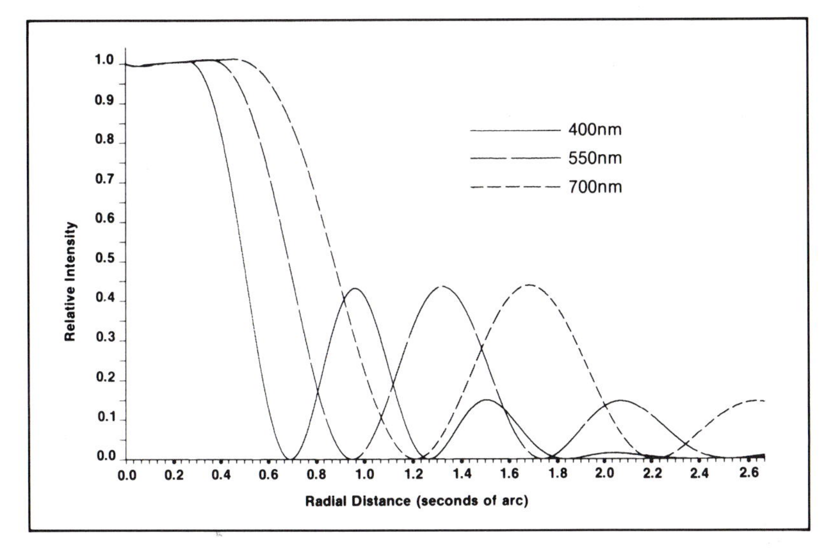

Fig. 4.4. The modified Airy Disk function is shown for wavelengths of 400 nm, 550 nm, and 700 nm for a 6 inch aperture system.

$$R_p = 3.062 \times f\ \mu m, \tag{4.6}$$

(where f is f/ratio). This implies that the smallest possible linear size of the faintest stars in any given imaging system (with reasonably good optics, using a panchromatic film, etc.) is entirely dependent on the f/ratio of the optical system.

f/4: > 12 μm f/6: > 18 μm f/10: > 30 μm
f/5: > 15 μm f/7: > 21 μm f/15: > 45 μm

This sets up a number of predictions which can be used to test this hypothesis; we hope to get some feedback from some readers on these tests. One prediction which should have been expected is that the resolution found on a plate exposed to blue light should exceed that of a plate exposed to red or white light, but we did not really look into the question of how much better the resolution would be. In any case, we must revise the earlier relationship to read:

$$R_p = \mathrm{AMAX}[(k/F), (24.8/A), (R_v)] \tag{4.7}$$

where R_p is in seconds, A and F are in inches, and k is a constant that is dependent on the film that is being used in the system ($k_{103a} = 103$, $k_{2415} = 26$). This also says that in optical systems where the f/ratio is greater than about f/4 that (for 103a-emulsions) the aperture controls the actual resolution.

Effects of seeing

Another condition that must be brought to bear on this growing relationship is that seeing at observatories using large-aperture telescopes is considered to be excellent when the diameter of the Airy disk is around 1 arc sec and that on very few occasions is seeing of 0.5 or 0.25 arc sec reported. So, it can be seen that if the seeing is only 1 arc sec, the term for R_{p2} is meaningless if the aperture exceeds 24.8 inches! Thus, the expression for

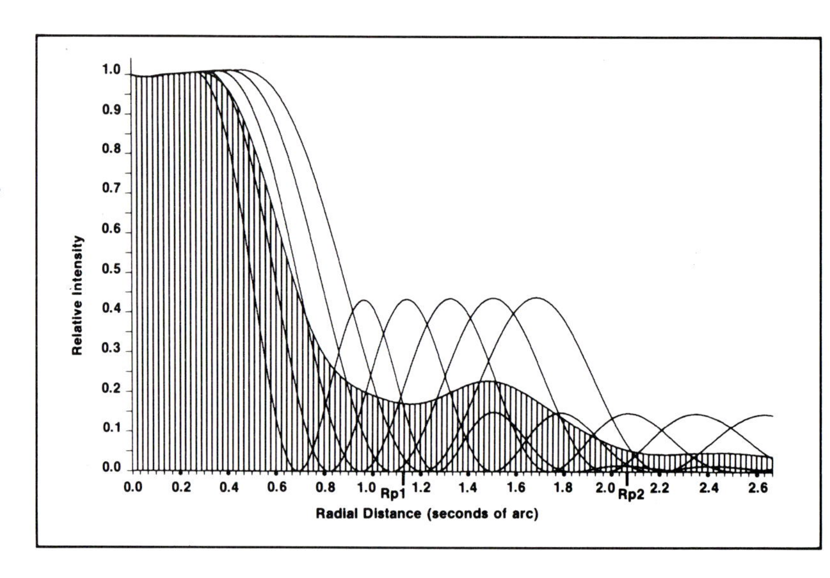

Fig. 4.5. Modified Airy Disk functions for a 6 inch aperture system are shown for 400 nm, 475 nm, 550 nm, 625 nm, and 700 nm. The mean of all curves is shown by the shaded curve and represents the net effect of light in the visual range. The points corresponding to Rp_1 and Rp_2 in the text are indicated.

photographic resolution must be further qualified to express the dependency upon seeing. We now rewrite the expression to read:

$$R_p = \mathrm{AMAX}[(k/F), (24.8/A), (S)] \qquad \textbf{(4.8)}$$

where (S) is the seeing in seconds of arc. This makes our expression capable of handling all the variables that we now know to be involved in the problem.*

Please note that this is based on a diffraction pattern in which the energy in the rings diminishes rapidly as one moves out from the central disk. If aberrations are present that cause more energy to be thrown into the outer rings it is probable that this relationship will be violated. At the very least, we would expect such aberrations to create the following symptoms:

1 star images with a poorly defined outer boundary ('soft' star images); this effect would also be expected when atmospheric scintillation causes the star image to jump around during the course of the exposure – the use of very thick emulsions might also produce this result;
2 star images whose sizes increase very rapidly with brightness when compared with images from less aberration plagued systems.

Ray-trace *v.* reality

As the capabilities of small computers increase and their costs decrease, the amateur has gained greater access to the capability of performing ray-trace studies on various optical systems. The ray-trace program simulates the sending of a number of rays of light into an optical system and traces the path of those light beams producing a graphic output that is a greatly magnified approximation of the pattern that would be produced when light from a star passes through the same system and is imaged at the focal plane. The process is very powerful when applied correctly and it has allowed a number of advanced amateurs to investigate thoroughly a variety of popular optical systems as well as to design new optical systems that enhance the performance of less-optimized variations.

Our interest in the ray-trace grew as we continued to see studies and diagrams of Newtonian 'star images' that were far larger than anything we had ever measured on any negatives that we had exposed. In fact, most comatic images that we have measured disagreed with the ray-trace diagrams and the predictions of geometrical optics by a factor of nearly three! We developed several theories about the reasons for this discrepancy, but were uncertain until finally acquiring a computer and several ray-trace

* While the theoretical curves and the hypothesis derived from those curves is based upon an unobstructed optical system, our empirical data is obtained from systems with central obstructions of nearly 40 % of the main mirror's diameter. This can be expected since the central obstruction does not shift the positions of the minima in the diffraction pattern. However, such an obstruction does throw energy into the first diffraction rings and a much smaller amount into the second diffraction ring. This much extra energy does not seem to affect significantly the actual resolution.

programs to test out some of our ideas. The answers were not long in coming as we tried to determine what could be learned from the ray-trace method, and what precautions should be taken in trying to interpret the information contained in the resulting spot diagrams. We found that if a few rays (30–100) were passed through a Newtonian system, using a regular array, one obtained the usual spot diagram and the disparity between ray-trace and reality remained. However, when a larger array (200–2000 +) of rays is passed through the system, the resulting spot diagrams contain considerably more information and the mystery is resolved.*

The ray-trace spot diagram must be viewed and interpreted very carefully in order to arrive at realistic conclusions about the optical configuration under study. This is especially true when a handful of points are tossed into the computer in a regular array and then bounced and scattered about to form the final image pattern. The use of a few points, combined with the use of a regular data array really destroys the information that potentially can be obtained from a ray-trace study. If one hopes to obtain more than the general shape of the final image, no fewer than 200 points can be used to obtain some idea of the actual image structure (see Fig. 4.6). Even this small a sample cannot be expected to give an image that matches reality, as there are many flaws in the entire procedure, especially when the photographic emulsion is involved. A brief listing of just a few well known factors would include:

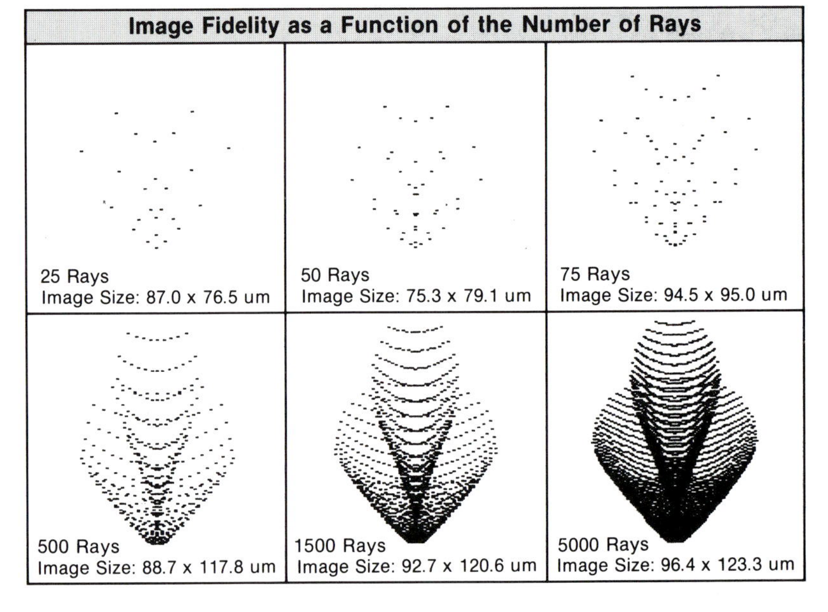

Fig. 4.6. Illustrates the effect of increasing the number of rays on the fidelity of ray trace images. The system shown is a 8 inch f/6.0 paraboloid. Note that the upper three spot diagrams not only give inaccurate information on the size of the image, but yield no useable information on the light intensity distribution within the image.

* Kingslake, Herzberger and others have studied and verified the validity of ray-trace spot diagrams and all agree that many hundreds of rays need to be traced in order to get an accurate representation of the appearance and size of an actual image. These conclusions are based on studies that have been conducted using many rays to produce a detailed spot diagram and then comparing the plot with microphotos of actual images produced by the optical system in question.

1 The Airy disk is considerably larger than the 2–4 μm image that the ray-trace shows for the on-axis images of some systems. What happens to the wave model of light? (Stolzman's articles address this issue as does our previous section on photographic resolution.)
2 The photographic emulsion can disperse the image even more than does the diffraction process.
3 The intensity of the image at various points is largely lost in the ray-trace diagram unless special steps are taken to obtain this information.
4 The faintest portions of the image are not recorded on the film until long after the image has grown considerably due to the general dispersion, reflection, and scattering of light in the photographic emulsion layer and its backing.
5 The longest comatic tail ever seen in a photographic image is considerably shorter than anything predicted by ray-trace by a factor of at least 2!

In order to address items 3, 4, and 5 above, the following ray-trace diagrams are presented for a 10 inch f/6.0 instrument and a star image 1.0 deg off-axis (Fig. 4.7). Some 15 000 random points were generated, and a histogram array of the frequency at which a single pixel was hit was formed. Using this array, slices where then taken through the three-dimensional ray-trace spot diagram (dimensions = x, y, intensity) and the following series of diagrams was made showing the effect of increasing intensity upon the size and shape of the spot diagram. The label 'Intensity: 32' means that only those points receiving *more* than 32 hits are plotted in that associated diagram. The 20 μm scale bar at the bottom is as small as the average amateur astrophotographer will see using such an instrument with 103a-films, but the real lower limit on star-image sizes that can be photographed using this instrument is around 18 μm. These diagrams present a great deal of information on the formation of a photographic star

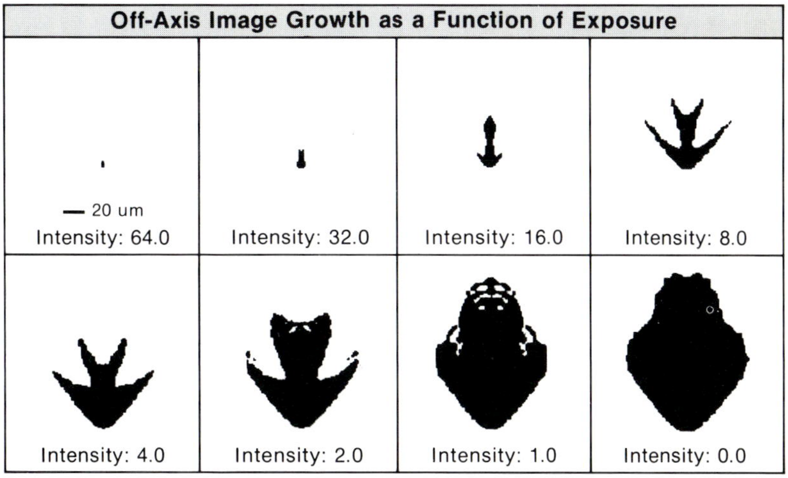

Fig. 4.7. Shows the growth of a star image formed by an 10 inch f/6.0 parabola 1 deg off-axis as a function of exposure (variations of either time or intensity). An image was traced using 15 000 rays and then slices were made through the three dimensional image that was formed. A label of 'Intensity: 16.0' means that only those points that were struck by 16 or more rays are plotted. Only one point showed the maximum intensity of 128 points.

image and we will explore some of the information that is available (see Color Plate 1).

Examining the series of spot diagrams produced in this manner, one can see that the most intense region of the star image formed by this Newtonian is quite small and well below the size that could be resolved photographically. In fact, it is not until one reaches an intensity level that is about one-fourth that of the most intense area that one sees an image that even matches the photographic resolution of the instrument. Then, the effects of astigmatism and coma build up the intensity of the resulting arrowhead-shaped image without significantly enlarging that image until the intensity grows by about another factor of four. Then, as light continues to fall on the film, the fan-shaped comatic tail (that is considered so characteristic of a Newtonian) builds up. The light that forms that large fantail is about 1/64 (0.0156) as bright, or nearly 5 magnitudes fainter than the intensity of the brightest portion of the image!

Additional information about the real appearance of an off-axis image can be gained by using a large number of rays and then plotting the maximum intensity of the image against the off-axis distance. This will show that the intensity drops slowly as the image grows and that the brightest region of an image produced by this same f/6 system 1 deg off-axis is about one-sixth the brightness of the brightest on-axis image. Thus, an off-axis image suffers from a built-in vignetting caused by the general growth, or spread, of the off-axis image. In addition, the edges of a Newtonian usually suffer from some vignetting due to secondary size, focuser construction, and camera design that may amount to a 20–40% light fall-off. Adding up the effects of mechanical vignetting, general image intensity fall-off, and the intensity structure of an off-axis image as shown in the above study, it becomes much easier to see why an f/6 system actually produces images 1 deg off-axis that are about 30–40 μm in diameter instead of images that are 60 μm by 130 μm in size. Only an extremely bright star at this distance off-axis is capable of producing an image that would appear anything like the image predicted by a ray-trace program.

Without such elaborate treatment, the information that can be derived from a spot diagram would seem to be accurate only when all of the points get concentrated into a small, symmetrical image that is uniformly illuminated. Thus, the ray-trace information from systems like the Concentric Schmidt–Cassegrain and the Ritchey–Chrétien would appear to approach reality. Therefore, if the ray-trace diagram gives some clue that the intensity is not uniform, one should perform some rather elaborate testing and be very careful about drawing the conclusions that the system is not suitable for use in astrophotography! There are very few systems that out-do the overall performance of a Newtonian for narrow-angle work; the Concentric Schmidt–Cassegrain (not to be confused with the Schmidt–Cassegrains designs that are being mass-produced currently by several telescope-making companies in the USA) is the only one that the present authors have seen.

Limiting photographic magnitudes

Any photographic instrument reaches its operational limit when the general sky illumination darkens the background of the photographic plate beyond a certain density level. The limiting photographic magnitude of any given instrument (at a given site) will be dependent upon the emulsion that is being used. Studies have shown that fine-grain, high-contrast emulsions will reach a much fainter limiting magnitude than grainy, low-contrast emulsions. The early improvements in astronomical emulsions gave us fast, high-contrast film like the 103-a emulsions (which were great improvements over normal high-speed film, which are generally low in contrast). Now, the latest emulsions are breaking down the grain barrier to provide even better films for use in astronomical photography. The new generation of emulsions began with the introduction of IIIa-J for professional astronomers and was soon followed by the introduction of 2415 Technical Pan film for amateur astrophotographers. The need for emulsions such as these has been recognized for over 40 years, but only recently has photographic technology reached the point where they have become a reality.

The limiting magnitude of a photographic instrument, refers to the faintest stellar magnitude that can be reached, regardless of the exposure that is used. In some cases, this same term is applied to mean the faintest stellar magnitude that can be reached in an exposure of a given duration. In fact, the latter is the most common use of this term even though, in this case, one is referring to the limiting magnitude of the exposure and not to the limiting magnitude of the instrument. We will discuss both of these situations in order to acquaint the reader with the factors involved in each case.

When dealing with either of the limiting-magnitude terms defined above, it is important to realize that the photographic intensity of an extended object (e.g. nebulae, distant galaxies, sky background, etc.) is inversely proportional to the square of the f/ratio while that of a point source is dependent upon the square of the aperture of the instrument. A very simple expression for the limiting magnitude of any given exposure with a 'fast plate', published in 1955, is:

$$mp = 5\log(D) + 2.15\log(E) + 6, \quad (1955) \tag{4.9}$$

where D is in inches and E is in minutes). When this same expression was evaluated for 103a-O emulsions in 1976, the increased speed and optimized reciprocity characteristics resulted in a formula given as:

$$mp = 5\log(D) + 2.50\log(E) + 9.5 \quad (1976). \tag{4.10}$$

Fig. 4.8 shows a plot of this function for a range of apertures. The reader

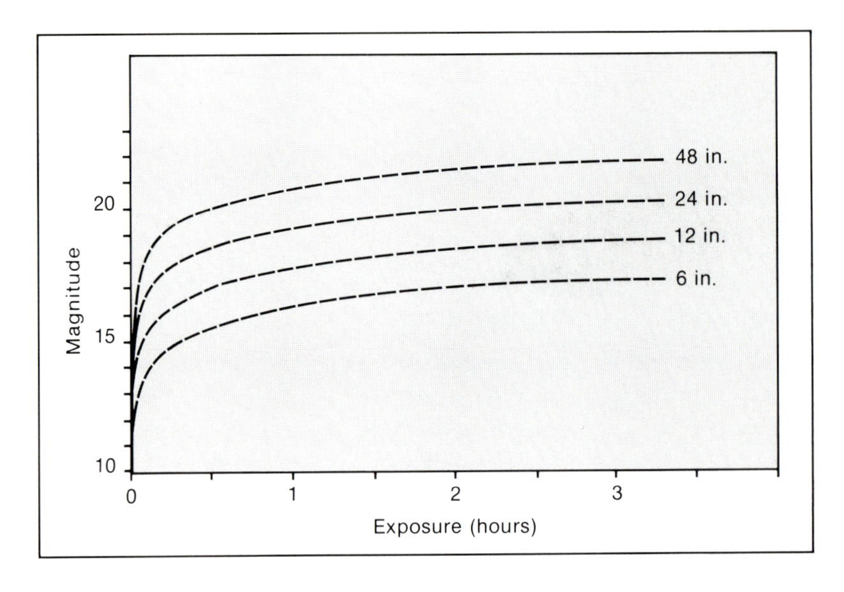

Fig. 4.8. Shows limiting magnitudes that can be obtained with various instruments using 103a-O as given by an expression derived in 1976. The values shown assume the use of D-19 with a 4 min development time at 20°C (68°F).

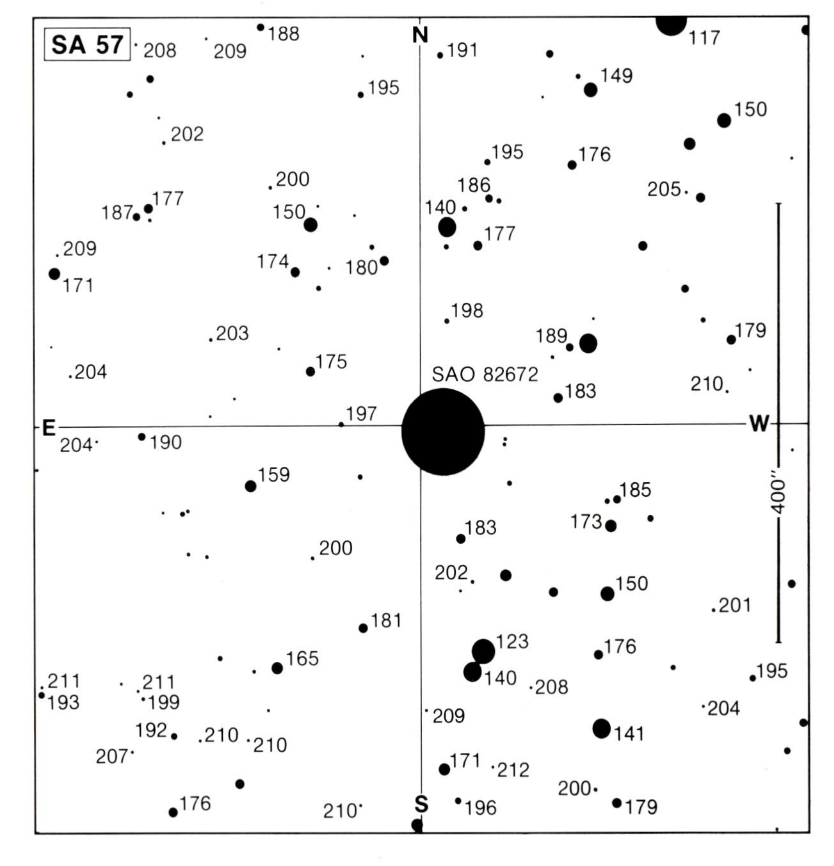

Fig. 4.9. The region around SAO 82672 is shown with photographic magnitudes of the stars in this 13 × 13 arc min region labeled. SAO 82672 is located in SA 57 and is at RA = 13 h 8.6 min, DEC = +29 d 23 min (Epoch 2000.0). (Chart prepared from photograph by Edgar Everhart, *Sky and Telescope*, Jan. 1984 with permission of the author.)

should realize that these expressions are approximations and that they must be evaluated for a given film and a given developer and that these expressions result from fitting large quantities of numerical data in order to derive a simple expression. Actual photographs must be taken in order to determine the limiting magnitude for a given exposure with any system (many variables may influence this value such as film, hypering technique, development, etc.). Figs. 4.9 and 4.10 provide charts of two areas that can be used to determine the limiting magnitude of moderate aperture instruments.

The equivalent expressions for the limiting exposure which can be used on any given instrument also show the effects of advances in photographic technology over this 21 year period. The limiting exposure for a fast plate of 1955 vintage is given as:

$$\log(E) = 0.6 + 2.325 \log(f), \quad (1955), \tag{4.11}$$

while the equivalent expression for 103a-O in 1976 is given by

$$\log(E) = 0.201 + 2.0 \log(f). \quad (1976). \tag{4.12}$$

Since these expressions also represent approximations, they should be used only as general guidelines and not as exact figures. There are numerous

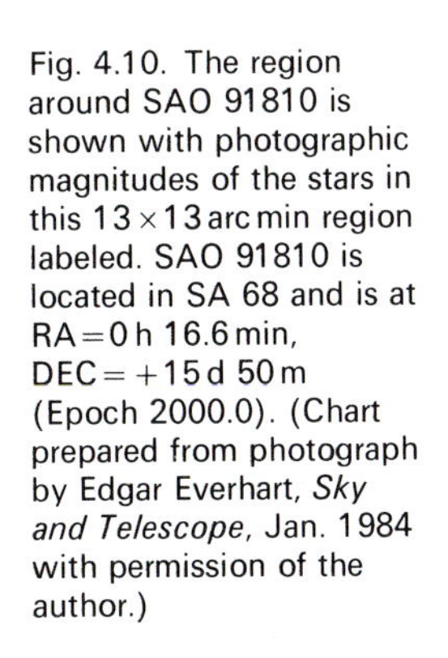

Fig. 4.10. The region around SAO 91810 is shown with photographic magnitudes of the stars in this 13×13 arc min region labeled. SAO 91810 is located in SA 68 and is at RA = 0 h 16.6 min, DEC = +15 d 50 m (Epoch 2000.0). (Chart prepared from photograph by Edgar Everhart, *Sky and Telescope*, Jan. 1984 with permission of the author.)

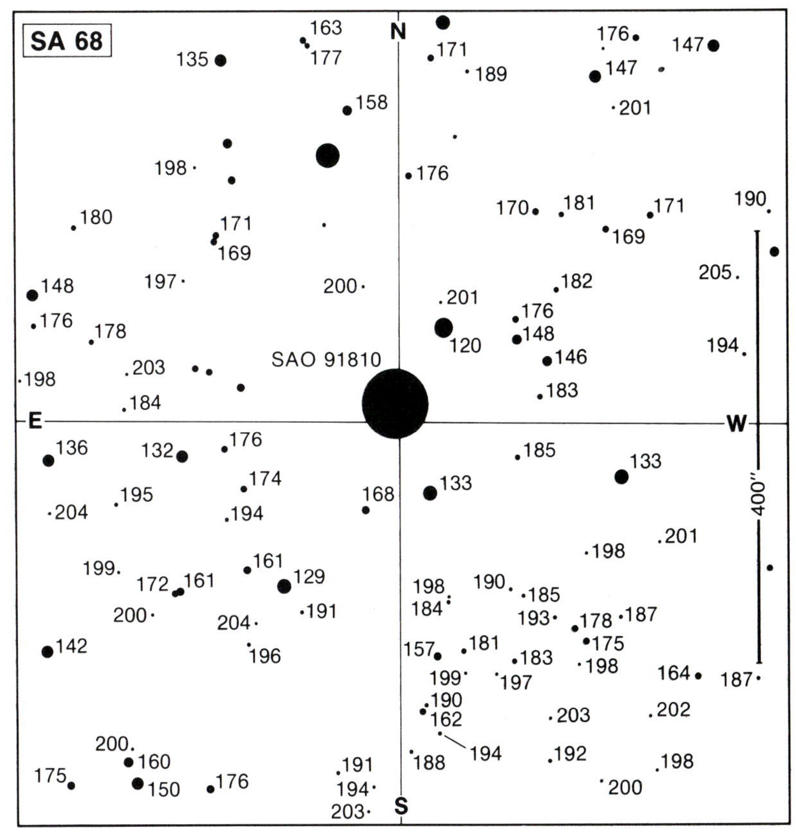

variables that can affect the limiting exposure of an instrument, including many that are outside of the control of the astrophotographer.

The factors that influence the limiting magnitude of any exposure are quite complex, but the investigation of these variables has yielded some very interesting information which has proved very enlightening to the present authors. A partial list of these non-instrumental variables would include:

1 distance of the sun below the horizon;
2 phase and position of the moon;
3 amount of sky light from extraterrestrial sources such as aurorae, zodiacal light, gegenschein (this is a variable that fluctuates considerably and is generally unknown at the time of exposure);
4 brightness of the skyglow contributed from local artificial sources (this too is a fluctuating variable that is dependent upon atmospheric conditions as well as the whims of local politicians);
5 local atmospheric conditions and the resulting atmospheric clarity (this can also affect factor 4 if there are weather conditions that result in changes in the brightness of the skyglow; for example, thick fog layers frequently cover much of Los Angeles during the months of May and June which result in dark skies for observers in the surrounding mountains);
6 seeing!

A detailed examination of the problem of determining an expression that would fit data from many sources was conducted by F.L. Whipple and published in 1942. This work built upon a detailed 1938 study by Ross which also investigated a number of factors that influence the limiting magnitude of a photographic instrument. The expression is out of date, but the information contained in this investigation is still pertinent today. Since any given expression is only good for a particular emulsion and must ignore many of the other variables that influence the actual results, the expressions derived from these studies are almost meaningless except as historical notes. The investigations that led to these expressions have aspects which have far-reaching implications. It is the factors that influence these formulas that are of importance and not the formulas themselves.

The relationship that was derived by Whipple can be expressed as:

$$ml = ms + 5\log(F) - 2.50\log(d) - 16.09. \quad \textbf{(4.13)}$$

Where ms is the brightness of the night sky (~22.0–22.25 in 1940 at very dark sites), F is the focal length of the instrument in inches, and d is the diameter of the faintest star images on the plate in millimeters. It is the expression for d which yields the information that is not apparent from any of the formulas that we have previously seen. The expression which Whipple derived for the diameter of the faintest stars on the plate brings many of these variables together into a single coherent expression (the reader is cautioned that much of Whipple's data was derived from data taken using large instruments with reasonably poor optics and coarse-grained films and that the exact figures given by this expression are not accurate when applied to smaller instruments using today's emulsions). The expression yielding the size of the faintest, and thus (according to Whipple) the smallest, star images on a given plate in millimeters is:

$$d = 0.002 + [F(s'' + 8.1''/D) \times 0.000123], \qquad \textbf{(4.14)}$$

where s'' is the diameter of the seeing disk in seconds of arc.

This single expression shows the effects of seeing, f/ratio and focal length upon the size of the star image and, when values for these quantities are substituted into eq. **(4.13)**, we can see the effects of these variables upon the limiting magnitude of the photographic instrument. While we feel that Whipple's lower limits were influenced by the instruments and films of the day, general conclusions can still be formed from this study, including the following:

1. as the f/ratio is increased (all other factors being constant), a fainter limiting magnitude can be reached;
2. larger apertures yield fainter limiting magnitudes;
3. in bad seeing conditions, the limiting magnitude will be decreased;
4. bad seeing will have a greater effect upon instruments of longer focal lengths.

And, from the previous equation:

5. sites with darker skies are capable of yielding fainter limiting magnitudes;
6. longer-focal-length instruments can produce fainter limiting magnitudes;
7. instruments that are capable of producing smaller star images will produce fainter limiting magnitudes (star-image size can be affected not only by seeing conditions but by focus and optical configurations as well).

With all of this, one begins to see why astronomers have turned to dark sites, large apertures, slower f/ratios, finer grained films, and better imaging systems as they have tried to reach fainter and fainter objects. In 1938, the possibilities of fine-grained films with high speeds did not enter into Whipple's work and the prospects of modern hypersensitizing methods were just beginning to dawn. Darker skies have become extremely difficult to find and observatories are now being located in extremely remote locations. An alternative way of getting a dark sky is to use filters to darken the sky in the region that one is trying to photograph. Even this is becoming more difficult as cities adopt high-pressure sodium lamps and the future will pose some difficult problems for professionals as well as for amateurs.

In addition to these considerations, there is the question of how to determine when the photographic limit of a given instrument is reached. Plotting the magnitude limit of a plate and the fog density of the plate against exposure time, will show that the magnitude limit increases rapidly at first and then climbs more slowly as the exposure progresses. At the same time, the fog density shows a similar pattern or rapid increase followed by slower increase. Many amateurs consider that they have reached the limiting exposure when the fog level reaches a density of 0.3–0.4. This looks pretty bad to an untrained eye but is far from a limiting background. If one were to take a negative exposed to produce a 0.4 background density and a negative exposed to produce a 0.6 background density and illuminate each of the negatives in such a way as to make them appear to have the same background density, one would see that the 0.6 density negative actually showed noticeably more detail and thus the 0.4 background negative was not given a limiting exposure (this same conclusion could be reached by printing the two negatives to produce prints with identical background densities).

In reality, studies have shown that an exposure which produces a negative with a sky density of 0.8–1.0 seems to be close to optimum for black-and-white emulsions. Beyond this point, the information gain slows as the exposure is extended and, at a point where the background density exceeds 1.0, finally becomes negligible for increased exposure times. With a color transparency film, one is interested less in information gained than in the aesthetic appearance which will be diminished at relatively low levels of sky fog. In addition, it requires a sophisticated filter to knock out skyglow while still preserving a reasonable color rendition of the object. With color negatives, one can print or copy the original to compensate for the effects of sky fog far better than one can with color transparency film. Thus, if one is interested in producing color photographs of faint objects and cannot get to sites with nice dark skies, the odds of obtaining a good rendition of the object would be more favorable if color negative materials were used (or if a transparency material were processed as a color negative).

There is a great deal of information available on the many factors that can affect an astronomical photograph and there is more to this field than just exposing film at the telescope, running it through a few miscellaneous chemicals in the darkroom as instructed, slapping the negative into an enlarger and making a quick print and then hanging the print on the wall and/or showing it to a few friends. There is more than can be accommodated by any single work and the *AAS Photo-Bulletin* and other journals are regularly filled with articles that are relevant to the topic. Many amateurs seem entirely intimidated by articles appearing in professional journals for reasons that are unclear to the present authors. We would urge amateurs (and professionals as well) to read the materials on photography and astrophotography (a list of journals of interest is included in our bibliography) and use that information to their advantage. It was due to the reliance on a single outdated source on astrophotography and a lack of new knowledge that the state of amateur astrophotography in 1970 was basically identical to that in 1950. In the remainder of this volume we will present a sample of topics that go beyond the usual telescope and darkroom techniques of astrophotography. Some of the following information will hopefully be found useful by many of our readers, while other sections will be found useful by only the most advanced and progressive astrophotographers. We hope that we will at least let many see the great scope and challenge of this field and will persuade others to take an interest in doing their astrophotography just a little better each time they go out into the field.

5

Sensitometry

There should be a Society for the Perpetuation of Plain Simple Facts.
Ansel Adams

Introduction to general sensitometry

The word 'sensitometry' used in conjunction with photography would seem to imply that it is concerned with the sensitivity of the photographic material. When the field initially developed, its focus was on just that single element. However, it now includes the study of many other general characteristics of the photographic emulsion, the study of the action of developers on photographic material and the study of methods of quantifying these characteristics. We will look into a few of the basic properties of photographic emulsions that are particularly relevant to astrophotography. For a more comprehensive look at this fascinating field, the works by C.E.K. Mees, C.B. Neblette, and G.L. Wakefield are highly recommended.

Photographic exposure (E) is defined as the product of the intensity (I) of the light falling on the emulsion and the duration (t) of the exposure to light (i.e. $E = It$). The instrument that is used to produce a series of controlled exposures on the photographic material for the purpose of measuring various film characteristics is called a sensitometer. A sensitometer consists of a standard light source and a mechanism for varying the amount of exposure reaching various portions of the film being tested. The amount of exposure can be regulated by either varying the duration of the exposure or by varying the intensity of the light making the exposure. For astronomical work, it is best to measure film characteristics for constant exposure durations and therefore to use sensitometers which vary the intensity of the light falling upon the emulsion. Such sensitometric exposures are termed 'illumination scale exposures'.

Since one is interested in measuring the properties of an emulsion under circumstances which are comparable to those under which the emulsion is to be used, the light source should be chosen to match normal working conditions. In normal photography, this is not too difficult a problem as one is generally familiar with the type of light that is illuminating the subject. In astronomical work, one is dealing with photographing the light sources themselves, and the characteristics of these sources vary widely. The approach that has been taken by a number of workers is to adopt the sun as a standard source for astronomical sensitometry. Thus, the source is taken as a light source having a 5500 K 'color temperature' if standard daylight is chosen as the reference; one could also choose the surface of the sun as the standard and adopt a 6100 K color temperature as the standard. Most workers have adopted the 5500 K standard daylight for astronomical sensitometry.

The term 'color temperature' refers to the characteristic spectral distribution that light would have if the light source was a 'black body' (a perfect thermal radiator) at that particular temperature, radiating light because of its ambient temperature (see Fig. 5.1). For our purposes, we will not address the 'black body' issue and proceed with the idea that an incandescent filament heated by a flowing electrical current is essentially the same as a 'black body'. When measuring the color temperature of various common tungsten light sources, it is found that those which we normally use fall quite short of this 5500 K mark (see Table 5.1). Fortunately, this problem faces all photographers and various filters have been designed to convert the light from a wide array of sources to the standard 5500 K color temperature.

When a photographic material is exposed to light, some of the silver halide grains are rendered developable and, upon development, are reduced to metallic silver. After development, this metallic silver deposit appears to

Table 5.1. *The color temperatures of some common incandescent lamps with their corresponding wattages*

Light (W)	Temperature (K)
15	2350
40	2650
100	2900
250	2980

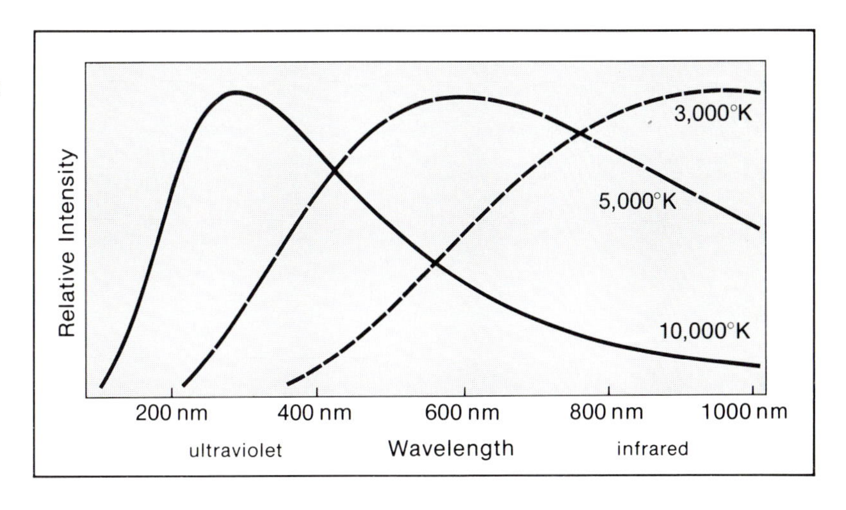

Fig. 5.1. The relative intensities of the light emitted by black bodies at three temperatures are shown as a function of wavelength.

the viewer as an increased optical density in the exposed area and the degree of optical density is an indication of the sensitivity of the photographic material to light. The optical density of the developed emulsion is directly proportional to the mass of the suspended metallic silver and can be easily measured. If I_0 is the amount of light incident on the film and I_t is the amount of light that is transmitted through the material, then I_0/I_t is the fraction of the incident light that is transmitted, this is known as the transmission or transparency (T). The inverse of T is known as the opacity ($O=1/T$). The optical density (D) of the light-absorbing silver layer is defined as the logarithm of the opacity (see Table 5.2)

$$D=\log_{10}O=\log_{10}(1/T).* \quad \textbf{(5.1)}$$

Thus, if a material transmits 1/10 of the incident light, it has a density of 1.0. If another material transmits 1/1000 of the incident light, it has a density of 3.0. If a beam of light is directed through a light-absorbing layer of material that transmits 1/10 of the light and then through another layer that transmits 1/100 of the light incident upon it, the net transparency is 1/1000 and the combined density is 3.0. The devices that are designed and built to measure the optical densities of light-absorbing materials are called densitometers.

The measurement of the optical density of film with a densitometer is complicated slightly by the fact that part of the light passing through the film is scattered by the metallic silver suspension. Thus, there are two mechanisms acting simultaneously which may affect the density measurements: absorption and scattering. If one measures only that light which passes through the film undeviated by scattering effects, the density will be greater than if all the light passing through the film is measured. Densities determined by measuring only the direct, unscattered are termed specular densities ($D\|$). The densities obtained by measuring both the direct light and the scattered light are termed diffuse densities ($D\#$). The difference between these two density values is dependent upon (*a*) the amount of light scattered by the silver deposit and (*b*) how much of that scattered light is included in each of the measurements.

In most printing situations, one is dealing with an apparent density that lies somewhere between $D\|$ and $D\#$ (contact printing is the exception to this where one is dealing exclusively with $D\#$). Thus it is desirable to measure $D\#$. The amount of light scattered by the emulsion is dependent upon the density of the material and upon the degree of non-homogeneity of the silver deposit in the film. In practice, the ratio $D\| : D\#$ is found to increase with the density and with the average grain size of the silver deposit. The divergence of these two measures of density caused by

* A similar relationship exists along with similar equations for defining 'reflection densities' for measuring the properties of photographic print materials. Our emphasis here will be on negative materials and the reader is referred to one of the references on sensitometry for more extensive treatment of this topic.

Table 5.2. *Values of density, opacity and transmission are tabulated for comparison*

Density	Transmission or reflectance (%)	Opacity
0.00	100.00	1.00
0.05	89.13	1.12
0.10	79.43	1.26
0.15	70.79	1.41
0.20	63.10	1.59
0.25	56.23	1.78
0.30	50.12	2.00
0.35	44.67	2.24
0.40	39.81	2.51
0.45	35.48	2.82
0.50	31.62	3.16
0.55	28.18	3.55
0.60	25.12	3.98
0.65	23.39	4.47
0.70	19.95	5.01
0.75	17.78	5.62
0.80	15.85	6.31
0.85	14.13	7.08
0.90	12.59	7.94
0.95	11.22	8.91
1.00	10.00	10.00
1.05	8.913	11.22
1.10	7.943	12.59
1.15	7.079	14.13
1.20	6.310	15.85
1.25	5.623	17.78
1.30	5.012	19.95
1.35	4.467	22.39
1.40	3.981	25.12
1.45	3.548	28.18
1.50	3.162	31.62
1.55	2.818	35.48
1.60	2.512	39.81
1.65	2.239	44.67
1.70	1.995	50.12
1.75	1.778	56.23
1.80	1.585	63.10
1.85	1.413	70.80
1.90	1.259	79.43
1.95	1.122	89.13
2.00	1.000	100.00
2.05	0.8913	112.2
2.10	0.7943	125.9
2.15	0.7079	141.3
2.20	0.6310	158.5
2.25	0.5623	177.8
2.30	0.5012	199.5
2.35	0.4467	223.9
2.40	0.3981	251.2
2.45	0.3548	281.8
2.50	0.3162	316.2
2.55	0.2818	354.8

Table 5.2. (*cont.*)

Density	Transmission or reflectance (%)	Opacity
2.60	0.2512	398.1
2.65	0.2239	446.7
2.70	0.1995	501.2
2.75	0.1778	562.3
2.80	0.1585	631.0
2.85	0.1413	708.0
2.90	0.1259	794.3
2.95	0.1122	891.3
3.00	0.1000	1000.0

emulsion characteristics is known as the Callier effect, and the ratio $D\|:D\#$ is known as the Callier factor.

One of the most useful methods used to illustrate the relationship between density and exposure in a given photographic emulsion is to plot the measured density (D) against the logarithm of the exposure ($\log E$) required to produce that density. This representation of the film's response to light has been almost universally adopted as the standard means of illustrating this relationship. This curve is known by a host of common names including: characteristic curve, Hurter–Driffield curve, H–D curve, and $D \log E$ curve. Hurter and Driffield were the first investigators to use this means of illustrating the response of a film to light in their classic work on photographic materials (see Fig. 5.2).

When the relationship between density and exposure is shown in this manner, a number of the features of the resulting curve are found to be common to virtually all photographic emulsions. A large portion of the curve is seen as a sloping straight line. Along this straight-line segment, the relationship between density and the log of the exposure is linear and proportional. The horizontal length of this straight-line segment is commonly known as the latitude (L) of the emulsion. Thus,

$$L = \log E_b - \log E_t. \tag{5.2}$$

The value of L for any given curve yields the exposure range over which the relationship between exposure (or $\log E$) and density remains linear.

Along this straight-line segment, the slope of the line yields the constant that relates a given change in exposure to the resulting change in density. The value of this constant is a measure of the film's ability to respond to luminosity changes in the subject; this is commonly known as the **contrast** of the material. The slope of that line, or the gradient (G) is found by dividing the produced density change ($D_t - D_b$) by the log exposure range required to produce that density change, or

$$G = (D_t - D_b)/(\log E_t - \log E_b). \tag{5.3}$$

The value of G is equal to the tangent of the angle (a) that the straight line portion makes with the horizontal axis of the D log E graph. It was found by Hurter and Driffield that, for any particular emulsion–developer combination, G was dependent upon the time and temperature of development. Hurter and Driffield adopted G as a measure of the degree of development and gave it the name **Gamma**. G can be calculated as given above, or it can be measured directly using a convenient device called a 'gammeter'. Thus,

$$\textbf{Gamma} = \gamma = G = \tan a. \qquad \textbf{(5.4)}$$

Many of today's photographic emulsions have characteristic curves which really show no true straight-line segment. Because of this, one may see the terms 'G-bar' (or '$\overline{G}$'), and 'contrast index' (or CI). G-bar is used in conjunction with Ilford materials and is Ilford's measure of film contrast, while 'contrast index' was created by Kodak for use on these new materials. G-bar is generally accepted as a valid measure of a films contrast, while contrast index is considered to be an arbitrary construct of somewhat less sensitometric value. Fig. 5.3 shows how the value of G-bar is obtained from the characteristic curve.

Below and above the straight-line portion, the relationship between density and exposure is non-linear and the amount of exposure required to produce a given change in density is greatly increased. Another means of stating this effect is to say that the 'contrast' is greatly reduced in these regions. It is important to remember that the given value of Gamma only applies to the 'straight-line' portion of the characteristic curve and that this value does not hold for any other region of the curve. The resulting effect of this compression is to make it difficult to distinguish luminance, or brightness differences which are present in the object being photographed. The region of the curve below the straight-line segment is known as the 'toe' while the region of the curve above the straight line segment is known as the 'shoulder'. Within the shoulder portion of the curve, a point is reached

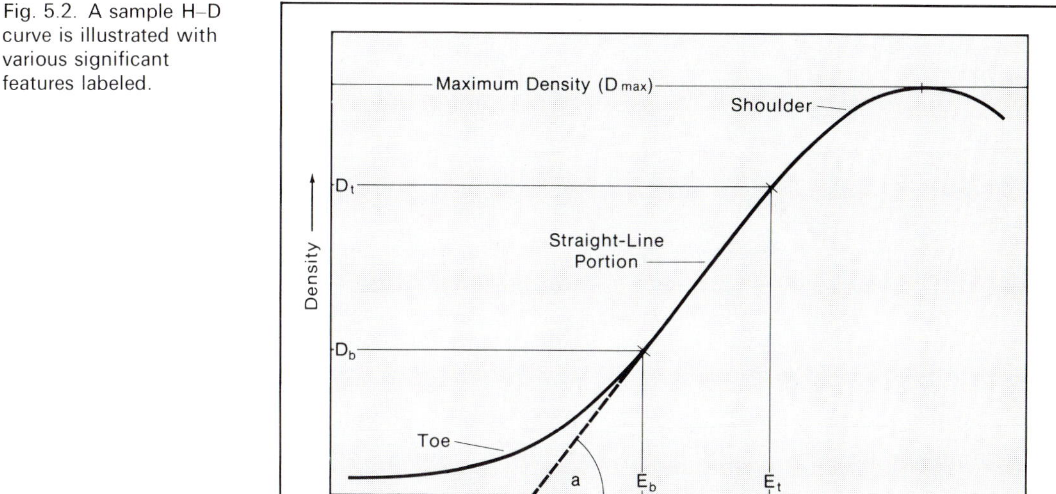

Fig. 5.2. A sample H–D curve is illustrated with various significant features labeled.

where it is not possible for the material to produce a density increase; the density at this point is known as D_{max}. Beyond this point, an increase in exposure will produce an actual decrease in the density of the photographic material.

Since Gamma is primarily a function of the degree of development to which any given emulsion is subjected, it is common for manufacturers to provide plots of the dependence of Gamma upon development time (Gamma is also dependent upon the wavelength of the light that is used for the exposure, see Chapter 7). These plots are exact for the carefully controlled temperature and agitation conditions specified for sensitometric work, but these curves will generally differ from what a normal user will encounter. Nevertheless, these curves still provide the user with guidelines that can be used to choose a development time which will yield the desired results. Examining the sample curve shown in Fig. 5.4, it can be seen that Gamma tends to increase rapidly for shorter processing times but that this rate of increase gradually slows and finally diminishes to zero. This situation will be found whenever a developer is used which has a very 'active' or highly energetic reducing effect on photographic emulsions. The Gamma measured at the point where no further increase can occur is known as 'Gamma infinity'. In choosing a development time, when using such developers, it is generally desirable to avoid times that fall in the region where Gamma is increasing rapidly; such processing times will make it difficult to obtain consistent results from processing run to processing run. Thus, it is a good habit to choose a processing time closer to the 'shoulder' of

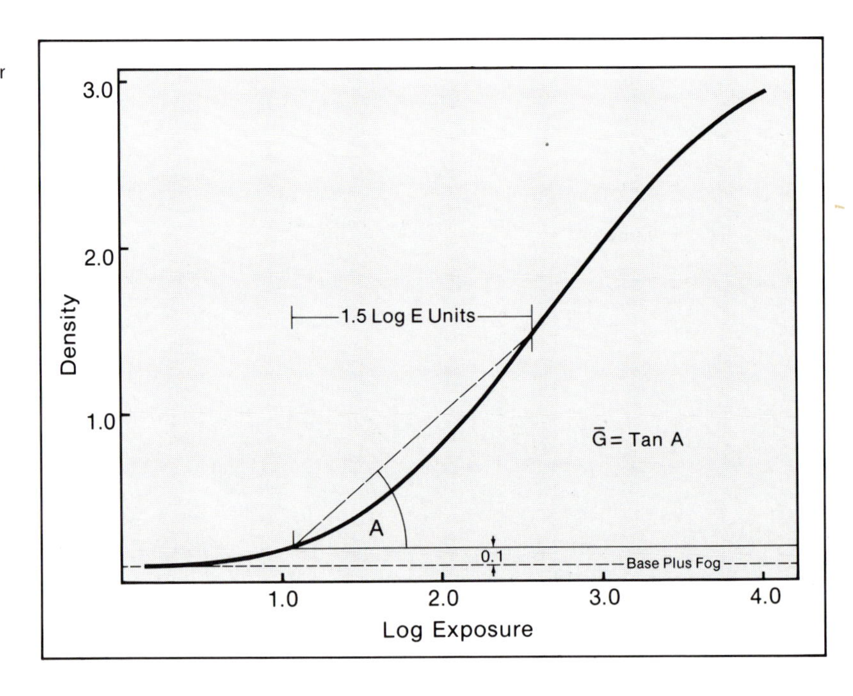

Fig. 5.3. The graphical method for deriving G-bar from an H–D curve is shown. G-bar is of particular use for films which exhibit only a very short straight-line segment.

this curve but not too close to the region where Gamma infinity is obtained. In the upper regions of these curves, near Gamma infinity, the development that occurs provides little information gain while continuing to increase the background fog level (and thus the 'noise' level).

Using a sensitometer and a densitometer one can measure Gamma and construct a 'time–Gamma curve' that is tailored to specific processing methods. In order to benefit from such an exercise, it is necessary to develop a set of 'darkroom habits' that will produce reasonably consistent results over a long period of time. This means finding a processing time, temperature, and agitation method that you feel comfortable with and that you will be able to stick to reasonably well. For example, if the cold water supply in your darkroom reaches a temperature of 73–5 °F in the summertime, it would be unwise to pick a standard processing temperature that is below 75 °F. Another important factor is the means of agitating the film while it is in the developer; some can consistently use the recommended methods, some roll the canister back and forth on a counter top, some place the processing canister on a mechanical drum roller, while others prefer to juggle the processing container. The actual processing method is not important as long as the agitation provided is uniform and your 'system' works for you. The important factor is consistency: once you have found a comfortable system, stick to it.

Another film property which is of particular interest in astrophotography is the speed of the film. For astronomical purposes, it is necessary to measure the film speed using exposures that are comparable to the exposures that are normally used at the telescope. There are several standard methods that are used for measuring film speeds. In each of these methods, the manner of exposure, the developer and the means of agitating

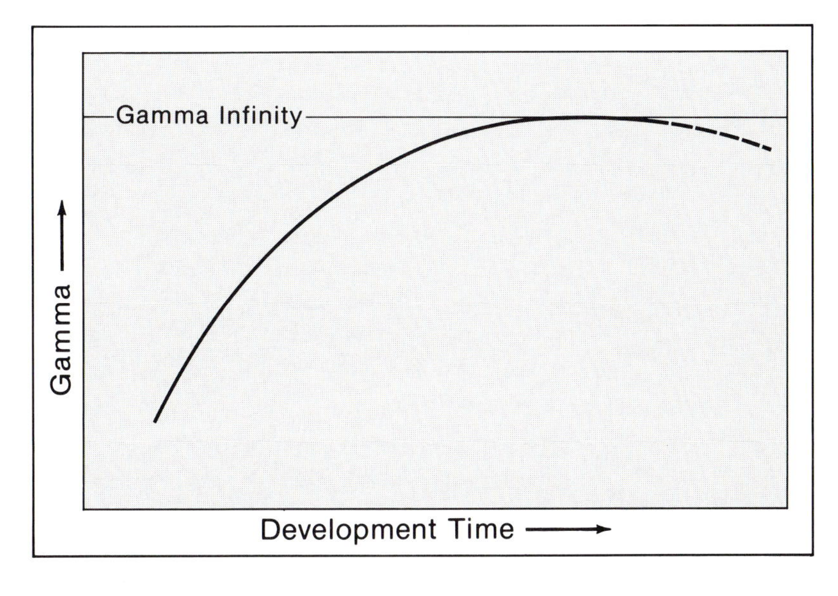

Fig. 5.4. This sample curve illustrates the variation of Gamma with development time. In this case, the film in question reaches a maximum Gamma and then slowly decreases as development time is extended.

the film during processing are described and followed in detail. Criteria which are used for measuring film speeds are:

1 the exposure required to produce the first visible density (this is the Weston criteria);
2 the exposure required to produce a density equal to Gamma;
3 the exposure required to produce a density of 0.1 above fog (this is the German Deutsche Industrie Norm (DIN) standard);
4 the exposure at the point of intersection of the straight-line portion of the characteristic curve when extended to the log E axis (this is the method used by Hurter and Driffield);
5 the exposure, E, at the point on the characteristic curve where the gradient is $0.3 \times$ the average gradient for a log E range of 1.5 when E is the minimum exposure (this was the method used by the American Standards Association – ASA – for the measurement of ASA values until 1962).

Of these methods, method 1 does not yield a speed that is a measure of the usefulness of the film, as such densities are virtually always on the 'toe' portion of the characteristic curve and it is not generally possible to print such low densities. Methods 2 and 4 are very dependent upon the contrast of the photographic material and upon the shape of the toe region of the characteristic curve. While method 5 is actually based upon the minimum exposure which will yield a useful print, it is somewhat dependent upon the contrast of the material and extremely difficult to calculate. Therefore, method 3, or some variant thereof (the ASA and ISO values that are quoted today are based upon a variant of this method), is the generally adopted means of measuring the speeds of photographic materials and is quite useful in astronomical applications. The density level of 0.1 above base plus fog is a density which can be printed by an experienced printer and it is relatively unaffected by either the shape of the toe portion of the curve or by the contrast of the material.

Using the method described above, it is possible to measure the relative speeds of many films by making exposures using a sensitometer and then measuring and plotting the densities caused by each exposure level. This will yield speeds that show exactly how much faster or slower different films are under any given conditions. This procedure can also be used to show the effects that different developers and different processing times have upon the speed of any given emulsion. In order to assign actual numbers to the speeds of various emulsions, it is necessary to assign a speed to one emulsion/developer/temperature/time combination that is adopted as a standard. When a 'standard' is chosen, the exact conditions needed to reproduce the standard speed must be reported and it is essential that the criteria used to measure the speeds be reported. In the following section we will describe the design, construction, and use of a sensitometer that can be used to make sensitometric exposures for astronomical applications.

Design and use of an inexpensive sensitometer

The issue of 'film speed' is a continuing topic of discussion and debate in astronomical photography. Many amateur articles have been dedicated to this topic but very few have reached conclusions that would stand up to any quantitative scrutiny. Most articles on this topic use methods that are highly suspect and utilize no controls to insure the validity of the test results. Even when such tests are carefully planned and executed, the final results can only conclude that Film *A* is faster than Film *B*. If the reported conclusions go beyond that point (for instance, by stating that Film *A* is twice as fast as Film *B*) the results can virtually never be supported due to the fact that no controls were used to conduct the test. Many of the test results reported end up raising more questions than they purport to answer. In order to measure accurately the effective 'speed' of any film used for long-exposure astronomical photography, a number of controls need to be incorporated into the test procedures and, in order for a reader of such a report to be able to interpret the report and to be able to judge the applicability of the results to any particular problem, the test controls need to be reported with the test results. The issue of controls and test procedures is of even greater importance when the film being tested is a color emulsion.

Among the numerous controls that need to be incorporated into any such tests of astronomical film speeds are the following:

1 the light source used must be of constant intensity;
2 the light source used must be of a constant color temperature which is known and which is appropriate for the planned test (this also implies an incandescent light source);
3 the intensity of the light source must be low enough to allow exposures that are typical of astronomical photography (i.e. 15 min to over 1 h);
4 the 'target' that is photographed must be constructed so as to provide accurate and measurable information on the sensitivity of the film;
5 the 'target' must provide many calibrated intensity levels in order to allow the measurement of various emulsion characteristics (e.g. speed, contrast, toe and shoulder characteristics);
6 the target used should be neutral grey in color, any studies directed at measuring color sensitivity should use filters to alter the color of the light falling on the film;
7 if the film under study is a color emulsion, the 'target' must be constructed to allow accurate measurements of the 'color balance' of the emulsion;
8 processing procedures must be standardized and carefully controlled;
9 the measurement sets that are produced in the course of testing any emulsion *must* be repeatable in order for the results to be of any value;
10 the methods used to achieve the above conditions must be reported with the results of the test.

These criteria are *all necessary* in order to guarantee the validity of any test results. If these criteria are met, accurate quantitative measurements can be made and the results of any tests would be repeatable by others interested in this area. In addition, readers of such test results will be able to relate the test conditions to the type of work that they are interested in doing. Going beyond the needs for accuracy and care in conducting such tests, the issues of cost and construction need to be considered. Therefore, the following points were added to the list above:

1 the materials used to construct such a device must be inexpensive;
2 the materials used must be readily available;
3 the device must be of simple design.

The design which was finally settled upon is shown in Fig. 5.5. It utilizes a 15 W incandescent light source, which was measured using a color temperature meter, to provide light at a 2350 K color temperature. This light is converted to a 5500 K color temperature (i.e. standard daylight color temperature) by use of a Kodak Wratten Color Conversion Filter (No. 78). This light is reflected off a neutral grey test card which is placed behind the test target. The target is a Kodak No. 3 step tablet, having 21 steps

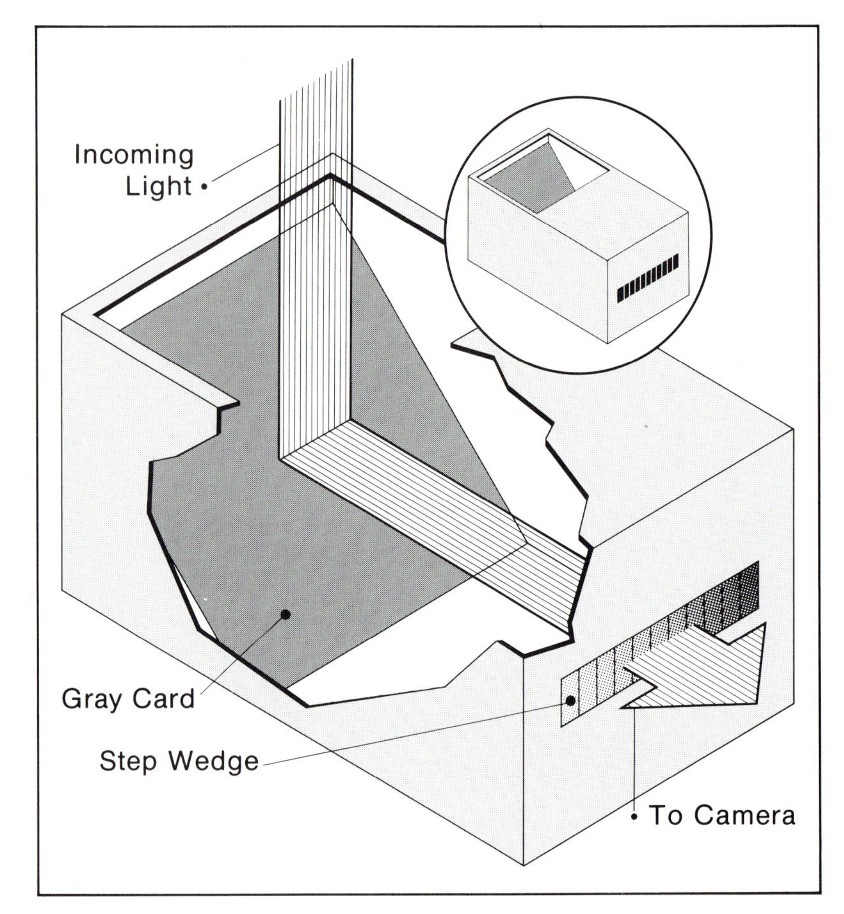

Fig. 5.5. A cut-away view of a simple and inexpensive sensitometer is shown to reveal its critical parts and the path of a beam of light through the sensitometer. The inset indicates the appearance of the working model.

differing by 0.15 density units (one-half stop). This illuminated target is then photographed using a camera equipped with a moderate telephoto lens (135 mm f.l.) set at an f-stop that will provide maximum useful information (determined by testing). In order to insure that tests can be repeated, the distance from the light source to the target must be constant from test to test. In addition, the tungsten light source must be allowed to warm-up for a period of at least 10 min prior to starting any sensitometric exposures since it has been determined that such a warm-up period is required in order to stabilize both the intensity and the color temperature of a tungsten source.

The sensitometer housing can be constructed from virtually any material. When we initially constructed the prototype of the device shown to evaluate it as a functional tool, it was constructed from cardboard and taped together. Our final model was constructed using thin (1/16 inch or 1.6 mm) plexiglass. Plywood would have served equally well, but the weight of this device was a prime consideration. In essence, all of the materials can be obtained from scraps left over from other projects. The only exceptions to this are the neutral grey test card and the calibrated step wedge. The step wedge can be precalibrated by the manufacturer, or an uncalibrated step wedge can be purchased and then calibrated by the user with a densitometer. If a densitometer is available to use for the calibration, the cost of the step wedge is greatly reduced. (Note: most color processing labs have color densitometers on the premises and it is usually possible to arrange to use these expensive devices. If such facilities are not available, surplus densitometers can be purchased at reasonable prices.)

Once this device has been constructed and an appropriate f-stop has been determined for the required exposure time, there are several pieces of information that can be extracted from the exposed film without the use of any additional equipment. The relative speeds of both black-and-white films and color films can be estimated by inspecting the comparison frames, and some relative judgments can be made regarding the overall color balance of color emulsions being tested. Without the use of a densitometer, no further information can credibly be derived from the test exposures. Thus, we can now make the following types of subjective statements based on the information obtained from the sensitometer test exposures:

1 Film *A* is faster than Film *B* when used in exposures of *X* min, and it appears to be about *N* stops faster (this statement can be made for both black-and-white emulsions and color emulsions);
2 color Film *A* yields a more neutral color balance when used in exposures of *X* min than does color Film *B*.

If a densitometer is available to examine the sensitometer exposures, quantitative measurements and statements can be made and more information can be obtained from each set of test exposures. However, in order to use the densitometer on these film strips, some more conditions must be imposed upon the testing procedures and standardized measuring procedures must be adopted and reported with the results. The conditions and conventions that must be adopted in order to insure the validity of the test results include:

1 the size of the individual step bars recorded on the film must be large enough to measure with the smallest densitometer aperture and allow extra margin space to insure that 'edge effects' do not impair the interpretation of the data;
2 standard filters must be adopted when measuring the sensitivity of individual color emulsion layers;
3 the 'speed' of all emulsions must be determined for particular exposure times;
4 the lens on the camera being used (if any) needs to be stopped down ~2 stops in order to reduce any vignetting effects;
5 the 'speed' of all emulsions must be measured at a particular density above the film (base density plus fog density) and this point must be a standardized value. (Note: The various conventions that can be adopted for measuring film speeds have been discussed earlier in this chapter.) The standard that we have adopted for measuring 'speed' is to measure at a density of 0.10 above base+fog density (this convention allows the speed to be measured at a point that almost always is free of the effects of contrast variations);
6 all standard and non-standard parameters must be reported with the results to allow for proper interpretation of the results by the readers.

Using the densitometer in conjunction with the above control procedures, the following parameters can be measured quantitatively from the test exposures made in this 'poor man's sensitometer':

1 actual numbers can be assigned to the speeds of the films that are being measured once a film/developer/time combination is adopted as a standard and a value is arbitrarily assigned to that reference film/developer/time combination (i.e. our standard reference is unhypered 103a-F processed in D-19 for 4 min at 20°C and we have assigned a value of 100 to the speed of this standard);
2 the H–D characteristic curves can be obtained by plotting film density against exposure (with this type of step wedge test target, one merely has to plot film density against step wedge density);
3 from the H–D curve, the contrast index can be accurately measured from the slope of the straight-line portion of the curve;
4 with color films, the relative speeds of all three emulsions can be accurately measured using tri-color filter sets in conjunction with the densitometer (this allows the color balance to be measured accurately and also allows one to determine the actual filtration that would be required to correct the color balance at any particular exposure time);
5 with color films, the contrast of all three emulsions can be plotted and measured to check on the severity of any 'cross-over' effect caused by reciprocity failure;

6 since actual relative speeds can now be measured, the effects of altering one or more parameters can be accurately evaluated (e.g. the effects of different developers, varying development time, intensifying a negative, pre-exposing film, hypering film, using cold cameras, etc. can be measured accurately);

It should be quite evident that the combination of this seemingly crude sensitometer and a reliable densitometer creates a very powerful tool that will allow numerous tests to be carried out independent of the numerous variables that normally affect the results of such tests when carried out at the telescope. Such 'in-telescope' testing is subject to too many variables that lie beyond the control of the experimenter. Seeing, transparency, temperature, viewing elevation all detract from the validity of film testing *in situ*. By carrying out these tests in the darkroom, stringent controls can be imposed upon the tests and valuable telescope time can be saved for actual photography. In addition, the imposition of these controls adds to the credibility and applicability of the test results.

In Fig. 5.6 we can see how the densitometer is used to transform the raw data (i.e. the test frames) into a format that can be measured to extract the information that we are actually seeking. The frames are measured and the test frame densities are plotted against the actual test target densities to prepare the characteristic curve. From this curve, we can extract the contrast and, by comparing the curve derived from any given test frame with the curve obtained from our reference standard, the relative speed can be obtained. Fig. 5.7 shows a number of plotted characteristic curves for some common astronomical emulsions and illustrates changes induced by different treatments. The speeds of a number of hypersensitized emulsion samples appears in Table 5.3 where the combination that resulted

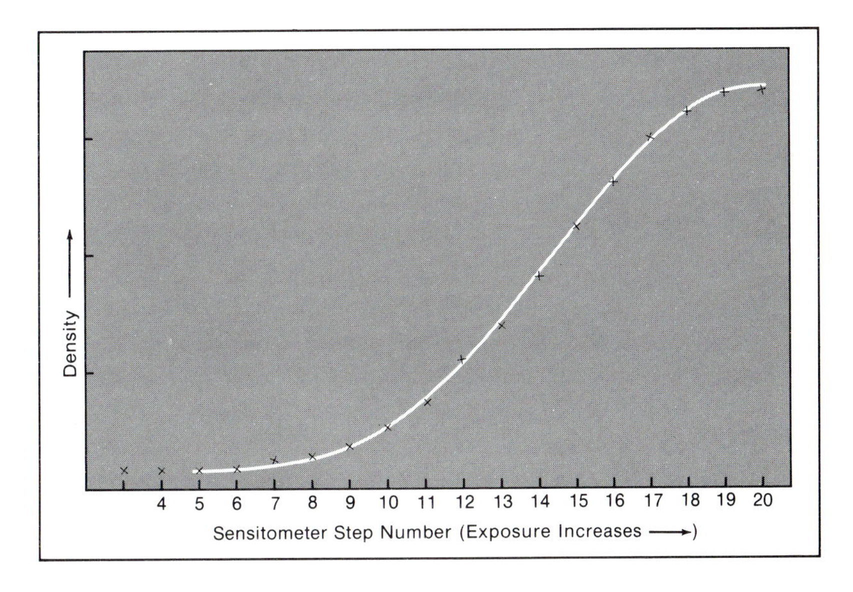

Fig. 5.6. This illustrates how the density measurements from a film sample exposed in the sensitometer can be plotted to form a characteristic curve.

in each of the curves is identified. By comparing the H–D curves and the tabulated speeds, it can be seen that higher relative speed values are assigned as the intersection of the 'normalized' characteristic curve (i.e. the density plotted is the density above fog + base density for all films) with the 0.10 density line moves farther to the left (i.e. the intersection moves in the direction of lower exposures. The interpretation of this effect is quite simple and straightforward: for a given film to be faster than some other film, it

Table 5.3. *Relative speeds of photographic emulsions during 1 h exposures*

Film	Developer	Processing time at 68 °F (min)	Speed[a]	Treatment
Plus-X	D-76(1 : 1)	8	13	None
103a-F	**D-19**	**4**	**100**[b]	**None**
103a-F	D-19	4	109	Continuous agitation
Tri-X	D-76(1 : 1)	8	115	None
Plus-X	D-76(1 : 1)	8	118	13 h hydrogen @ 70 °F
Tri-X	D-76(1 : 1)	8	144	13 h hydrogen @ 70 °F
Plus-X	MWP-2	10	162	16 h hydrogen @ 70 °F
103a-F	MWP-2	10	165	None
Tri-X	D-76	8	316	13 h hydrogen @ 70 °F
Tri-X	D-76	8	437	16 h hydrogen @ 70 °F

[a] The speeds listed above should only be used for general comparisons as film speeds (particularly astronomical films) vary from emulsion batch to emulsion batch.
[b] This was chosen as the standard and assigned a speed of 100.

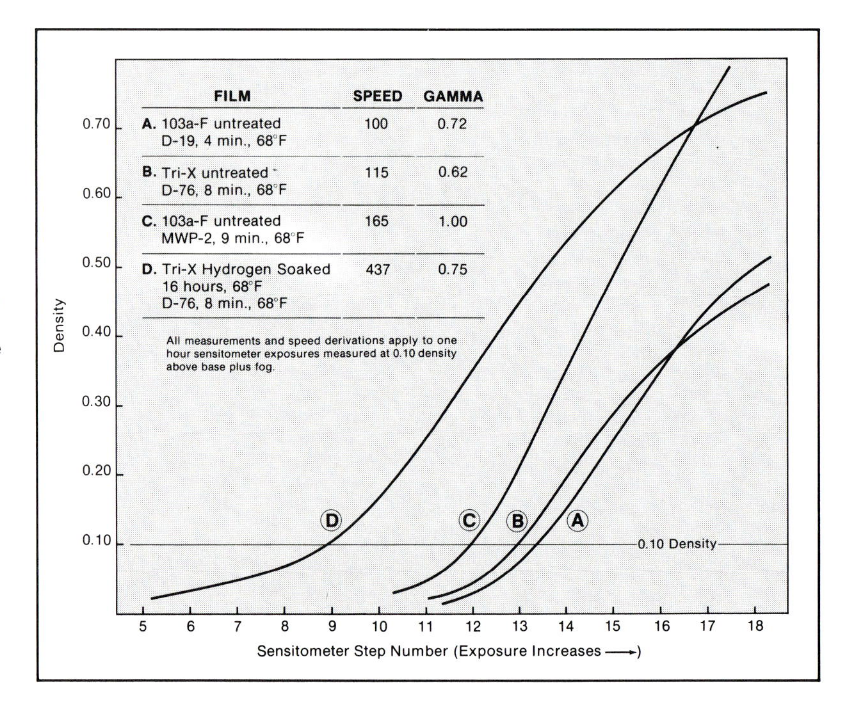

Fig. 5.7. The characteristic curves of two well-known Kodak films are plotted before and after they are given treatments that yield significant speed increases. The curves and speeds are all plotted for 1 h sensitometer exposures and the speeds have all been measured at a level of 0.10 ANSI diffuse density above base plus fog.

must require less light to produce the same film density (see Figs 5.8 and 5.9).

It cannot be overemphasized that all conditions surrounding any tests of emulsion speeds or emulsion color shifts must be reported with the final results in order for such tests to be of value to others interested in utilizing these test results in their own work. When variables are left out of test reports, the results are virtually invalidated and the time spent in conducting the test and in preparing a report of the results has been wasted. By reporting all facts surrounding the test, it also allows others to study your results and to suggest changes that might improve upon the results that were observed.

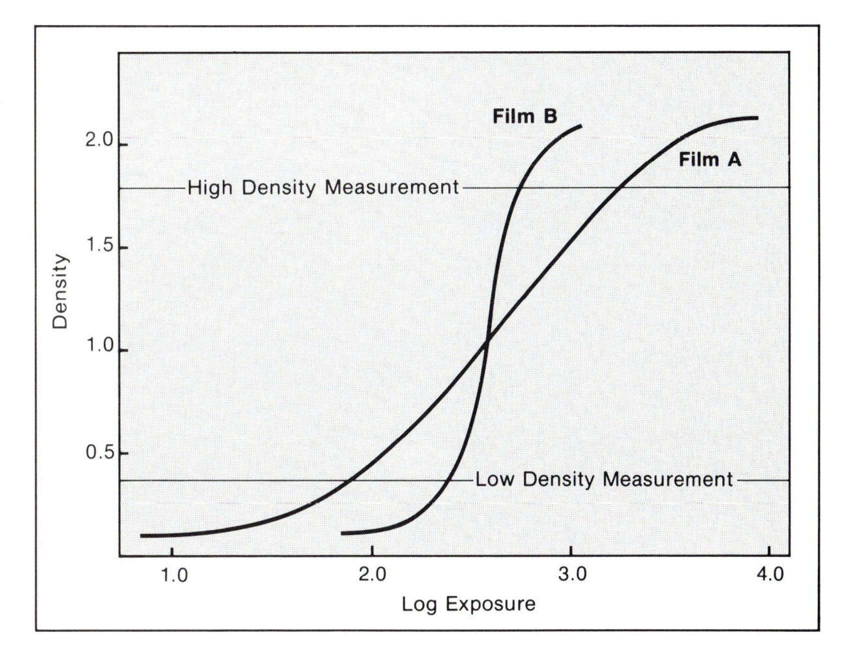

Fig. 5.8. Correct and incorrect methods for measuring film speeds are illustrated. While Film B is clearly more contrasty than Film A it is obvious that Film A is considerably faster than Film B. Only if the film speed is measured in the higher densities does one get the false impression that Film B is faster.

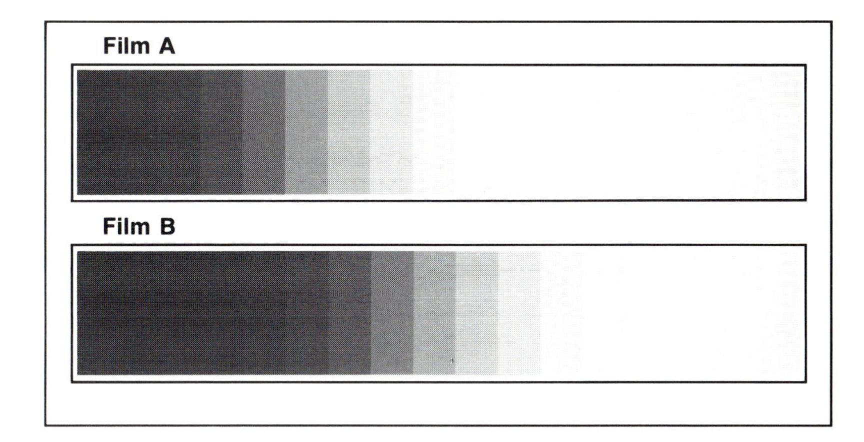

Fig. 5.9. This set of photos shows the appearance sensitometer strips from two films of significantly differing speeds.

The densitometer

While we have mentioned numerous things that one can do with a densitometer and the advantages of using a densitometer instead of making an educated estimate, the nature of this laboratory beast has as yet not been revealed. There are several models available and different applications will require different capabilities. We think it prudent to discuss the general design of a densitometer along with some of the features available and the uses of these in a number of astronomical studies.

A transmission densitometer used for measuring film densities is a device that forms an image of a small area of film, that is illuminated from behind, which is then directed onto the surface of a photosensitive cell. If the light illuminating the film is diffuse, one is measuring diffuse density, while measurements of density produced using a collimated beam of light are termed specular density. The density values produced by each of these methods are quite different so it is necessary to state the type of density measurements for interpretative purposes. Diffuse density measurements are found much more frequently than are specular density readings. The current output by the photo cell is then measured in some manner so that actual numerical values can be attached to the amount of light that falls upon the photocell. It is obviously important that the densitometer be calibrated regularly in order to make these numerical readouts meaningful and, if color work is being done, it is important to keep the light source color temperature constant. A densitometer used for color work has a number of added features including a standard set of color filters for measuring the film density in red, blue, and green light. The readout system can consist of a meter as on older models, and a few of the less expensive newer models, or it can be a digital readout. The digital readout form is more accurate, easier to use, and less subject to user errors and is to be preferred whenever possible. A densitometer designed for use on prints is known as a reflection densitometer and has the same basic elements as a transmission densitometer but with a markedly different structure. In a reflection densitometer, the light must be directed down onto the surface being measured and the light reflected off this surface is then measured.

A high quality color transmission/reflection densitometer with a digital readout cost approximately US$3000 in 1982. The cost is generally prohibitive to the individual but several alternatives are available. Since one cannot honestly say anything quantitative about film speeds, developer effects, or hypering effects without the use of a densitometer, we strongly urge all who are interested in such studies to investigate methods of gaining access to a quality densitometer. One alternative to buying a very expensive densitometer is to investigate the possibility of locating a surplus model at a significantly lower price. In the USA in 1982, a quality black-and-white densitometer could be found for as little as $100 while a good color densitometer could be had for $700–$1000 (it should also be noted that a black-and-white densitometer can be used for color work by using some relatively inexpensive filters). A second alternative that is available is to locate a densitometer that is not in constant use and try to obtain permission to use it (provided that it is found to be in good condition). Many quality custom photographic laboratories have excellent densitometers that sit unused for much of the time and one can often persuade the owner of such laboratories to permit the use of the densitometer for some photographic studies (the owners are often also devotees of quality in photography and interested in promoting conscientious photography enthusiasts).

The final alternative is to build a densitometer of one's own. With a primitive knowledge of electronics and a reasonable degree of mechanical skills, one can build a model that is close to the accuracy of a good commercial densitometer. The crude schematic that we show in Fig. 5.10, gives the basic details of the bare necessities of a densitometer but, for more on the requirements of a precision model, we refer the reader to G.L. Wakefield's work on sensitometry. If one uses a commercial photographic spot meter in place of the upper stage of the densitometer in Fig. 5.10, a reasonably accurate model can be built with a minimal amount of

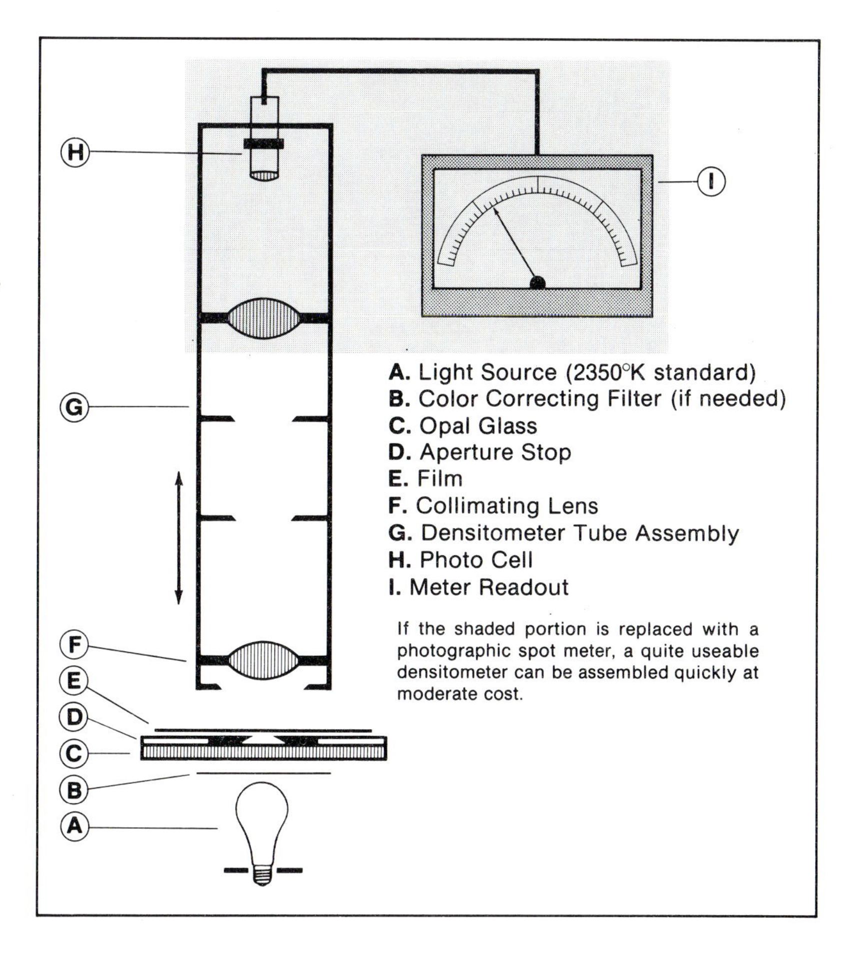

Fig. 5.10. A schematic design for a simple densitometer is shown. The upper tube assembly is moved down into contact with the lower light stage in order to make density measurements. The upper portion of the tube assembly (shaded) can be replaced with a commercial photographic spot meter to produce an accurate instrument at moderate cost.

mechanical and electrical skills. The use of a photographic spot meter saves one the trouble of wiring a photo cell and a readout device but it also severely limits the precision of the work that can be performed with the densitometer. A commercial spot meter is seldom readable to any greater accuracy than one-third of a stop, which translates out to about 0.10 density units. This would allow one to place approximate numbers on film speeds if the speeds were significantly different, but it would become less useful when measuring very similar film speeds. However, even this degree of accuracy is an order of magnitude better than casually comparing two strips of film and then declaring to interested friends that 'Film A is at least three or four times faster than Film B'! If one cannot buy or borrow a densitometer, it is reasonably easy to build such a simple instrument and if one is interested in testing film speeds and the effects of things like developers and hypering methods on film speeds, the densitometer is indispensable.

6

Black-and-white photography

Now, we are faced with an absolutely impossible task: describing the aspects of black-and-white photography in a single chapter. It is barely possible to give a reasonable overview in a single book, much less a meager chapter. We will not even try! Instead, we will provide some information that is generally not available in the amateur literature: an introduction to the physics and processes that allow photography to exist. There are many reasons for learning the basics of photochemistry and emulsion structure but the main reason is quite simple: with some background knowledge, one is able to understand what is happening in various processes and thus it is often possible to evaluate possible applications of 'new' techniques without investing large amounts of telescope or darkroom time. We hope that our readers will share our interest in this area of photography and be willing to investigate some of the sources that we have included in the extensive bibliography contained in this volume.

Film characteristics

The emulsion: physical composition

The photographic emulsion and the process which permits light falling on the emulsion to be captured and converted into a form which can later be developed into a visible image that can be reproduced is extremely complex. Although the photographic process has been known for over 150 years, carefully documented and studied in private and commercial laboratories, and emulsions have been dramatically improved through such research efforts, the process is not fully understood. Many interesting photographic effects are known and documented, the literature on photographic research fills libraries. New and improved films and developers are continually being introduced into the commercial market, yet the exact mechanism still eludes researchers and the list of factors influencing the performance of a photographic emulsion continues to grow. As data are accumulated and various photographic effects are studied in greater and greater detail, the complex nature of this process is slowly being unraveled.

The silver halide suspension. The photographic emulsion consists of extremely minute crystals of silver halide suspended in a gelatin medium which acts to 'solidify' the suspension into a 'stable' mass which can be spread thinly upon some supporting structure, such as glass or the Estar bases that are commonly encountered in modern commercial films. Halides are those elements which are members of the halogen family, sharing various common attributes. Halides include: chlorine, bromine, iodine, fluorine, etc. In negative emulsions, the emulsion is normally made up of crystals of silver bromide with a small amount of silver iodide. In virtually all commercial photographic papers, the emulsion is composed of a mixture of silver chloride and silver bromide.

The silver halide crystals in negative emulsions range in size from submicroscopic to a maximum of 3–5 μm and make up around 40 % of the mass of the emulsion. The size range of the crystals varies significantly depending upon the average speed of the emulsion and upon the exact chemical composition of the halide suspension (i.e. bromo-iodide v. bromo-chloride emulsions). These crystals vary in form from triangular and hexagonal platelets to long needles and minute spherical shapes. In addition to the light-sensitive halide crystals and halide ions, various contaminents, such as silver sulfide and metallic silver, are also suspended in the emulsion. Some of these contaminants play an important role in the process of converting light into a developable image.

The gelatin which supports the photographic emulsion serves a number of purposes which go far beyond its role as a substrate. The gelatin itself is not a sharply defined chemical, it is produced by processing animal hides and thus is composed of proteins and chains of amino acids. At ~40 °C, it becomes a solution when placed in water, but at 20 °C it is a 'rigid jelly' which can then be dried out to 'set' it into an even more stable form. Once set and dried, the gelatin is still permeable enough to allow developers and other processing solutions freely to penetrate the emulsion. The gelatin also acts to regulate the size of the silver halide crystals as they are formed in the suspension. In addition to all this, some of the impurities present in gelatin are known to increase the sensitivity of the emulsion to light. The gelatin also forms a barrier around the silver halide crystals which causes a differential in the rate at which exposed and unexposed crystals are affected by the developer.

The silver and halide ions are bound together in a lattice having a cubic structure. If the crystal is formed in the presence of an excess of halide ions, it will be negatively charged with respect to the rest of the solution. This condition is a prerequisite for the formation of a developable latent image, as the formation of positively charged crystals would result in grains which would be developable even if unexposed. The excess halide ions are usually supplied by the addition of a salt, like potassium bromide, to the emulsion mixture. These excess halide ions adhere to the silver ions at the surface of each of the crystals and thus the entire crystal surface is covered with halide ions. In addition, the normal crystalline structure of each grain contains defects created by the incorporation of minute contaminant particles in the crystal lattice. These contaminants appear to play a major role in the formation of the latent image, their loci are termed sensitivity specks.

A 1917 study gave the first clue to the existence of sensitivity specks and raised another round of questions about the formation of the latent image. A number of exposed emulsion samples were studied by stopping the development process at various stages and then studying the grains under the microscope. The first traces of any developed image appeared as tiny specks showing that the development had commenced at a small number of sites on the surface of the grain. As development proceeded these specks grew in size and the grains developed to completion by the growth of these same specks and not by the formation and growth of any new specks.

The photographic mechanism

Latent image formation. Decades of studies on the formation of the developable latent image and the contributions of quantum chemistry have yielded a theory which explains many of the observed phenomena and provides a fairly good picture of how the radiant energy is transformed into a form which the developing agent can render into a visible image. This theory was proposed by Gurney and Mott who utilized the fact that metallic silver is present in the crystal lattice and combined this with the knowledge that crystal structures are not perfect but usually contain flaws in the form of missing ions and/or ions out of their normal positions. These missing ions create 'holes' in the lattice structure which are probably created by the thermal motions of the ions within the crystal.

Using this information, Gurney and Mott proposed the following explanation of the photographic mechanism: when a silver halide crystal is exposed to light, some of the energy is captured by electrons around the ions which causes some of those electrons to be elevated to a higher energy state; this, in turn, allows the electron to wander about freely in the crystal. These freed electrons wander until they come in contact with a speck of metallic silver and are then trapped. The speck of metallic silver is thus given a negative charge and this causes it to attract a positive silver ion which neutralizes the charged center and adds another atom to the mass of metallic silver. As this exposure to light continues, this process repeats itself again and again. The silver ions migrate from hole to hole through the lattice to neutralize the clump of negatively charged silver and the mass of metallic silver continues to grow. Thus, the effect of light which is uniformly absorbed by a silver halide grain is concentrated into just a few small specks. As the exposure increases, these specks grow and form increasingly larger clumps of metallic silver.

Spectral sensitivity of photographic emulsions. The basic emulsion formed by coating a suspension of silver bromide on some type of support layer is only sensitive to the blue, violet and ultraviolet portions of the spectrum. It was not until 1873 that Hermann W. Vogel discovered a means of altering the spectral sensitivity of the photographic emulsion. Vogel was apparently following the work that was being done on the study of the halation effect and noticed that the addition of dyes to the emulsion layer not only helped to eliminate some of the effects of halation, but also helped to eliminate the 'blindness' of the photographic emulsion to the longer wavelengths of the spectrum. After making this observation and studying the process employed in the halation work of Wortley, Vogel reasoned, quite correctly, that the addition of a dye to the emulsion will sensitize the emulsion to the regions of the spectrum which are absorbed by the dye being used. Vogel tested and described the effects of a number of dyes and found that cyanine was the most efficient sensitizing agent for photography of the orange and red regions of the spectrum. Today, various members of the cyanine family dominate the group of dyes that are used to alter the sensitivity of photographic emulsions to suit a wide variety of applications. Photographic emulsions can now be sensitized to light from the red, and infrared regions of the spectrum out to around 1300 nm.

To expand on this principle, one can alter the sensitivity of an emulsion to make it sensitive to some region of the spectrum by adding a dye to the emulsion that will absorb the light from the region of interest (see Fig. 6.1). Thus, the addition of a red dye will increase the emulsion's sensitivity to the cyan and green regions of the spectrum and the addition of a cyan dye will increase the sensitivity to the red area of the spectrum. In addition, mixtures of dyes can be added to the emulsion to provide a broad sensitivity band that covers a large portion of the spectrum from the ultraviolet (UV) to the near infrared (IR). Over the past century, numerous volumes have been written on the properties of dyes and classes of dyes that have qualities which can affect the photographic sensitivity in some desirable manner.

For most practicing astrophotographers, the molecular properties of the various dyes and the quantum mechanics of the energy-transfer process to the halide centers is of little importance. However, since most of the emulsions in use (the 'O' emulsions being basically a raw emulsion that has

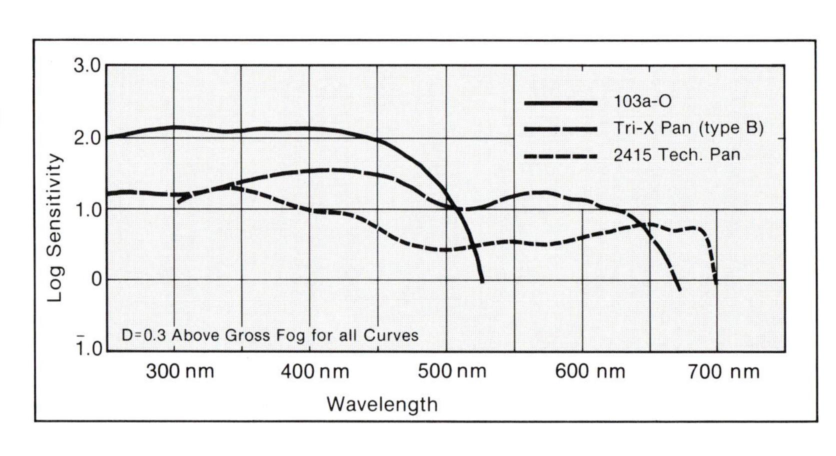

Fig. 6.1. The spectral sensitivities of three well-known Kodak emulsions. The logarithmic scaling of the sensitivity axis produces a compressed appearance as each division of the vertical axis represents a difference of a factor of ten in sensitivity.

not been dye sensitized) have been spectrally sensitized in some manner, it is important to realize that there are some side effects of the dye sensitizing process. The primary effect that has a significant impact in astronomical applications is known as 'dye desensitization'.

When a dye is added to the raw emulsion, the resulting emulsion has an added sensitivity to some new region of the spectrum. However, it is generally found that the sensitivity in the region of the spectrum that could be recorded using the raw emulsion has been decreased by the effects of the dye agent. Only by observing the speeds of various emulsions at a variety of wavelengths can one determine readily what effects the dye sensitizing process have had on a number of similar emulsions. However, most photographers fall into the trap of looking at the spectral sensitivity curves that are published by Kodak and others and see that the sensitivity is the same over most of the spectrum except in area X where they would like to get some extra sensitivity. However, the families of spectral sensitivity curves commonly published have been plotted on logarithmic scales which greatly conceal sensitivity differences. For instance, a sensitivity difference of a factor of 2.0 results in a change of 0.3 when plotted on a logarithmic scale and a sensitivity difference of a factor of 3.0 results in a change of just 0.48 on a logarithmic scale.

This situation is commonly seen in astrophotography when one is selecting one of the 103a-emulsions for some application. Many will just go by the time-hardened concrete rule that one uses 103a-E for H-alpha nebulae and uses 103a-O for galaxies and most reflection nebulae and will never even bother to look at the curves. Most others will look at the curves and, having no idea of what has caused the sensitivity of 103a-O to be altered to form 103a-F and 103a-E (and a host of others), simply see that 103a-E has a nice little sensitivity hump near the 656.3 nm H-alpha (Hα) line and conclude that the old rule is correct. However, a few letters to Kodak, some lab tests, and some browsing through the literature will turn up hard facts (which should have been noticed while working at the scope over a few years) which will show that, when exposed to blue light, 103a-O is actually almost twice as fast as 103a-F and that 103a-F is again some 25 %+ faster than 103a-E. If the three sensitivity curves are then readjusted and plotted to show actual sensitivities, a substantial difference will be seen.

Finding that the suspicions which we had accrued after several years' of work with 103a-O, 103a-F, and 103a-E were justified led us to adopt a whole new policy about these films. We now use 103a-O for galaxy work, star-cluster work, and shots where we know that we are working with a blue reflection nebula. We have dropped the use of 103a-E for any applications and have replaced its role in nebular photography with 103a-F. While some still argue that 103a-E yields better results when using a No. 29 filter to photograph Hα nebulosity, we have not been able to see any significant differences in contrast but can see the 25 % speed loss caused by using 103a-E. We will also point out that when photographing star fields containing nebulosity without the use of a red filter 103a-F helps to shorten exposure times in two ways: since the film does not have a sensitivity dip in the green, the integrated light from the background stars is accumulated much more quickly than with 103a-E and the increased sensitivity of 103a-F in the green also helps to record the nebulosity by picking up the light from the 500.7 nm OIII light.

Image structure: granularity v. graininess. Once light begins to strike the film and the photochemical reactions start within the emulsion, the processes that take place are clearly quite complex and generally beyond the interest of the average photographer. Yet knowledge of some of the processes is of use in understanding some of the effects that can be measured and observed in the negative that results from the process. For instance, the ultra-microscopic nature of the structure of a small portion of an image is of little concern to one who only makes 5 × to 10 × prints, yet this is how one learns of the growth of a photographic image and some very useful applications have come from this knowledge.

The image of a point source of 2–5 μm diameter does not just produce a small circular image of 2–5 μm on the emulsion in question. Normally, the image of such a source will be at least 10 μm in diameter, and the size of the image will depend upon the duration of the exposure (see Chapter 4 for a thorough discussion of factors that influence the limits of photographic resolution.) In fact the relationship between exposure and image size for a given imaging system can be so well expressed that it is generally possible to use a photographic image for making photometric measurements of all stars over a range of about 8 magnitudes on any given plate. Besides this well-known effect, it has been known since the early 1920s that the image of a star that is lightly exposed only penetrates a few molecules deep into the emulsion, while a heavily exposed image can penetrate the entire thickness of the emulsion layer. However, unlike the applications of radial-image growth to the field of photographic photometry, the knowledge of the third dimension of the image was only lightly used in a few special applications until the late 1970s when David Malin began to exploit this characteristic to enhance faint objects lying near the limits of detection of the plate. In the process that he has now standardized, the exposed plate is placed face down on either high-contrast printing paper, or high-contrast film and a contact exposure is made using a diffused light source. The use of the diffused light source causes the exposed grains near the surface to have a greater influence on the final image than exposed grains that are uniformly distributed through the emulsion and it tends to reduce the effects of existing plate blemishes since these should have an even chance of occurring anywhere within the emulsion layer.

When an image is examined microscopically, it is seen that small strands and lumps of silver tend to clump together to make larger (but not well bounded) clumps of silver. These clumps then appear to blend together when viewed at even lower magnifications to make what is finally perceived as a continuous tone image. These silver clumps and the degree to which these clumps blend to make up a continuous tone image give rise to the concept of the 'granularity' of a given emulsion which when seen in the final print or transparency is perceived by the viewer as a sensation that is termed 'graininess'. When studies of perceived graininess have been performed, it is seen that the middle-density regions of the final print or transparency are the areas where graininess is most readily detected by the viewer. Also, since graininess is a subjective sensation obtained while viewing an image, it should be noted that the graininess of the final image is not simply a function of the negative characteristics. If a low-contrast printing medium is used, the graininess will be perceived as being less than on a print from the same negative made on a high-contrast material.

While graininess is largely determined by the combinations of materials that are used to make up the final print or transparency, the granularity of the negative is largely determined by the makeup of the emulsion layer of the film. Slower films are generally composed of a rather uniform distribution of smaller grains in the 2 μm size range. Faster emulsions are composed of a far less uniform distribution of grains in the 2–6 μm size range with a high frequency of the larger grains. The only means of altering the granularity that will ultimately be seen or measured in a given emulsion is through the choice of developers but the effects of different developers on granularity are small. A 'fine-grain developer' can slightly alter the size distribution of the grains in a fast emulsion in favor of the smaller grains, while a 'fast' high-contrast developer can push the distribution of grains towards the larger end of the size spectrum and even aid in the growth of the larger silver clumps. However, once given a fast grainy film, the choice of a developer will have only a small effect upon the granularity of the final image.

Halation and antihalation measures. The emulsion layer is quite thin when compared with the total thickness of a sheet of film and the light from a bright source is generally able to penetrate the entire thickness of the emulsion and then pass on into the film support base. When this light strikes the back of the film base, most of the light that strikes the base at near normal angles will generally be transmitted through the back of the base, while a larger portion of the light that strikes the back of the base at more oblique angles will be reflected back and will then re-expose the film surrounding the original image. As the angle of incidence increases, more and more of the stray light will be reflected back to the emulsion until an angle is reached where all of the incident light is internally reflected. The effect of all this when a very small illuminated disk is penetrating the emulsion is that the reflected light tends to form a well-defined ring around a bright star image. This ring has been called a halation ring over the years and is essentially the simplest form of halation that is encountered in various photographic applications.

Two methods are now commonly employed in order to diminish or eliminate the effects of halation. In many of the more common photographic emulsions, a dye is incorporated in the base that is meant to absorb any light that should manage to penetrate the base through the emulsion layer. This dye is then removed during the processing stages, leaving the film almost perfectly clear. In astronomical applications, more effective measures are required and a thin absorbing layer is normally applied to the back of the film base (or glass plate) that helps to prevent light from being reflected back at the rear surface of the film base and thus serves to absorb virtually all of the light that does completely penetrate the film. This absorbing layer must also be removed during the processing steps but it usually requires a conscious effort to do so and can sometimes require some effort to remove this layer without damaging the emulsion.

Low-intensity reciprocity failure. The 'Reciprocity Law' that was formulated by R.W. Bunsen and H.E. Roscoe in 1862 stated that the density of the silver emulsion was dependent upon the total energy that was absorbed by the emulsion and thus was independent of both the duration of the exposure and the intensity of the exposure as long as the product of the two remained constant. In other words, as long as the equation:

$$E = I \times t$$

remains constant, the emulsion will produce identical densities no matter how much the intensity and the exposure are varied. W. de W. Abney was the first seriously to examine this 'law' and show that the relationship did not hold during extremely long exposures to faint sources. Abney also showed that this relationship broke down with extremely short exposures. In fact this 'law' is only close to being valid over a rather narrow range of exposures from about 0.001 s to something less than 10 s. It is the departure from the relationship expressed by the 'Reciprocity Law' that has spawned the term reciprocity failure. In reality, it was the law that failed to describe the real world correctly.

Most astronomical exposures involve the long exposure range where the effect is properly referred to as 'low-intensity reciprocity failure' (or 'LIRF'). With films designed for 'normal' photography, the sensitivity of the emulsion drops rapidly as the exposure goes beyond the minute mark. By the time one has reached a 5 min exposure, the film has lost all but a small trace of its sensitivity to light and exposures beyond 20 min show very little increase in density or in limiting magnitude (see Fig. 6.2). In the early days of astrophotography the only available options were to expose for extremely long times (exposures exceeding 8 h were not uncommon), to build faster telescopes and live with the aberrations, or to wait for the emulsions to be improved. Today, emulsions have been designed specifically for the problems of astronomical photography and yet the astronomer is still searching for methods of eliminating the last traces of LIRF in order to reach fainter threshold limits. Several hypering techniques work to diminish greatly the effects of LIRF in addition to increasing the speed of the photographic emulsion, cooled-emulsion photography is popular among some amateurs, and new emulsions are still being introduced to improve the efficiency of the photographic emulsion for capturing photons (hypering techniques and cooled-emulsion photography are discussed in greater detail in Chapter 10).

The mechanism responsible for low-intensity reciprocity failure appears to be related to the energy needed to form stable (developable) silver centers as an exposure increases. With enough energy, a sensitive speck of silver is inundated with a burst of photons that provide a large amount of energy, and many neutral silver atoms form yielding a stable mass of silver very quickly. With a low level of incoming photons, energy is accumulated slowly and the few silver atoms that have been formed tend to be unstable and lose electrons reverting back to undevelopable silver ions. At some threshold level of intensity, the rate of decay is very close to the rate of growth and an image is basically unable to form.

Black-and-white films for astrophotography

Among the questions that you should ask when doing astrophotography is, 'Why?'. When you pick a film to use, this question should enter your mind and when you choose a developer for your film that same fundamental question should be pondered. You need to examine all the available choices and the pros and cons of each before making the decisions

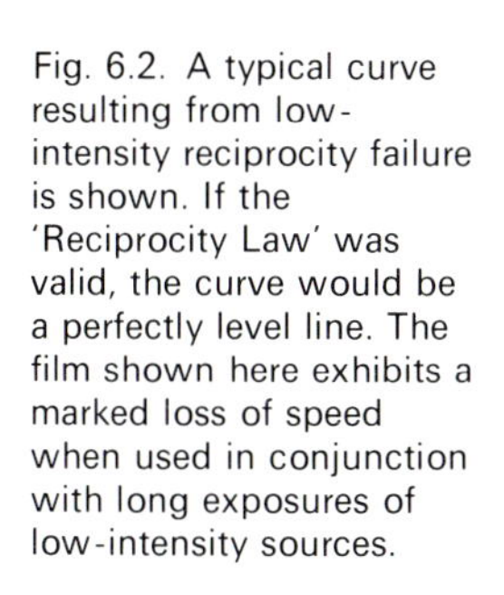

Fig. 6.2. A typical curve resulting from low-intensity reciprocity failure is shown. If the 'Reciprocity Law' was valid, the curve would be a perfectly level line. The film shown here exhibits a marked loss of speed when used in conjunction with long exposures of low-intensity sources.

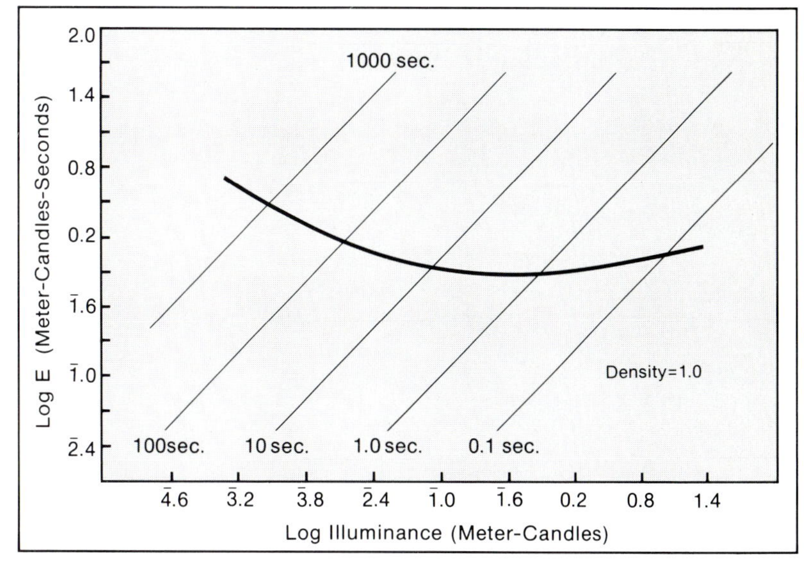

about film and developer. You should also seek information from several sources, including the manufacturer, as errors in the amateur literature are frequently encountered owing to a lack of basic understanding of the technical literature. If you are unsure about some point(s), a letter to the manufacturer can save many woes.

The choice of a film is an important one that is often done automatically by the starting astrophotographer. We feel that more thought should be given to this problem and we will present some of the information that is available on a few of the films that are around today. Kodak's publication: *Films and Plates for Science and Industry* is an excellent source of information and a few hours' of reading will certainly not hurt anyone.

103a-O

103a-O is a blue-sensitive emulsion, with coarse grain, and high contrast (Gamma = 1.0–1.2). This is the fastest of the three spectroscopic emulsions to be discussed here (but is sensitive to blue light only!) and is the worst of the three emulsions as far as sky fog is concerned (using a UV-blocking filter or a nebular filter will be of use to some for obtaining maximum image contrast). This emulsion is most useful for recording galaxies, blue reflection nebulae and loose star clusters.

103a-E

103a-E is a 'semi-panchromatic' emulsion with lowered sensitivity in the green and yellow regions and a red sensitivity peak near the Hα line (656.3 nm). It has coarse grain and high contrast and about half the blue sensitivity of 103a–O. Most useful when filters are being used to isolate the Hα region.

103a-F

103a-F is a panchromatic emulsion with coarse grain and high contrast. This emulsion is actually slightly faster in the Hα region than 103a–E which is the emulsion that is usually recommended for Hα photography. Sky fog has less of an effect on this film than on 103a–O. This film is most useful for recording red nebulosity and loose star clusters. If you can only afford to purchase one type of spectroscopic emulsion, this is the one to choose.

IIIa-emulsion

IIIa-emulsion is Kodak's latest entry into the astronomical film array, and is presently available in IIIa–J and IIIa–F plates. It is finer grained and slower than 103a-emulsions but when hypered it is just about as fast as its unhypered 103a counterpart and is capable of reaching fainter limiting magnitudes.

Tri-X

Tri-X is Kodak's all around panchromatic emulsion for normal photography. It has medium–coarse grain and low contrast (Gamma = 0.4–0.6).

This film *has virtually no sensitivity in the* Hα *region!* It can be treated in hydrogen (or forming gas) or used in cold cameras to greatly increase its sensitivity during long exposures. Useful for galaxies or tight star clusters but its lack of red sensitivity makes it a poor choice for Hα nebulae. Fortunately for those that persist in using this film, hydrogen clouds also emits lots of blue light from the H-beta (Hβ) line.

HP-5
HP-5 is Ilford's answer to Tri-X, it is panchromatic but with an Hα sensitivity that is about 2–4 times that of Tri-X. This can also be hypered in hydrogen or used in a cold camera to dramatically increase sensitivity during long exposures. Useful for galaxies, tight star clusters, blue Nebulae and has more sensitivity to Hα nebulosity than similar Kodak films.

Kodak 2415 Technical Pan
Kodak 2415 Technical Pan is a remarkable panchromatic film with extended red sensitivity beyond the Ha region. This film has extremely fine grain, normally has high contrast but that can be widely varied by the use of special developers (gamma =0.4–1.4). This film is not useful for astrophotography unless hypered in hydrogen, forming gas or silver nitrate. When hydrogen hypered, this film can achieve a speed that is about equal to that of 103a-E (hypered speed will vary depending on technique used). This film (when hypered) has the widest range of uses of any of the films discussed.

When you plan a series of photographs (i.e. one evening's work), the nature of the objects should be considered before selecting a film. If it is a nebula, the type and color would eliminate several films from consideration. If the object is faint and has a fairly uniform surface brightness, then a fast and contrasty emulsion would be useful. However, if the object has a very non-uniform surface brightness (i.e. very bright and very dim areas in the same object), it is better to avoid high contrast films or else to lower the contrast by selecting an appropriate developer.

Developers

Theory of development

In order for the latent image in an emulsion to be rendered visible, the silver ions that are formed during the exposure must be reduced to metallic silver by the developing agent. Many reducing agents are known, but the real trick is to find an agent that is capable of discriminating between the exposed and unexposed areas of the emulsion. In fact, even with common developing agents, it is a matter of the reducing rate of the agent since prolonged development can eventually reduce all of the silver in the emulsion. Again, we are in an area where the exact nature of the process is still not understood. The large number of variables that affect the development process add numerous bits of information that can be studied to help solve the complex problem but the sheer volume of available data has simply overwhelmed the community of photographic scientists. However, in the literature studying the development process one can find many useful facts that can be employed in dealing with unusual problems. In theory, there are two basic types of development processes, 'physical' and 'chemical'. However, studies tend to show that the two processes seldom ever occur in a pure form.

Common developing agents

In order to be of any practical use, a developing agent must have a good ability to differentiate between exposed and unexposed silver halide centers (i.e. have a low fogging rate), it must be soluble in water and reasonably stable in solution. In addition, it should not soften the gelatin layer enough to produce any damage to the image and, for commercial developers, it should preferably be non-toxic. For the past century developing agents have been sought, tested, and some have found use in various special applications. Out of the hundreds of agents that have been discovered, about two dozen remain in use to some extent today. Of these developing agents, about half are used only for special applications or to recreate effects seen in old photographs that cannot be duplicated using modern chemicals.

Hydroquinone. Hydroquinone was discovered in 1880 and is generally found with Metol in most developers in use today. When used on its own, it must be mixed with a strong alkali in order to produce a reasonably active developer. A pure hydroquinone developer would produce high contrast negatives of low speed. The properties of hydroquinone–Metol and hydroquinone–Phenidone developers are discussed with Metol and Phenidone. The fine white crystals need to be stored in a container that can be well sealed as they will slowly turn brown with exposure to air.

Metol. Metol was introduced in 1891 and is now just about the most common developing agent in use. Metol (or 'Elon') is generally combined with hydroquinone owing to the superadditive properties of these two compounds. A pure Metol developer would be characterized by a high emulsion speed (measured at the 'toe' of the characteristic curve), low constrast and fine grain. Useful 'soft' developers may be mixed from Metol and sodium sulfite (D-23 developer) or sodium carbonate can be added to obtain a faster-working solution (D-25 developer).

When Metol is combined with hydroquinone to form a Metol–hydroquinone (MQ) developer, a developer is formed that has the short induction period and high speed produced by a Metol developer but with the contrast characteristic of a hydroquinone developer. Thus, the new mixture has properties that could not be obtained by the use of either or both developers by themselves.

Phenidone. Phenidone is a relatively new developing compound that was discovered by the Ilford laboratories in 1950. It is similar to Metol in many ways, but different in several significant properties. When used alone, it produces high emulsion speeds with low contrast but has a tendency to produce fog. It, like Metol, also has the property of being able to activate hydroquinone but considerably less Phenidone is required (about 1/40 the weight of Metol that would be required for a similar developer). MWP-2 is a Phenidone-based developer.

When mixed with hydroquinone, Phenidone produces a developer similar to Metol–hydroquinone mixes but with the following differences:

1 a phenidone–hydroquinone (PQ) developer can function at a lower pH than an MQ developer;
2 two processes related to the oxidation of both Phenidone and hydroquinone work together to prolong the working life and the shelf life of the PQ developer;
3 as MQ developers oxidize they eventually give rise to a very unpleasant odor, PQ developers do not;
4 PQ developers are less likely to produce dermatitis;
5 when mixing a PQ developer, the Phenidone should not be added until both the sulphite and the alkali have been added (this will help to dissolve the Phenidone quickly);
6 PQ developers are less affected by the bromide by-products that are created as the developer is exhausted.

Other developer ingredients

Virtually none of the known developing agents are used by themselves to formulate a developer. A developer always contains one or more additional elements which are necessary for the solution to function properly. These additional constituents are: a preservative, an alkali and a restrainer. The purpose of each will be described briefly in order to see what each does and to give the reader some degree of understanding about how a developer functions and why developer 'A' has ingredients X, Y, and W.

The preservative. This compound (normally sodium sulfite) is used to prevent wasteful oxidation of the developing agent and also to prevent the discoloration of the used developer (which could, in turn, stain negatives and prints). Variations of the amount of preservative only alter the keeping qualities of the developer and, since one is already dealing with an almost saturated solution, the only real change is towards a shorter shelf life.

The alkali. This is sometimes called the 'accelerator' as it serves to make the developing agent sufficiently active. The alkali is used to raise the pH of the developer to a point that raises the activity of the developing agent to a practical level. Today, the most commonly used alkali is the anhydrous form of sodium carbonate.* Other alkaline agents that are sometimes encountered include sodium hydroxide, potassium hydroxide, and potassium carbonate. An increase of the amount of alkali in the developer will serve to shorten the development time or, for an equal development period, increase the contrast of the negative. A decrease in the amount of alkali will have the opposite effect.

The restrainer. The function of the restrainer is to delay or to stop the development of the unexposed grains in the emulsion (i.e. to prevent fog). Restrainers also affect the exposed grains to some extent and thus can also affect the speed of the film. Of the organic and inorganic restrainers that are available, two are found in much greater frequency than any other agents: potassium bromide (inorganic) and Benzotriazole (organic).

Potassium bromide is almost universally used because of the fact that bromides are produced as a by-product of the development process. Inclusion of bromide compounds in the original solution helps to minimize the impact of these by-products as the developer is used. Phenidone developers are less affected by bromide than are Metol developers.

Benzotriazole is a very powerful organic restrainer and is of particular interest when Phenidone is used and is especially useful when high-speed emulsions are to be developed using Phenidone. In this latter case, the amount of bromide restrainer required would be excessive and there could even be a danger of staining the film. Benzotriazole and other organic 'anti-fog agents' are more capable of restraining fog *without affecting the film speed significantly* than are their inorganic counterparts. However, they are quite powerful and must be measured carefully when compounding a developer! We have found that a moderate increase in the amount of Benzotriazole can bring astronomical film speeds down drastically. However, if one is willing to endure a little extra fog, the amount of Benzotriazole in the formula for MWP-2 could be decreased by 10–20 % in order to gain a small amount of film speed. Consequently if one would like to get rid of some of the fog that comes with hydrogen hypering of color emulsions, one can add a *very small* amount of Benzotriazole to the first developer in a color chemistry package and trade fog for a little less film speed. (However, experiment on regular photographs before just adding a pinch of Benzotriazole to your color chemistry – a little goes a long way!)

Miscellaneous variables

The primary variables that effect film development are the time, temperature, and agitation during the development period. Another variable which can drastically affect film development is the pH of the

* If your developer formula calls for sodium carbonate it usually means adhydrous sodium carbonate. The three forms of sodium carbonate are not equally interchangeable; use the form that is called for or look up the proportions for each of the three forms.

developer. However, this is generally of little concern unless a small amount of stop bath has been spilled into the developer. If such an accident occurs, the developer should be discarded immediately. The basic effect of these other variables is to change the rate of development. If any of these other variables are increased, the rate of development will be increased and vice versa.

The basic effect of *small* changes in any of these variables is to alter the contrast of the negative, although the speed of the emulsion will also be affected to a small degree (see Fig. 6.3). Many claims are made about how these variables can affect dramatically the speed of the film. If one measures film speeds incorrectly, in the densest regions of the negative, then large 'speed shifts' will be seen which are largely due to contrast changes. However, if the speed is correctly measured in the less dense regions (0.10 density is a point where film speed is often measured) of the negative, only slight changes will be recorded (see Fig. 5.8).

When any of these variables is changed by more than just a small amount, other effects are seen and each of these variables produces its own set of characteristic symptoms. Too low a temperature results in no development at all, while too high a temperature creates a great deal of chemical fogging unless special developers are used. Too long in the developer results in a fogged contrasty negative, while an extremely short development time creates a negative where most of the details appear only faintly, and where almost all of the details appear at nearly the same density. An extreme amount of agitation during development creates a contrasty, fogged negative similar in appearance to one formed by prolonged development, but too little agitation creates a low-contrast image

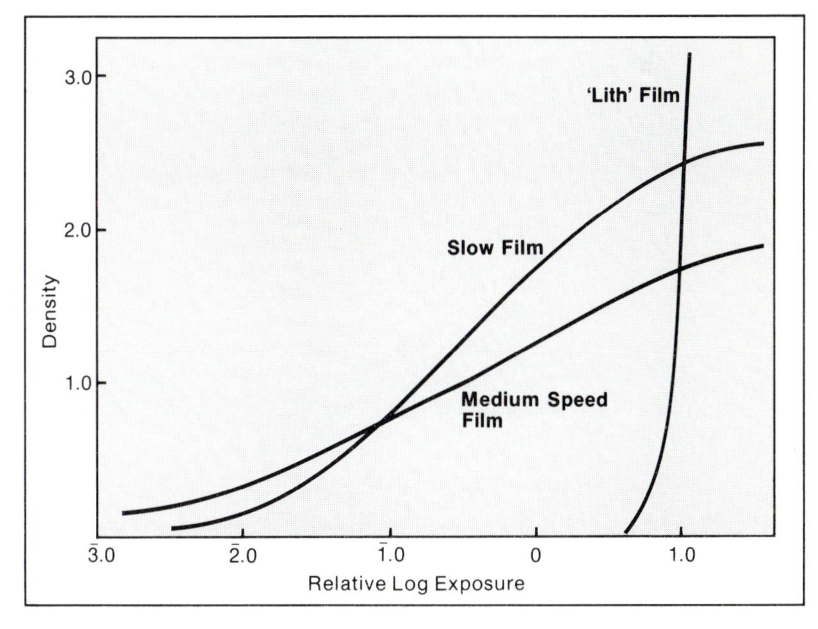

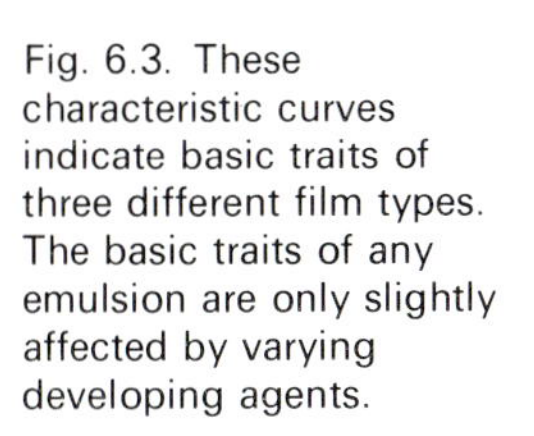
Fig. 6.3. These characteristic curves indicate basic traits of three different film types. The basic traits of any emulsion are only slightly affected by varying developing agents.

which may show blotches and streaks from the effects of exhausted developer.

In addition to the above effects, changes in any of these variables will also affect the granularity of the negative. While decreases in any of these variables will show only a very slight decrease in granularity, a marked increase in any of these variables will result in a noticeable increase in the granularity of the negative image and thus in the graininess of the final print.

Again, we will emphasize the fact that increasing any of these variables will not bring about a great speed change in the less dense regions of the emulsion. Increasing any or all of these factors will bring about a contrast increase that will dramatically affect the density in the denser regions of the negative (see Figs 6.4 & 6.5). This density increase can be

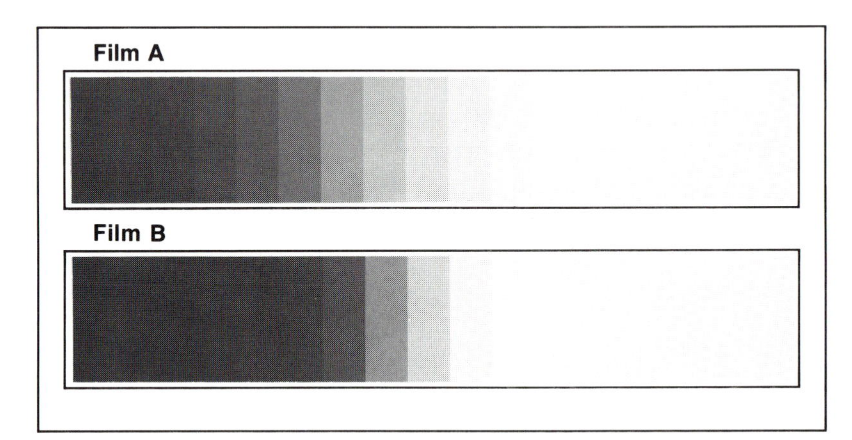

Fig. 6.4. This set of photographs illustrates the appearance of sensitometer strips from two films of extremely different contrast. Film B is considerably more contrasty than Film A.

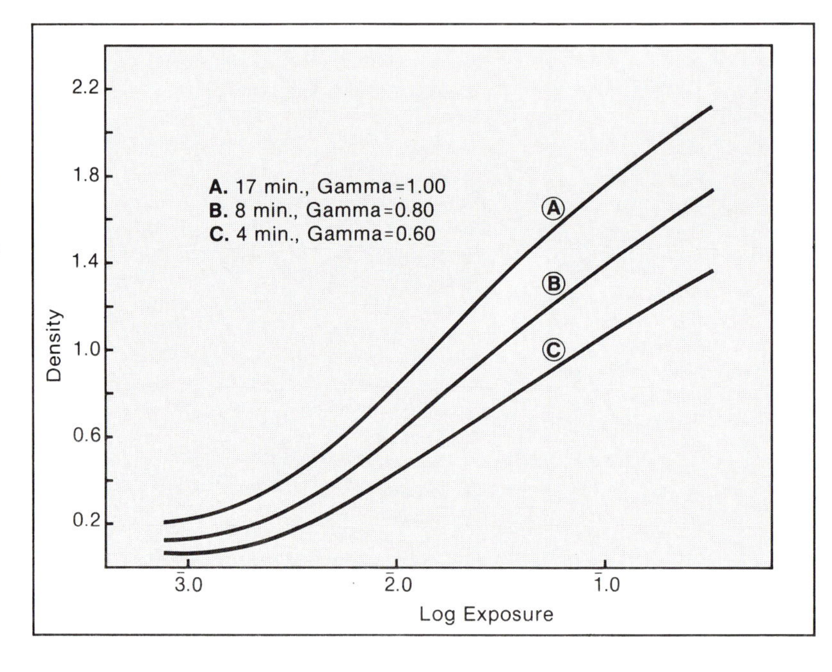

Fig. 6.5. Sample H–D curves illustrate the effects of prolonged development on the photographic emulsion. Notice that the contrast increases and that the fog level increases but that the 'speed' of the film, if measured correctly, shows very little increase.

misinterpreted as a speed increase, but the dense negative regions are not where a speed increase is needed and the increased contrast is achieved at the expense of increased grain and increased fog. In addition, there is also a good possibility that the more dense 'highlight areas' will be pushed up onto the shoulder of the H–D curve and will become 'blocked up' and details will then be lost in these blocked highlights.

Some useful developers

There are many developers available and their actions on films are varied. Most people select a developer based on a recommendation made by the manufacturer or even on a friend's advice. In all probability, fewer than 1 % of all photographers ever test a series of developers to determine which one is best suited to their own needs. A developer that works for one person's photographic needs, with his processing techniques, his enlarger's characteristics, and his print-processing methods, may not work well for another person's needs when transplanted. We would like to describe a few of the developers that are commonly used in astrophotographic work and list some of the properties of each developer. If you mix your own developers, you should always use distilled water as municipal water supply quality varies dramatically, making it difficult to achieve consistent results.

D-19. D-19 is the common developer for spectroscopic films. It is a fast developer that produces high contrast and moderately coarse grain. Typical developing times for 103a-type films are 4–5 min at 68 °F (20 °C). This developer has been shown to yield excellent results on Kodak 2415 Technical Pan when processed for times of 10–15 min at 68 °F (20 °C).

MWP-2. Mount Wilson–Palomar formula 2 (MWP-2) is a greatly improved D-19 formula. MWP–2 yields about twice the speed of D-19 on 103a-type films with about the same grain. One feature that is noticed readily when comparing prints from D-19 and MWP-2 negatives is that the grain from an MWP–2 negative shows very few 'giant' grains and that the grains are very uniform in size. MWP-2 also suppresses fog on the negatives. The only real drawback of MWP-2 is that it is not available on the market in small quantities and must be mixed from its basic chemical ingredients. Developing times for 103a-emulsions and 2415 Technical Pan are 11–16 min at 68 °F (20 °C). The formula for MWP-2 appears in Table 6.1.

D-76 (or ID-11). D-76 produces less grain and less contrast than either MWP-2 or D-19, but it also produces lower speed negatives than either of those developers. This developer is very good if you want to reduce contrast slightly when photographing inherently contrasty objects, but you must remember to increase exposures by at least 60 % (a 100 % increase would not hurt). Typical developing times range from 9 to 12 min.

POTA. POTA was created for a very specific application, specifically the drastic reduction of contrast. This developer was designed for processing films of nuclear bomb tests. It also produces a speed loss of about 50 % and a reduction in grain. It is particularly useful when photographing inherently contrasty objects on high-contrast films like 103a-emulsions and Kodak 2415 Technical Pan film . This developer MUST be mixed from chemicals and used immediately after it is mixed. Typical processing time for 103a-films is 20 min; for films like HP-5 and 2415, 8–15 min is typical. The formula for POTA appears in Table 6.1.

Technidol LC. In late 1982, Kodak responded to the growing demands for their 2415 Technical Pan by introducing Technidol LC. This is essentially Kodak's equivalent of POTA.

Other special developers. Developers for maximum emulsion speed: There are many developers on the market, and some advertise very attractive promises regarding speed gains. These promises are difficult to pass by as many customers believe the speed promises of less-than-honest advertising campaigns. One class of developer that does usually produce a noticeable speed gain is that which consists of the 'fogging developers'. The best known of the developers in this class is Edwal FG-7 developer. The problem with using fogging developers on astronomical films is that fog is one of the biggest nuisances in astronomical photography, especially when hypering

Table 6.1. *Developer formulas*

MWP-2[a]		POTA[b]	
Water at about 125 °F	750 ml	Sodium sulfite, anhydrous	30.0 g
Sodium sulfite, anhydrous	105.0 g	Phenidone	1.5 g
Hydroquinone	10.0 g	Water at 100–125 °F to make	1000 ml
Phenidone	0.40 g (6.2 grains)		
Benzotriazole	0.60 g (10.8 grains)		
Potassium bromide	2.0 g (30.6 grains)		
Potassium carbonate ($2K_2CO_3 \cdot 3H_2O$)	30.0 g		
Cold water to make	1000 ml		
Developing time	7–16 min at 68 °F		

[a] Normal times for 103a-emulsions and 2415 Technical Pan range from 9 to 16 min at 68 °F. It is of particular importance that the sodium sulfite and the hydroquinone are completely dissolved prior to the addition of the Phenidone.
[b] Normal processing times for 2415 Technical Pan range from 8 to 15 min at 68 °F. In this formula, the Phenidone will be more difficult to dissolve than in the MWP-2 formula. Keeping the water temperature above 100 °F will be of some assistance.

techniques have been used to boost the effective speed of the film. One should avoid fogging developers because of their undesirable side effects. The modest speed gains obtained by using these developers do not come close to the gains that can result from the various hypering processes.

Other developer variants that actually do provide speed gains include decreasing the amount of anti-foggant in the developer formula, adding iodide compounds to the developer, adding hydrazine compounds to the developer, and compounding special two-solution developers. Of these variants, the last is probably the easiest and safest to try since it has the least number of potentially disastrous side effects. With a two-solution developer, the developing agents (and some preservative) are mixed in the first solution and the alkali and restrainer (and more preservative) are placed in the second solution. The film is soaked in the first solution until it is saturated with the developer solution and then placed in the second solution where the alkali solution accelerates the developing process and builds up density. Since the film can only soak up a limited amount of developer, when the alkali accelerates the process, the heavily exposed areas exhaust the limited developer quite rapidly and do not build up the density that they would in a one-solution processing bath. This type of developer could also be considered a 'compensating' developer since it tends to work more on the low-density regions and retard an overall contrast increase.

Another claim that is often seen is that of the so-called fine-grain developers. While there are several strategies that can be used for obtaining negatives having less granularity than those processed normally, we would like to emphasize the fact that the main determinant in this area is the film itself. Once a particular emulsion is chosen, the variations that one can impose upon the innate grain structure are minimal. The reader should also be aware that while some 'fine-grain' developers (like D-76 or ID-11) can be used on astronomical emulsions without any problems, others can bring about extremely disappointing results. Any new developer should be tested before dropping a roll of film containing many hours' of effort into the solution!

This represents just a small sample of the wide array of developers that are now available; new formulas are appearing continually. Many photography books and darkroom manuals contain numerous interesting formulas and the reader is urged to seek out such sources. There are many developers that will produce satisfactory results and you are encouraged to test newly recommended formulas to determine which will produce results that you like. In addition it is recommended that you always test for an optimum developing time. It is a poor gamble to waste film that has been exposed at the telescope in an untried developer. Tests should always be run in the darkroom (using a sensitometer such as the one described in Chapter 5) to check out the effects of any new processes before risking the waste of long hours at the telescope. The most important thing to remember at this point is that the manufacturer's recommended developer is not the only developer that will produce excellent results! If you are interested in learning more general information about films and developers, we recommend Neblette (1952, 1962) as an excellent starting point. For more information on specific films and developers, the manufacturer should be contacted.

7

Color photography

We have been cocksure of many things that were not so.
Oliver Wendell Holmes

In the past decade there has been an increasing interest in color astrophotography in both the amateur and professional communities. While the primary interest among the great majority of amateurs is an aesthetic one, professionals are interested in accurate color photographs as a means of quickly conveying a great deal of information. An accurate color photograph presents information in a form that is far more striking than viewing a set of black-and-white plates taken in different wavelengths of light. In addition, for the professional, there is the public-relations value to consider when it comes to obtaining financial support in this era when esoteric scientific budgeting is being sacrificed in favor of funding less aspiring military projects.

It is imperative that the reader realize the problems that are inherent in writing a single chapter on the topic of color photography. It is probably equivalent to comparing the act of viewing a postcard of the Grand Canyon to the experience of walking up to the edge and seeing the Canyon for the first time! Just as the history of photography goes back at least to the Roman Empire (the dye which provided the royal purple robes that were reserved for the emperor, was obtained through a photochemical process from a specific variety of Mediterranean snail), the dream of obtaining exact color representations of objects and scenes also covers the same timespan. We have included numerous classical references in our bibliography and urge interested readers to further pursue the topic by taking time to read other in-depth works. The works by Neblette (introductory), Evans, Itten, Hunt, and Evans, Hanson, & Brewer are all highly recommended.

While some thought that the path to a method of making color photographs had been revealed as early as 1810 when it was found that exposing wet silver chloride to a spectrum produced faint traces of the colors, the origin of the history of practical color processes must be placed in the year 1861 when James Clerk Maxwell presented a paper before the Royal Institution in London. Maxwell showed that colors could be reproduced by mixing red, blue, and green light in the proper proportions. Maxwell even projected three black-and-white positive images through the same red, blue and green filters through which they had originally been exposed in order to produce a color reproduction at this meeting. This three-color process using red, blue and green colors is known as tri-color photography; it is the basic principle upon which all further developments in color photography have evolved. Maxwell's work was based on the findings of Thomas Young who, after extensive investigations, had hypothesized in 1801 that human vision was based on some type of three-color analysis. Investigations of Maxwell's fundamental process have continued and have shown that most colors normally encountered in everyday situations can be reproduced with a reasonable degree of accuracy using the additive process.* However, we will be discussing some of the situations that occur in astronomical applications where the process cannot produce accurate results.

Maxwell had constructed his demonstration by exposing three negatives through red, blue, and green filters. He then turned the black-and-white negatives into black-and-white positives and projected these images in register using three projectors through the same red, blue, and green filters to produce a final colored image. The red, blue, and green images were thus 'added together' to form the final color reproduction of the original scene. This form of color synthesis is referred to as the additive color process. The alternate form of color synthesis is known as the subtractive color process and it is based on using the colors cyan, yellow, and magenta to remove color from a beam of 'white' light. The most common form of the subtractive process would be implemented as follows. First, the same three negatives would be exposed using the same red, blue, and green filters that were used for the additive process. However, one would then make the three positives of these negatives and use a process that would dye the red filter positive to form a cyan positive, dye the blue filter positive to form a yellow positive and dye the green filter positive to form a magenta positive. Now, the cyan, yellow, and magenta positives can be stacked in register and viewed using a white light source to reveal a color reproduction of the original scene.

If you noticed that the additive process requires three images to be projected separately to form the final image, while the subtractive process only requires a single 'projector', you can easily see why the subtractive process has become the more popular alternative for color synthesis. It is the subtractive process that is the basis of most color photographic processes in use today and it is the process that is used in all commercial color films currently available.

* The additive primaries (red, blue and green) have the property that when combined, or added, they produce white (the combination of all the colors of the visible spectrum). The subtractive primaries (magenta, cyan and yellow) are each composed of all colors of the spectrum minus one of the additive primaries. Thus magenta represents all colors with green subtracted (magenta is often referred to as 'minus green'). Similarly cyan is 'minus red' and yellow is 'minus blue'. When all the subtractive primaries are combined the result is black since all of the additive primaries have been subtracted.

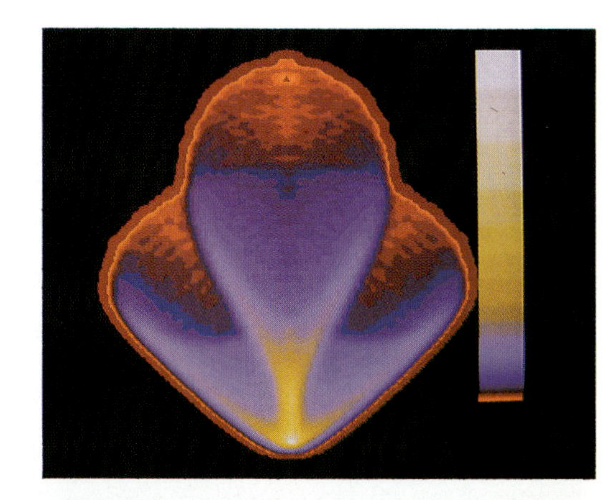

Plate 1. The intensity of an off-axis Newtonian star image is shown in this false color display. The image was formed using 15 000 rays and a VAX 11/980 was used to do additional image processing. The ray-trace was done for an 8-inch f/6.0 Parabola 1 deg off-axis. The image itself is 97 by 123 um and the intensity runs from 0 to 255. The intensity scale at the right can be used to determine the intensity distributions through the image.

Plate 2. Tri-color photo of the Horsehead Nebula made by combining three separate exposures taken in red, blue and green light. An 8 inch F/6.0 Newtonian was used. Photo by Astrophoto Laboratory.

Plate 3. Tri-color photo of the Lagoon Nebula (M8). Photo by Astrophoto Laborarory.

Plates 4, 5. Color film comparisons four photographs of Jupiter (Ektachrome 160, Kodachrome 64 and Ektachrome 2483); two photographs of Saturn (Ektachrome 64 and Ektachrome 160). Jupiter: Top left – Ektachrome 160, 8 inch f/7, 5 s at f/105. Top right – previous photograph duplicated onto Kodachrome 25 for increased contrast. Bottom right – Kodachrome 64, 8 inch f/7, 7 s at f/105. Bottom left – Ektachrome 2483, 8 inch f/7, 8 s at f/80. Saturn: Left photograph – Ektachrome 64, 6 inch f/6.3, 5 s at f/95. Right photograph – Ektachrome 160, 8 inch f/7, 5 s at f/105. Photos by James K. Rouse.

Plate 6. Color photograph of lunar crescent showing earthshine and color photograph of 1975 lunar eclipse. Lunar crescents and lunar eclipses require fast color film. Crescent: Ektachrome 160, 8 inch f/7, 30 s. Eclipse: GAF 500, 8 inch f/7, 90 s. Lunar rate drive was used for both photographs, utilizing a drive corrector. Photos by James K. Rouse.

Plate 7. Two high magnification color photographs of the moon. Crater Copernicus: Celestron C–14, Ektachrome 160, 4 s at f/80. Lunar craters: Celestron C–14, Ektachrome 160, 2 s at f/80. Photos by James K. Rouse.

Plate 8. **Lee Coombs**. Jupiter occultation, 0.5 s on Ektachrome 400 with 10 inch Newtonian at f/90. 2 April 1983.

Plate 9. **Jack Marling**. NGC253, 25 min on gas-hypered Fujichrome 400 processed as a negative. 17.5 inch f/4.5 Newtonian without filtration.

Plate 10. **Jack Marling**. M83, 25 min on gas-hypered Fujichrome 400 processed as a negative. 17.5 inch f/4.5 Newtonian without filtration.

Plate 11. **Jack Marling**. NGC2174, 40 min on gas-hypered Fujichrome 400 processed as a negative. 17.5 inch f/4.5 Newtonian with Lumicon deep-sky filter.

Plate 12. **Jack Marling**. M16, 45 min on gas-hypered Fujichrome 400 processed as a negative. 17.5 inch f/4.5 Newtonian with Lumicon deep-sky filter.

Plate 13. **Jack Marling**. NGC7293, 53 min on gas-hypered Fujichrome 400 processed as a negative. 17.5 inch f/4.5 Newtonian with Lumicon deep-sky filter.

Plate 14. **Jack Newton**. M51, 17 min on cooled Fujichrome 400 with a 20 inch f/5 Newtonian.

Plate 15. **Jack Newton**. M1, 20 min on cooled Kodak VR 1000 film with a 20 inch f/5 Newtonian.

Plate 16. **Jack Newton**. NGC2024 and NGC2023, 20 min on cooled Kodak VR 1000 film with a 20 inch f/5 Newtonian.

Plate 17. **Jack Newton**. M42, 15 min on cooled Kodak VR 1000 film with a 20 inch f/5 Newtonian.

Plate 18. **Brad Wallis** and **Robert Provin**. M20, 60 min on hydrogen-hypered Kodak High-Speed Ektachrome with an 8 inch f/10 Schmidt–Cassegrain.

Plate 19. **Brad Wallis** and **Robert Provin**. M22, 60 min on hydrogen-hypered Kodak High Speed Ektachrome with an 8 inch f/10 Schmidt–Cassegrain.

Plate 20. **Brad Wallis** and **Robert Provin**. M42 and NGC1973–75–77, 60 min on Fujichrome 100 with a 6 inch f/6 Newtonian.

Plate 21. **Brad Wallis** and **Robert Provin**. Region around Scutum star cloud, 2.5 h on Kodak Ektachrome 400 (pushed to 800) with 90 mm lens at f/4.5 (Pentax 6 × 7 cm camera).

Plate 22. **Brad Wallis** and **Robert Provin**. Comet IRAS–Araki–Alcock (1983) with M44 and M67, 15 min on Fujichrome 400 with 135 mm lens at f/2.8. 12 May 1983. Camera tracked stars so comet is trailed.

Plate 23. **Brad Wallis** and **Robert Provin**. Aurora photographed from 34° N Latitude, 119° W longitude, 2 min on Fujichrome 400 (pushed to 800) with 24 mm lens at f/2.8. Sharp eyes and preparedness can sometimes capture a rare event.

General information on color films

How color films work

In this section, our primary accent will be on color transparency films which are also referred to as color slide films (i.e. Kodachrome 64, Ektachrome 200, Fujichrome 400, etc.). The general statements will also apply to color negative films which are commonly known as color print films (i.e. Vericolor, Kodacolor, etc.) and some attention will be given to specific differences. We will not be addressing the issues of which color films are best to use since the whole array of available films is changing at a very rapid pace. Few of the common films that were around six years ago are still on the market. Even Kodachrome has been improved over the years. Instead, we will concentrate on general principles and specialized aspects of color photography. The reader can see film comparisons quite regularly in the various popular publications and can evaluate the top two or three using the equipment that will be employed for 'normal' applications. It is important to realize that films should be tested for both wide-angle and narrow-angle photography and that the exact processing conditions need to be considered when evaluating or publishing any emulsion test. Similarly, we will avoid dealing with the specifics of color processing since it too is undergoing tremendous changes at this time. Again, the reader can find the latest information on new processing methods in popular photography magazines.

Today's color films are multilayer films consisting of three separate emulsion layers that are sensitive to blue, green, and red light. The first emulsion of this type was introduced in 1935 and was known as Kodachrome. In most of these emulsions (except Kodachrome) a 'dye coupler' is suspended in the emulsion that will, when activated during development, create the dye image that will make up the final image. In these films, an extremely thin layer of gelatin is applied between the three emulsions to isolate them from one another. Because it is virtually impossible to make a green- or red-sensitive emulsion that is not also sensitive to blue light, the blue-sensitive emulsion layer is placed on top and a yellow filter is incorporated into a layer immediately beneath. This yellow filter removes the blue light from the incoming light which continues on to expose the green and red emulsion layers (see Fig. 7.1). If this filter was not incorporated into the emulsion at this point, it would be impossible to obtain an accurate color rendition of any subject. The material and the method is

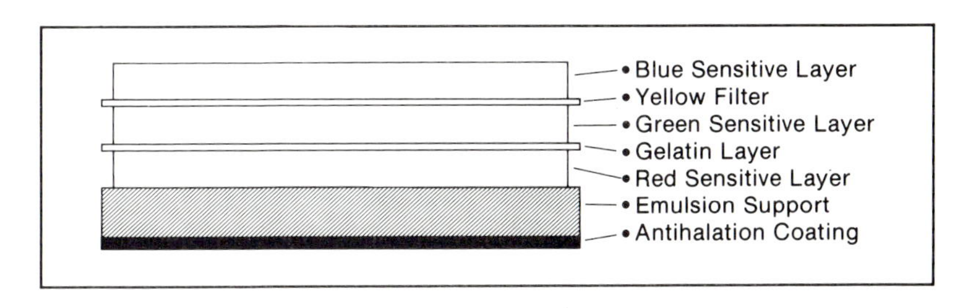

Fig. 7.1. The structure of a typical color film. The relative thickness of the emulsion support layer has been greatly reduced in this diagram.

complex but the system works extremely well when the film is exposed as the manufacturer meant it to be. These films were meant to be exposed to a subject illuminated by a continuum light source of a particular color temperature using exposures from ~1/2000 s to 5 or 10 s. Unfortunately these conditions eliminate nearly all astronomical subjects except the sun, moon, and the brighter planets.

When most transparency films are processed (Kodachrome being the outstanding exception to this process), the images are first developed in a developer that is quite similar to most common black-and-white developers, the image is bleached, then 'redeveloped' in a solution which activates dye couplers that are suspended in the emulsion to form the dye layers that will form the final image. The final product of this is a positive color image which is a close approximation to the original scene that was photographed. During the processing stages the blue-sensitive layer is transformed into a yellow positive, the green-sensitive layer is changed into a magenta positive, and the red-sensitive layer becomes a cyan positive. When light is passed through the transparency, the yellow positive absorbs blue light, the cyan layer absorbs red light and the magenta positive absorbs green light. Thus, one can see that the 'subtractive process' is used to subtract colors selectively from areas of the beam of white light that passes through the transparency to the viewer's eye or to a projection screen.

Owing to the fact that the image in color films that is present after processing is (in most cases) composed of organic dyes, and no longer contains a silver component, the image no longer has the archival qualities that characterize black-and-white emulsions. The exceptions to this statement include Polaroid instant print films (including the SX-70 films) the relatively new instant 35 mm films from Polaroid and the Cibachrome glossy printing paper.* In these cases, the dyes are not organic, but are a type of metallic dye that allows the film (or paper) to possess archival properties. Beyond these three examples, no other color processes can truly be considered to be archival; the photographic image will slowly fade away over a period ranging from 10 to 20 years for films like Ektachrome and Fujichrome to perhaps 50 to 100 years in the case of Kodachrome. If one were to list a small sample of the available color emulsions in approximate order from most archival to least archival, the list would run as follows: Kodachrome (by a long lead), Kodacolor, Vericolor, Ektacolor, Ektachrome (and most of the other 'chrome' films like Fujichrome, Agfachrome, GAF Color Slide Film, etc.). It is interesting to note here that the films used most commonly by amateurs are the ones that are the least archival and have a life time (before fading becomes noticeable) of as little as 10 years. The useful lifetimes of color materials can be extended somewhat by storage in refrigerated dark areas and by not projecting slides (or printing from them) more than is absolutely necessary.

* Cibachrome pearl surface paper is not archival.

Reciprocity failure and color emulsions

Color emulsions are designed by their manufacturer to produce satisfactory renditions of 'normal' subjects (i.e. trees, sky, clouds, flesh, etc.) when exposures remain within certain limits. Within this range of exposures, the 'Reciprocity Law' accurately expresses the relationship between the intensity of the light striking the film, exposure, and time exactly the same as with black-and-white films. These limits vary, but they usually fall in the range from 1/1000 or 1/2000 s to around 5 or 10 s. Within these limits, the film's speed is constant but, once outside these limits, the speed begins to decrease in a manner that is not readily predictable. The manufacturer usually provides some information to aid in predicting the behavior of a film outside of these limits, but these guidelines usually do not extend beyond exposures of 100 s.

As we discussed earlier, color films consist of three separate layered emulsions. For short exposures, these emulsions are designed so as to appear to have almost identical characteristics, where speed and contrast are the characteristics of prime concern to the photographer. Once outside of the designed exposure range, the effect of reciprocity failure poses a complex problem. Since we are dealing with three different emulsions, it is possible for each emulsion to change one or both of these characteristics in a totally independent manner.

A change in speed of one or more of the emulsions results in a color shift that is uniform throughout the entire range of usable densities (see Fig. 7.2). For example, if a particular film has a red-sensitive layer which does not lose speed as rapidly as either of the other emulsions, the resulting image will be too red. Such a uniform color shift (if minor) could be corrected by projecting or viewing the slide through a cyan filter of an appropriate density. A preferred method of color correcting, which will allow one to correct larger color shifts than with the above method, is to place an appropriate cyan filter over the camera lens while taking the

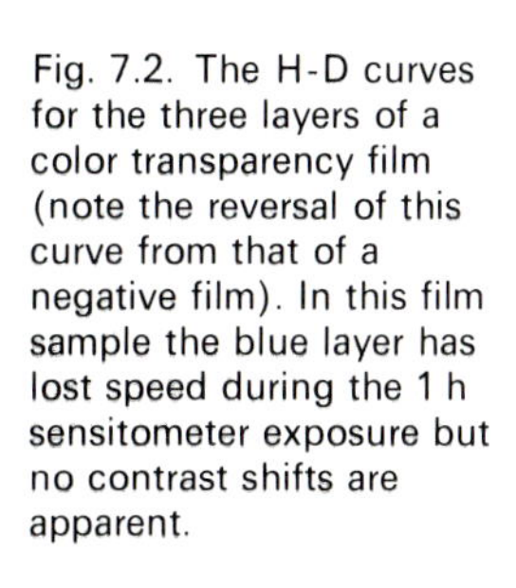
Fig. 7.2. The H-D curves for the three layers of a color transparency film (note the reversal of this curve from that of a negative film). In this film sample the blue layer has lost speed during the 1 h sensitometer exposure but no contrast shifts are apparent.

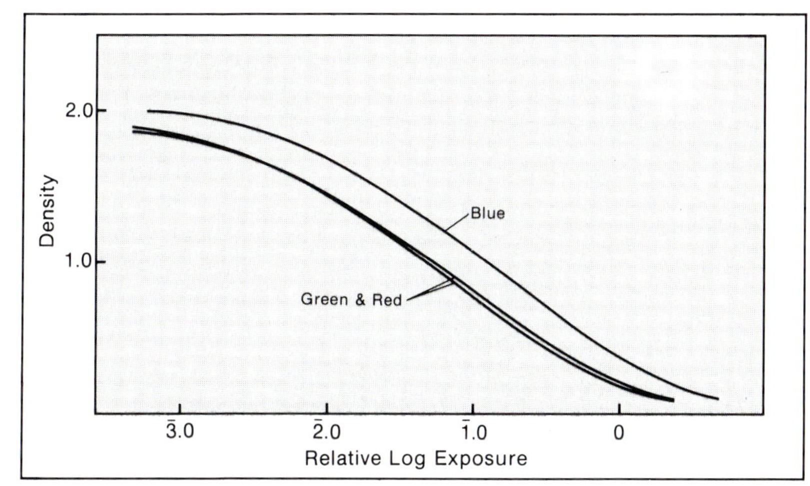

picture. If all three emulsions change speed at different rates, it would be necessary to use two filters to absorb some of the light to which each of the two faster emulsions is sensitive. However, the required filtration will usually be different for exposures of different durations and, as pointed out earlier, the manufacturer does not have information available for exposures greater than ~ 100 s.* Therefore, the burden of generating filter data for exposures from 10 min to 2 + hours falls upon the astronomer, assuming that there is interest in obtaining color photographs which bear some resemblance to the actual colors in the nebula or galaxy being photographed. We will discuss how one can determine the appropriate combination of filters to correct color films during long exposures, after we discuss another aspect of reciprocity failure in color emulsions.

When photographic emulsions are exposed for periods of time outside of their designed operating range, the inherent contrast of an emulsion may change as may its effective speed. Since each of the three emulsions in a color film is inherently different, there is a very good chance that one or more of these layers will change contrast at a rate which differs significantly from that of the other emulsion layers. While different emulsion speeds can be compensated for by using filters during or after the exposure, the problem created by emulsions of differing contrasts cannot be corrected in such a simple manner.

In order to see the effect of a change in contrast in one emulsion layer relative to the other two layers, consider Figs 7.2 and 7.3. As we discussed before, Fig. 7.2 shows the effect of a simple speed change in one emulsion layer which is seen by the viewer as a uniform color shift over the entire range of usable densities. In Fig. 7.3 the H–D curves have been plotted for a color film in which one layer exhibits an increase in contrast relative to the others. By comparing these two figures, it should become clear that the effect of a contrast shift is to produce a color shift which will vary over the entire density range. A sensitometer test will reveal a film which seems to exhibit too much of one particular color in the lower density range and too much of that color's complement in the higher density range. In Fig. 7.3, the blue layer exhibits a decrease in contrast combined with a decrease in speed. The figure shows that, if this film is used, one will find that the higher light level regions will appear to be too blue while the lower light level areas will appear to be too yellow (or, stated in other terms, they will not exhibit enough blue). This defect is known as 'color cross-over', and the effect cannot be corrected with the use of filters. Some compensation for the distortion introduced by this contrast effect may be obtained through the use of careful color-masking techniques. The easiest solution is to choose a film which does not exhibit this effect.

* Please understand that astronomical applications of any manufacturer's particular emulsion amounts to a tiny portion of the manufacturers' demands; to blame the manufacturer for not supplying this type of information is totally unrealistic.

Overview of the calibration process

(warning: this section could drive you to drink!)

The only way to determine how reciprocity failure affects any particular film is to photograph a grey scale and see how neutral the color appears in the processed transparency. If this sounds easy, you may be underestimating the nature of the task. The problems which arise in performing this 'simple' task include:

1. the grey scale must be illuminated by the light with which the film was designed to be used;
2. the exposure must be of the same duration as the exposure planned at the telescope (if one uses a wide range of exposures, the filter data must be determined for a series of exposures);
3. what is grey, how is it defined, how do you measure it? A color transmission/reflection densitometer is extremely useful here, but cruder equipment can be improvised for less accurate but still satisfactory work. One could approximate a grey, within certain limits, as a color change of 0.10 density is usually noticeable but the exact color shift may be difficult to define for non-color experts. The densitometer makes this an easy task as a perfectly neutral grey should have the same density when measured in any color.

In order to satisfy condition 1, the grey scale must be illuminated by a light source having a color temperature of 5500 K (degrees Kelvin) for daylight films or 3200 K for tungsten films. Color temperature meters and Kodak Wratten filters can be used to measure and adjust the color temperature of any particular light source.

Condition 2 means that it will be necessary to vary the illumination intensity. This can most easily be accomplished using a neutral step wedge as the photographic 'target' but watch out! Many so called 'neutral' density filters are not very neutral (i.e. Wratten No. 96 ND filters). The best are, in order of preference, Inconel-coated quartz filters, Inconel-coated glass and

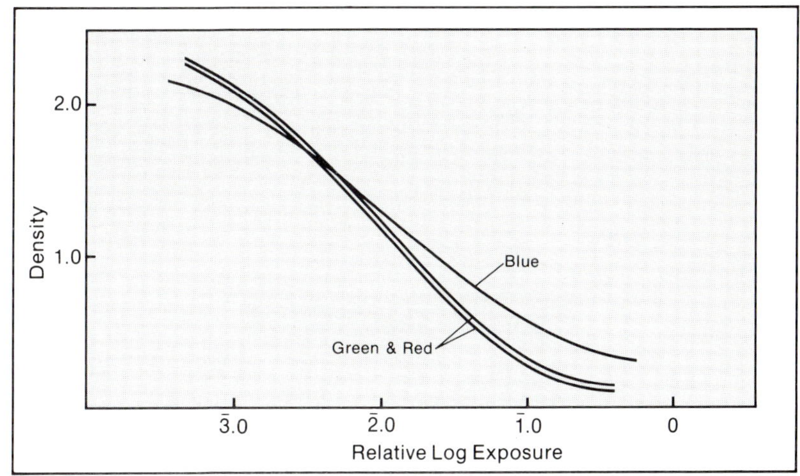

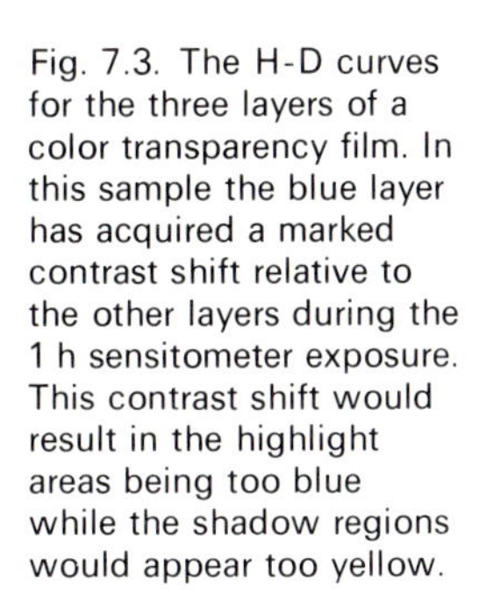
Fig. 7.3. The H-D curves for the three layers of a color transparency film. In this sample the blue layer has acquired a marked contrast shift relative to the other layers during the 1 h sensitometer exposure. This contrast shift would result in the highlight areas being too blue while the shadow regions would appear too yellow.

some filters made from photographic emulsion. Inconel filters are fairly expensive, but Kodak's photographic step tablets offer a reasonably priced and reasonably neutral alternative. The alternative to using a step wedge as a target is to vary the distance of the light source from the target or to use a series of sources of varying aperture.

The problems created by 3 can best be overcome by buying a new or surplus densitometer. Surplus black-and-white densitometers can be obtained for very reasonable prices and Wratten filters (92, 93, 94A) can be used to measure the color densities. Alternatively, one could build a densitometer from one of the kits advertised in many photographic magazines; some of these units are quite excellent. A final alternative is to get to know the people at a local custom color lab. These companies almost always own a good color densitometer, and it is not in use for much of each day. The authors have gained access to several densitometers in this manner. If this overview has not dulled your enthusiasm, the second section in Chapter 5 contains a detailed account of the requirements and construction of a sensitometer for astronomical applications.

Inherent difficulties with color emulsions

Before starting on the construction of a sensitometer, you should know that there are a number of problems relating to the use of color films in astronomical applications which cannot be corrected by any means. We do not want to discourage the use of color films, but it is important for the serious worker to be aware of these problems. If you find, through the use of the sensitometer, a film which exhibits one of these traits to an unacceptable degree, you can test others to find one that is more acceptable. The benefit of building a simple sensitometer is that tests can all be done in the darkroom eliminating the need to waste telescope time running disappointing, and often indeterminate, emulsion tests! If you choose to go ahead and use films recommended by other amateurs in the astronomy magazines, without further testing you will at least understand why some films behave as they do.

Color films were designed for the purpose of recording objects which are illuminated by a light source radiating a wide range of the visible spectrum and which exhibit colors by reflecting or transmitting a relatively large portion of the spectrum of the incident light. The light emitted from such sources and the light reflected by objects illuminated in this manner is termed 'continuum' light. All the objects that we see around us in everyday life display such spectral characteristics, even stars and most galaxies (being mostly stars) exhibit such light. Fig. 7.4 shows the reflection spectra of several everyday objects.

In contrast to these, are 'objects' (atoms or molecules) which radiate light only at specific wavelengths, caused by the transition of electrons from a higher energy state to a lower energy state. If these 'objects' emit light at numerous discrete wavelengths, the combination of all the visible emission lines of various intensities is perceived by the eye as a single color. Common examples of this are neon lights, mercury vapor lights and sodium vapor lights, to name just a few emission objects with which we are all familiar. The fact that our new improved street lights emit at just a few discrete wavelengths has made it possible to construct 'nebular filters' to allow amateur astronomy to continue in a world where light pollution permeates a great portion of the northern hemisphere. A more detailed discussion of such light sources can be found in Chapter 11.

Color films were not designed to photograph sources emitting light at such discrete wavelengths. They can yield an approximate color rendition, but cannot distinguish the colors of, say, the blue of 400.0 nm and the blue of 405.0 nm since all blues over a wide range of wavelengths will be recorded as a single color. Thus, it follows that the colors seen in a photograph of the spectrum of an emission source are probably not true to those that the eye would perceive if it could be used to view the colors of the spectrum directly. Colors, other than the 'pure' red, blue, and green characteristic of the individual dye layers, can only be produced when light is transmitted by more than one emulsion layer. Stars, galaxies and reflection nebulae are all continuum sources, but planetary nebulae and emission nebulae radiate in numerous discrete wavelengths, making the task of recording their 'true' color an impossible one for the simple system employed in color emulsions.

This problem is further intensified by the positions of the more intense nebular emission lines relative to the sensitivity curves of the three emulsion layers in color films. The most intense line in emission nebulae like that in Orion are at 656.3 nm (Hα) and 500.7 nm (OIII), two other intense emission lines (although considerably less intense than the Hα or OIII lines) are found at 486.1 nm (Hβ) and 495.9 nm (O) (Fig. 7.5). In Fig. 7.6, we see these nebular emission lines superimposed upon a curve representing the net summed sensitivities of the three filters normally used for tricolor photography. As one can readily see, the Hα line falls very close to the sensitivity peak of the red emulsion, while the other intense lines (the OIII

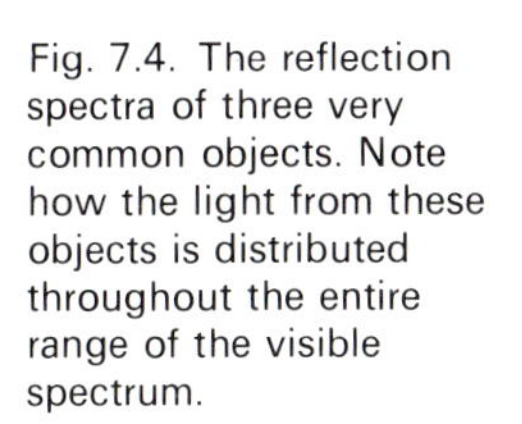
Fig. 7.4. The reflection spectra of three very common objects. Note how the light from these objects is distributed throughout the entire range of the visible spectrum.

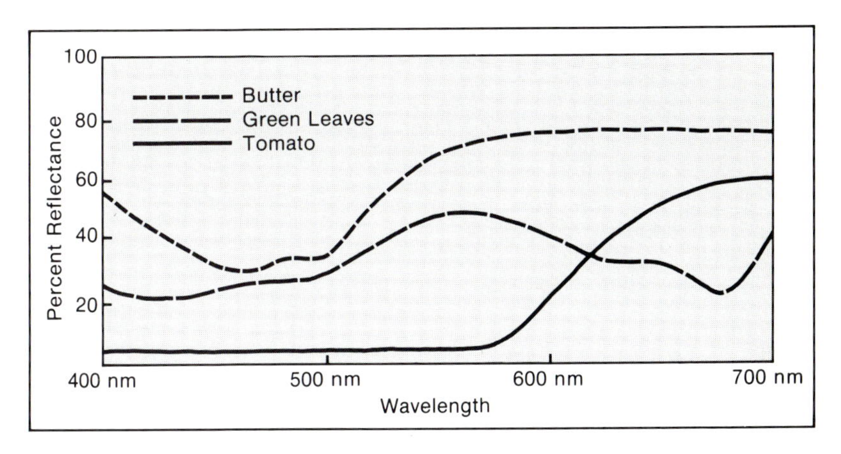

line at 500.7 nm in particular) fall in an area of the spectrum where the blue and green emulsions are losing sensitivity quite rapidly.

Even if the net sensitivity curve (the curve formed by summing the curves of the red, blue and green emulsions) is considered, there is a significant dip in the response of color films near this greenish region of the visible light spectrum. Many observational astronomers complain that the Orion nebula appears greenish blue when viewed through large telescopes while it photographs in various shades of red. This effect is often pointed to as a flaw in the nature of color emulsions while some use this as an argument against the use of color films in astronomical photography. In reality, this discrepancy is the product of three faults viewed simultaneously. The human eye at night is relatively insensitive to light at the Hα wavelength and near its peak sensitivity in the spectral region containing the blue-green emission lines while the color emulsion is less sensitive in this greenish-blue region than near the Hα line. The spectral sensitivity differences between the human eye and the photographic emulsion (and photo-electric detectors) are vastly different and one should avoid mixing apples and watermelons in this already complicated field.

Another problem which affects the fidelity of reproduction in color emulsions is related to the cyan, magenta and yellow dyes which are used in the emulsion to reconstruct the final image. These dyes do not behave as the so-called ideal dyes that would be incorporated in the ideal color emulsion. In theory, the ideal situation for color reproduction would be to have a yellow dye which absorbs blue light only, a magenta dye which absorbs green light only and a cyan dye which absorbs red light only. In reality, the cyan dye absorbs some green light, and some absorb blue light also, and the magenta layer absorbs quite a bit of blue light and some absorb a small amount of red light as well (Fig. 7.7). By using some rather advanced color-masking techniques, some of the color distortion created by these dye imperfections can be removed. This is a highly technical process and

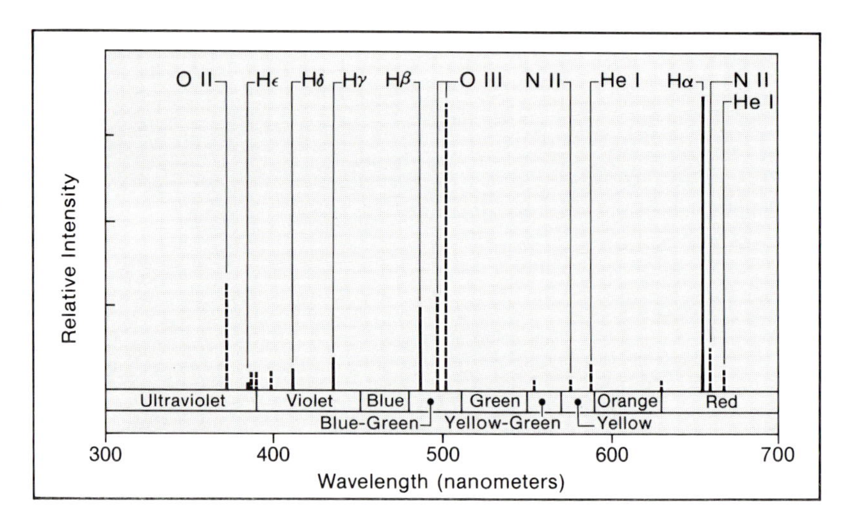

Fig. 7.5. The spectra of a typical emission nebula, with spectral lines identified. The regions of the spectra corresponding to commonly used color names are labeled in accordance with accepted color standards. The nebula used for this example is the Trapezium region of M42. Other nebulae spectra vary widely from this example; the lines that generally vary in intensity from nebula to nebula are shown with dashed lines.

significant effort is required to correct what are often very subtle color shifts. The interested reader is referred to the works of Neblette, Hunt, and Evans, Hanson, & Brewer for further information on this topic.

Tri-color photography

Another method of obtaining color images is the process known as 'tri-color photography' which actually duplicates the method used by Maxwell to obtain and reconstruct the first color images. In this process, the scene or object is photographed, in black-and-white, three times on three separate sheets of film using red, blue, and green filters. These images are then recombined, by one of several methods, to reconstruct the original scene. This method requires at least as much care to obtain 'accurate' colors as the the more common method we have discussed above, and requires considerably more work to produce a final positive image, but it allows the user a degree of image control that is not possible using standard tri-pack color emulsions.

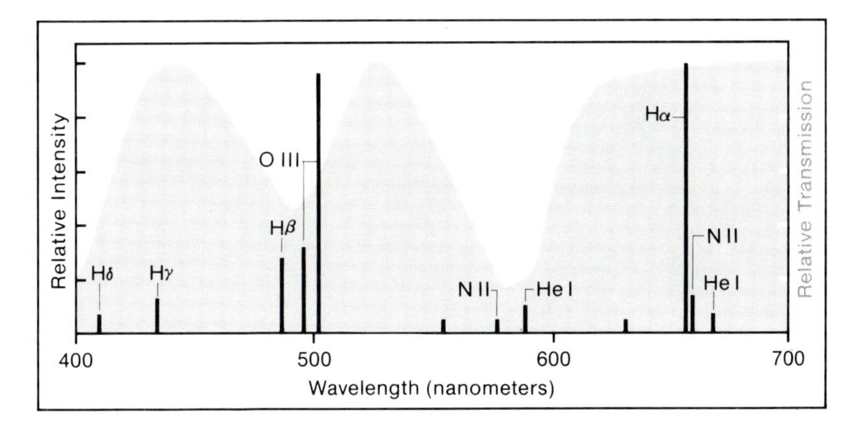

Fig. 7.6. The spectra of a typical emission nebula, along with the net transmission curve from a standard tri-color filter pack. Note the severe sensitivity dips in the blue–green region and in the yellow region of the spectrum.

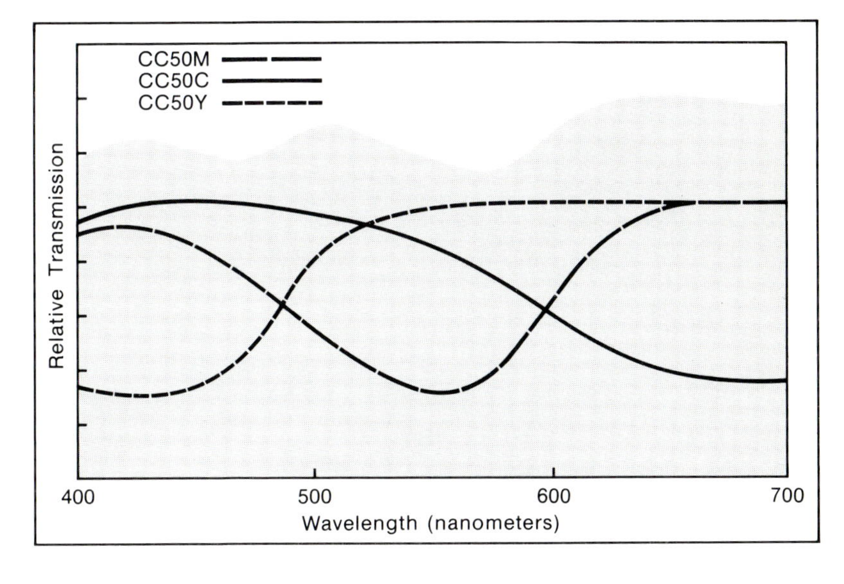

Fig. 7.7. The relative transmission curves of cyan, magenta, and yellow filters similar to those used in color emulsions and for tri-color reconstruction. In addition, the net transmission curve of the three filters is shown. Note how the overlap regions of these filters produce a relatively smooth net transmission curve.

There are several variations on the basic tri-color technique and all three have been applied in various astronomical applications. The technique which is usually adopted is to use three emulsions of different spectral sensitivities. An O-sensitive emulsion is used for the blue exposure with a cut-off filter that eliminates light below ~390 nm, while a G emulsion is used for the green exposure with a cut-off filter to eliminate light below ~480 nm, and an E emulsion is used for the red exposure with a cut-off filter that absorbs light below ~600 nm. A second variation on this theme uses an O emulsion plus a cut-off filter for the blue exposure and a panchromatic F emulsion for the other two exposures. In this variation, a broad-band green filter is used for the green exposure, and a red cut-off filter is used for the red exposure. The final variation, which we prefer, uses a single panchromatic emulsion for all three exposures in combination with blue and green broad-band filters and a red cut-off filter.

This latter method is the first method used to obtain color photographs and there are several reasons for choosing it over either of the other approaches. The most obvious reason is the financial aspect; the user only needs to obtain a single emulsion type when using this method, which produces the second additional advantage. The fact that only a single panchromatic emulsion is needed allows the user to select from a wide range of films and match the film to the subject being photographed. For most astronomical applications, a high-contrast emulsion like 103a–F will be quite suitable, while other objects may be better suited to a wide-latitude panchromatic film such as Ilford's HP–5. The introduction of 2415 Technical Pan offers another interesting panchromatic emulsion as a choice for tri-color work. This film can be processed for either high or low contrast and its fine grain and exceptional resolving power make it an attractive choice for this work.

A critical point to consider when choosing filters (and films) for tri-color work is that accurate color reproduction over the entire range of visible light is only possible if there is some overlap of the transmission bands of the three filters used. With continuum radiation, this requirement is not as critical as it is with emission objects because of the fact that the light from the source is spread over a wide range of the spectrum. If a small portion of the light from the object is not recorded with the three exposures, a close approximation to its actual color will still be obtained because portions of the light from the object were recorded on two or more of the exposures. If the object was only recorded on one of the exposures, it would be reproduced as being of exactly the same color as the filter with which the exposure was made (assuming that the same filter is used for exposing the final print). In the case of emission objects, the possible impact of this problem should be obvious. The light from all lines that are recorded on only a single exposure will be reproduced as a single color. However, if some of the emission lines from the object fall in a region where none of the three filters transmits light, that light will not be recorded. This latter situation accounts for the inability of most color tri-pack emulsions to record the blue–green light from the OIII line at 500.7 nm which is so intense near the trapezium in the Orion nebula.

When the transmission curves of the normal Kodak tri-color filters (Nos 47, 58, and 25) are plotted and the net light transmission is summed (Fig. 7.8) it is apparent that there are two regions of the spectrum where the net transmission for tri-color work is very low: around 470 nm and 580 nm. By searching through the transmission curves of Kodak's various filters, a set of tri-color filters can be found that improves upon these low-sensitivity regions. This filter set consists of filters 47, 57A, and 23A and the summed sensitivity curve for this filter set is also shown in Fig. 7.8. This alternate tri-color filter set provides a far better mix of colors across the visible spectrum and also provides for a flatter net sensitivity curve. However, the net sensitivity still drops steeply in the violet region of the spectrum and provides virtually no UV sensitivity. At present, there is no readily available filter set known to the authors that can remedy this fault.

The best solution for the photography of emission line sources would be to find a set of filters which would, when the individual transmission curves were summed, result in a uniform sensitivity curve across the entire visible spectrum while still maintaining overlaps between the three

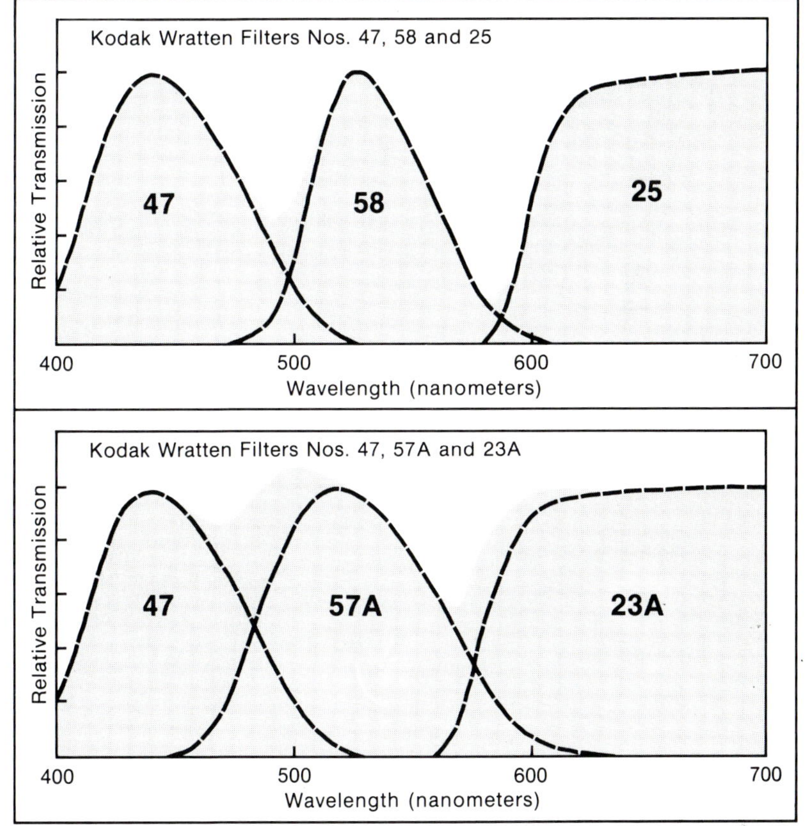

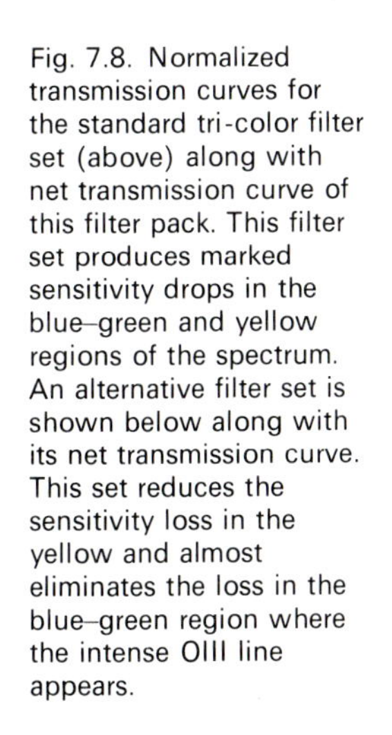
Fig. 7.8. Normalized transmission curves for the standard tri-color filter set (above) along with net transmission curve of this filter pack. This filter set produces marked sensitivity drops in the blue–green and yellow regions of the spectrum. An alternative filter set is shown below along with its net transmission curve. This set reduces the sensitivity loss in the yellow and almost eliminates the loss in the blue–green region where the intense OIII line appears.

transmission curves. Unfortunately, a filter set which fulfills this requirement does not appear to be available at this time. The tri-color results from the above method are much more likely to produce faithful color renditions of emission sources than the results from using three emulsions along with three cut-off filters that have non-overlapping transmission curves. This has been demonstrated by Spencer and a number of other researchers as early as 1935 yet many of those working in astronomy have continued to use the 'block-filter' approach to tri-color photography using three emulsion–filter combinations that have virtually no spectral overlap.

Once a film and filter combination has been chosen, the next problem is to choose the proper set of exposures required for each film–filter combination. In order to do this, one needs to construct the sensitometer that we have discussed earlier (the acquisition of a densitometer is desirable for this procedure, but close approximations can be obtained without the use of such an instrument). In order to begin the calibration process for tri-color work, a series of exposures of the grey scale in the sensitometer needs to be made using each of the three film + filter combinations. A series of exposures of 30, 45, 60, 90, and 120 min will usually be enough to provide the necessary data set for the calibration process.

Once the series of exposures has been made and the film processed, the density of each step is measured and the density of several of the steps between D = 0.4 and D = 0.9 on each piece of film is then plotted above the exposure given to that sheet of film. Finally, a line is drawn through all of the points showing the density changes of the same sensitometer step as a function of exposure time. When this is done, one must plot this density variance for a number of steps on each sensitometer exposure in order to establish how the density of each step has changed as the exposure is increased. Once this is done, an exposure value is chosen for the least sensitive film–filter combination, and a horizontal line across the graph will determine the exposure for each of the other two film–filter combinations, as shown in Fig. 7.9. It is important to realize that a set of exposure values must be determined using this method for each change in the longest exposure value that is planned. It is not possible just to multiply through by an exposure factor and still maintain accurate color balance. Normally, sets of exposures can be chosen so as to make the longest exposure of the set something on the order of 60 min, 90 min and 120 min, in order to photograph progressively fainter objects. As one example, using hypered Tri-X, we have found that an exposure of 90 min through a Wratten No. 47 and 59 min through a Wratten No. 57A and 49 min through a Wratten No. 23A will all produce equal densities when exposed to the illuminated grey scale in the sensitometer.

One might think that it is now time to go out to the telescope and expose some film with the newly determined exposure data. However, there is yet another step remaining before such exposures are of use. Once you have a set of exposures, how do you re-create a color photograph of the object? The problem should be clear. With a color film, the film structure is planned so that, once development is completed, one has a finished image that is ready to be projected or printed for viewing. With tri-color work, planning needs to be done in order to minimize the work needed to reconstruct the image and to maximize the accuracy of the final image reconstruction.

There are a number of ways to proceed with the problem of reconstructing an accurate final image. One could photograph a field of stars using the exposures determined above and then work to re-create the image by trying to match the colors of stars in the field (O stars being blue, G and K stars being pale yellow, and M and N stars being orange to red). This method poses a number of troublesome problems, including the difficulty of finding tables of the spectral types of stars down to magnitudes 12 to 18 and then locating the correct stars in the photographic field. In addition, the color of the star will vary as the exposure changes and it will require a number of exposures in order to insure an accurate reconstruction. This process, in our view, requires too much effort and does not guarantee an accurate reconstruction even after the effort has been expended.

The process that we have settled on is to use the sensitometer in order to determine a means of balancing the final color reconstruction. Using one of the exposure sets previously determined for the film + filter combinations that are to be used, photograph the sensitometer grey scale once again and process the negatives. The next problem concerns the method to be used to perform the final image synthesis and to form a positive color image that can be viewed and reproduced in a conventional manner. We will go through the details of the simplest and most direct method and refer our readers to the bibliography for additional information. The problem that

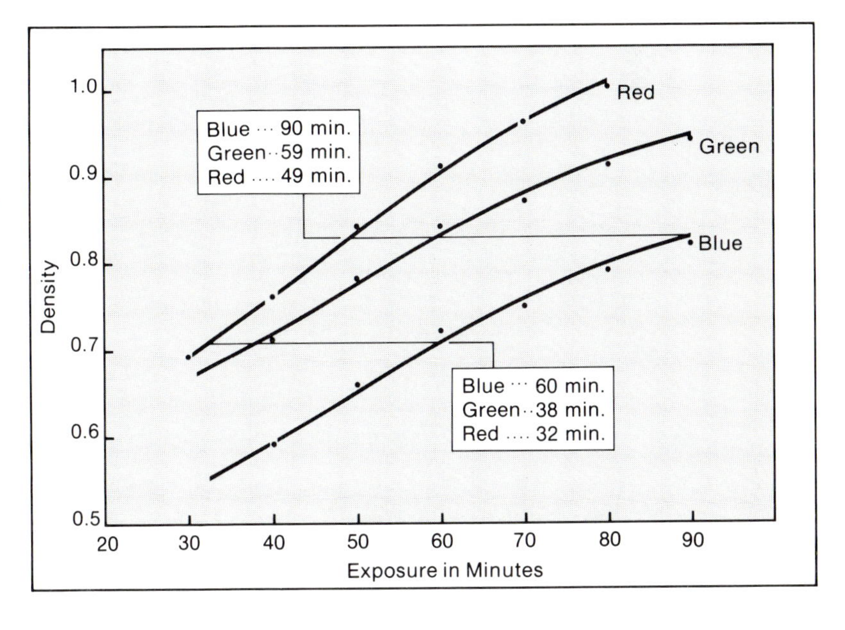

Fig. 7.9. Sensitivity curves for Tri-X, plotted as a function of exposure using the modified tri-color filter set shown in Fig. 7.8. Once the exposures have been made (six in each filter for this example) and the curves of density *v.* exposure have been plotted, sets of exposures for tri-color work can be derived graphically, as this example shows.

arises in this process lies in the fact that you now have three black-and-white negatives that must be turned into a color transparency or print. You could produce red, blue, and green positives and project them one by one onto a sheet of color transparency film or color reversal print paper, or you could produce cyan, magenta, and yellow dye positives and stack them (in register) on a sheet of color transparency film or color reversal print paper and make a contact print to produce a positive color image. Both of these processes can work and both have been used successfully although the additive process gives the user more control than does the subtractive process. However, both involve extra steps that can introduce additional variables that need to be controlled and one process requires the use of products that are not generally handled by most advanced professional photographers unless they are involved in lithography.

We prefer to use a more direct, single-step technique (initially used for this purpose by Alt, Rihm, Brodkorb & Rusche) to produce the final transparency. In this process, the three negatives are projected sequentially onto a sheet of Kodak Ektacolor print film 6109 to produce a final color positive. However, in this step it is important to insure that the red negative only registers on the red-sensitive layer of the film and that the same is true for the blue and green negative and the blue- and green-sensitive layers of the color positive film. If these conditions are not met, the final image will be contaminated and color accuracy will be sacrificed. In order to insure that no contamination occurs, each of the images is projected through a medium-band filter instead of the broad-band filter that was used to take the negative originally. The red light negative is projected through a Wratten No. 70 filter, the green light negative is projected through a Wratten No. 99 filter, and the blue light negative is projected through a Wratten No. 98 filter. Each of the three negatives is projected, in sequence and in register (see pp. 195–202 on integration printing for further detailed information), onto a sheet of 6109 film and is then processed. This first run will have virtually no chance of showing a color-balanced result, but it will serve to provide some information (once one is familiar with the process, this step will produce a color balanced reproduction).

The three negatives from the sensitometer should, when reproduced correctly, produce an image showing a neutral grey step wedge running from white to black. If a densitometer is used to read any step on a 'perfect' grey scale, it will theoretically yield equal density readings in the blue, green, and red (this is the same theory that was used to determine the proper tri-color exposures earlier). Now, we can use this same piece of information to get a close idea of the proper exposures to use to recombine our three negatives and produce a balanced color positive. In order to use this principle each of the three negatives must be printed through the appropriate filter side by side onto a single sheet of 6109 film. When this sheet is processed it can be examined and one can estimate (or measure with a densitometer) which steps on each colored image are of the same density.

If this is measured, one can calculate the exposures necessary to produce a print using all three images that will show a nicely balanced grey scale (barring problems with reciprocity). In order to get an even closer approximation, a second print of the individual negatives can be made using the calculated exposures and the steps (now more closely matched) can be measured again and a better set of exposure values can once again be calculated. Now, it is time to produce a print in which all three negatives are combined to produce a color print that should show a balanced color image of a grey scale. It is well to spend some time on this since it is this step that will determine how well the system performs from this point forward.

In all likelihood, the grey scale will not be perfectly matched over its entire density range. Instead, it is likely that it will be balanced over some density range and that the color will be off at one or both of the extremes of the scale. This cross-over effect has been discussed in the earlier section on color reciprocity failure but, in this instance, the photographer has the option of correcting the color shift if one is willing to expend some effort. Quite probably, the color cross-over that is now seen is due to the fact that black-and-white films increase in contrast when exposed to light of longer and longer wavelengths. Thus, a blue light negative is lower in contrast than a green light negative and a green light negative is, in turn, lower in contrast than a red light negative. In a multilayer color film it is impossible to correct for the cross-over effects created by reciprocity failure since all three emulsions must be processed identically. However, in this tri-color application, the three images can be processed separately and thus the contrast of each emulsion can be carefully controlled. Since the wavelength effect will probably result in only a small shift over the range from black to white, it is unlikely that any but the most fanatical amateurs will wish to pursue the problem of performing the final stages of color balance adjustment.

Once the exposures for printing have been determined for the sensitometer strips, it is finally time to go to the telescope. The same set of negative exposures is used at the telescope and then the set of printing exposures that was previously determined is used at the enlarger. If the positive requires exposure adjustment in order to lighten or darken the image, it is important to remember that all three exposures must be increased or reduced in proportion in order to maintain the correct color balance. Thus if one wants to shorten the exposure by 10 %, each of the three exposures must be shortened by 10 % of their respective durations.

Tri-color photography has some advantages that help to offset the disadvantages of three long exposures and some relatively complex darkroom stages during the setup and calibration process. One advantage is that the grain structure of each emulsion is considerably sharper than that of a color image in a fast color emulsion. This is due to a number of factors including the fact that a dye image is more diffuse than a silver image and the fact that the light is not being dispersed by its passage through a stack of three individual emulsions. In addition, since the three images are being printed in register, one is effectively doing integration printing and the graininess of the resulting image will actually be reduced by a factor of ~ 1.7 over that present in a single image using the same black-and-white emulsion (see Chapter 9).

While color photography, as applied to astronomical situations, certainly has its faults, we would not want to discourage its application. The photographs that can be obtained using even the smallest instruments can be breathtaking and certainly are inspiring for the beginning astrophotographer. The information that we have presented should be seen as a knowledge base from which to draw in order to understand some of the interesting effects that can be encountered in this area. With a reasonable background, one knows if and how some problems can be solved and even has an awareness of some of the alternative processes.

8

Introductory darkroom techniques

Over the years, we have met a number of amateurs who claimed that they simply did not have the time, space, patience or money to do their own darkroom work. Their work reflected this decision! It is impossible to turn out quality photographs without doing the processing of both film and prints. The requirements for processing and printing astronomical subjects are so radically different from those of 'normal' photography that only an ultracustom lab would be able to produce satisfactory results. The rates for a high-quality black-and-white printer run in the range from $10–$50 for a single print! The caution that we will issue is that one needs to do processing and printing right and the information needed to do it right will not fit into a 1 page information sheet or a 20 page booklet or even a single book. One needs to do considerable reading, take a few classes, and work in the darkroom in order to develop the needed skills.

Anyone who uses film and does not possess an understanding of that medium beyond the 1 page cookbook-information sheet that is provided by the manufacturer will more than likely end up wasting a lot of film and time and will not be able to recover from abnormal situations because of that lack of knowledge. In order to produce photographs that have a 'professional quality' some understanding of photographic media is required, along with some care in the processing and printing procedures. Basic knowledge of films, developers, photographic papers and chemical properties is required in order to understand the cause(s) of errors that are encountered and to allow one to go beyond the limits of the *status quo*. We hope to be able to convey some understanding of what happens in the darkroom, and of why some processes will work while others will not.

Darkroom equipment*

We will not spend a lot of time detailing a long list of equipment that one can collect and use in the task of producing better results and analyzing astrophotographs. Such a list could expand almost endlessly and overwhelm the darkroom budget of all but the richest enthusiasts. Instead, we

*The statements made in this chapter regarding equipment and processing techniques are generally directed to users of 35 mm roll film. While some equipment changes will be necessary, the principles discussed will apply equally well to plates and/or sheet film.

will outline some of the basic needs and discuss some points which should be kept in mind when adding to the arsenal of equipment.

We will first provide a list of equipment that should be considered as basic for darkroom work, and divide the list into sections for film processing and printing.

Processing	*Printing*
Film-processing reels	Enlarger
Film-processing tank	High-quality enlarging lens
Accurate timer	Processing trays
Accurate thermometer	Enlarger timer
Graduates	Drying screens
Accurate scale	Focusing magnifier
Drying clips	Print easel
	Safelights
	'Dust off'
	Camel hair brush

There are many types of processing tanks and reels on the market today and each variety will have its own proponents. Our preference, from long experience, is for the stainless steel variety. However, we have found that the stainless steel tanks with metal lids tend to leak chemicals, so it is preferable to have plastic lids for the processing tanks. The trays used for processing prints need to be larger than the size of the prints. If you are making 5 × 7 inch prints, you will need 8 × 10 inch trays and 8 × 10 inch prints will require trays that are at least 1 inch larger in each dimension. The thermometers for black-and-white (B&W) processing need to be accurate to about ± 0.5 °F and they need to be able to respond quickly to temperature changes. As the current boom in electronics technology grows, digital thermometers will undoubtedly become almost as inexpensive as mercury thermometers. Our advice would be to avoid inexpensive models, as one needs to be certain of accuracy and repeatability when trying to achieve the consistency required for really good-quality astrophotographs.

The timers that are used for timing enlarger exposures need to be accurate to about ± 0.1 s while the timers required for timing processing times need to be only within about ± 5 s (although greater accuracy is desirable). Again, the electronics boom is rapidly providing more and more accurate darkroom timers which are able to resist some of the corrosive chemicals that are commonly found in photographic chemistries. Timers for use in the darkroom need to be readable in the dark, but they do not need to be very bright (you would be amazed at how bright things become after 10–20 min in a totally darkened room).

The enlarger and its lens is a vital tool for the darkroom; the enlarger needs to be very solidly built, so that vibration is held to a minimum. When it is set at a particular height and focused, it should stay at that setting and not slip unless force is purposely applied to change its position. The enlarger can be either a diffusion enlarger or a condenser enlarger; each has its own merits and both produce excellent results when properly used. The diffusion enlargers are generally preferred by photographers who do fine printing, for the print quality that they can provide. The diffusion enlarger produces lower-contrast prints from a given negative and is able to penetrate very dense regions of the negative while still maintaining details in the 'shadow' areas. The most critical item in the darkroom inventory is the enlarging lens: a poor enlarging lens will never produce an excellent print! When it comes to purchasing the enlarging lens, the best is just barely good enough. Today, Rodenstock, El Nikkor, Fujinon and Schneider stand out as the best lenses available. If a used lens by a reputable manufacturer is to be purchased, the coatings on the lens should be inspected and should be in excellent condition. The f/ratio of the lens is not of any real importance for B&W work but, if one is planning to print from color transparencies (i.e. slides), the faster enlarging lenses provide some extra light for focusing and setting up the print composition. Table 8.1 summarizes the focal lengths generally recommended for various film formats, shorter focal length lenses can be used for greater enlargements of the central portions of a frame.

Once prints are finished and washed, they need to be dried. The easiest way (and just about the best way) is to lay them out on a nylon window screen, face up, and to let them air dry. The use of blotter books is to be avoided as any contamination of the blotter pages will result in the contamination of prints that are later dried in the book. The same comment is true of mechanical dryers with cloth belts. When hypo* residue is left in any print, it will eventually yellow and the image will fade very rapidly. With a nylon drying screen, the prints made on glossy paper dry to a semi-glossy finish and, if the screen should get dirty, it can be rinsed off and restored to a virtually sterile condition.

Darkroom materials

The well-equipped darkroom contains more than just equipment. The expendable materials must also be stocked and kept available. There are

Table 8.1. *Generally recommended enlarger lens focal lengths for various film formats*

Film format	Focal length
35 mm	50 mm
6 × 6 cm	75–85 mm
6 × 7 cm	90–105 mm
4 × 5 in	135–150 mm

* The term 'hypo' is derived from 'hyposulfite' which is known today as sodium thiosulfate. The excellent fixative properties of this compound were discovered by Sir John Herschel in 1819.

very few materials which need to be purchased outside of a well-stocked photographic store and these can be obtained on special order or purchased in a chemical supply shop. In general, most of the less-available chemicals will be required for mixing special developers like MWP-2 and POTA. A reasonably complete list of expendable materials would include:

Film developer mixes	Paper developer mixes
Concentrated acetic acid	Fixer (with and without hardener)
Hypo eliminator	Print hardener (potassium alumate)
Wetting agent	Selenium toner
Film cleaner	High-quality grade 3, 4 and 5 papers
Cotton gloves	Plastic bottles (1 qt or litre, 1 gal or 4 litres)

From this list of materials, the user has a very wide latitude in making choices due to the high degree of standardization that has evolved in the field of photography. Different manufacturers make very similar products to achieve the same result, and most products are of very high quality. The one area where we will offer guidelines for product selection is in the area of selecting printing papers for this is where the outcome of all the work at the telescope and in the darkroom is determined.

For fine prints, you should select a high-quality brand of printing paper (i.e. a paper containing lots of silver in the emulsion, allowing printing of full tonal range images) and purchase boxes of grades 3, 4, and 5 paper with a white glossy surface finish (quality brands in 1986: Ilford, Ilford Gallerie, Agfa Brovira 111, Oriental Seagull). These high-quality papers generally contain more silver than their lesser counterparts and thus provide richer tonal ranges in the finished prints. The use of multi-grade papers is not recommended as these papers will only span contrast grades in the range from about 1.5 to 2.5 and such low-contrast papers generally prove unsuitable for most astronomical subjects. At present, the use of resin-coated (RC) papers is severely discouraged, as is the use of stabilization paper processors. While comparisons of densitometer readings from prints made from the same negative on the different papers show no major differences, the eye sees a substantial difference when comparing prints. In addition, RC papers are not archival and the plastic surface cracks and splits as the print grows older. RC papers and stabilization processors may one day acquire the quality level that is needed for fine printing, but that level of refinement has not yet been reached.

Processing

We will now proceed on the assumption that you have chosen a film and developer combination based on information obtained through careful testing or from a reliable source and are ready to process some film. We are also going to assume that you do not have access to a fancy darkroom or any sophisticated equipment. We will discuss equipment and procedures that will produce negatives of very high quality. (Kodak also publishes excellent booklets that describe most of the procedures that we will discuss.)

The easiest and most efficient way to process 35 mm and 120 film is to spool the film onto a reel and load the reel into a light-tight tank for processing. Our personal preference is for stainless steel reels because the construction of plastic reels restricts the flow of chemicals over the film surface and may cause surge markings. Therefore, a set of stainless steel reels and a stainless steel tank should be used. We have found that stainless steel tanks with plastic lids are the best combination to help stop leakage. When stainless steel reels are first used, the loading procedure can be difficult to master, but a little practice with some scrap film in the light can make the learning process go a lot faster. If you continue to have problems, a simple device known as an 'auto loader' can make the process into a 30 s task requiring only minimal manual dexterity.

You should begin by mixing and assembling all of the required chemicals and bringing all of them to the desired temperature. For black-and-white processing, you should try and maintain the temperature of all solutions to within ± 1 °F. You can maintain the temperatures of the mixed solutions by placing the containers into a tray containing a water bath at the required temperature. The water bath can be static if the air temperature does not differ from the required processing temperature by much, but a slow-running water bath is preferred if no other means of maintaining temperature is available. A stable-temperature water bath can be constructed quite easily and inexpensively by using a submersible aquarium heater (of either 25 or 50 W) and an aquarium-water-circulating pump in a water bath of 1–2 gallon (4–8 l) capacity. In areas where the summers are relatively hot, is is very useful to standardize processing using a temperature of 70–5 °F to avoid the problem of having to cool the processing chemicals.

When mixing the various solutions, there are some procedures that we use that are not standard for most photographers. One of these was recommended by Wm C. Miller of Hale Observatories and it is a standard at many observatories. When mixing the stop bath, you should only use a 2 % acetic acid solution i.e. less than half the concentration that is usually recommended. This is done to prevent bubbles from rupturing the film as the acid bath is poured into the basic developer residue to stop the developing process. We also mix-up our fixer at half strength and then frequently test for exhaustion, as it is used. In addition, we prepare two fixing baths and split the fixing time up between these two baths. This provides much better fixing than is achieved using a single bath and extends the life of the fixer. Finally, you should also prepare a solution of hypo eliminator to cut wash times in order to save water. There is no sacrifice in the final image or archival quality of the negative and only a third as much water is needed for washing.

When all the solutions are at the proper temperature, you are ready to start. If you have a darkroom timer, you should set it for the developing time that you plan to use. If you do not, a clock will be fine, as long as there is some way of seeing seconds count by during the processing period. After setting the timer, quickly pour the developer into the film container and then start the timer.* Now, agitate the tank vigorously (by inverting repeatedly and twisting simultaneously) for at least 15 s and then, gently, for the remainder of the first minute of development. At this point, gently (but not too gently) bump the container on a firm surface to release any bubbles that are trying to stick to the film and then set the tank down in the water bath and wait 30 s. Then, agitate the tank gently for 5–10 s of every minute throughout the entire processing period. It is extremely important to adopt a consistent method of film agitation and stick to it. About 15 s before the developing time is finished, begin pouring out the developer. When the time is up, pour in the stop bath unless you are using a film with a rem-jet antihalation backing (like 103a-emulsions). If you are processing such films, you should use a 30 s water rinse prior to using the stop bath in order to help remove the antihalation backing. When you pour out this water rinse, it will be loaded with a soot-like suspension of carbon particles that will adhere to virtually everything, so pour it carefully down the drain. Now, pour in the 2 % stop bath and agitate continuously for 30 s and you are ready to fix the film.

Again, we use two half-strength fixing baths for more thorough fixing. The first bath can be a slightly used, but still active, solution and the film should be agitated every minute for 4 min. The second bath should be an absolutely fresh hypo solution and it should also be used for a 4 min period with agitation every minute. When you finish with the second bath, pour the used solution into the bottle for used hypo to avoid contamination of the fresh fixing solution.

Now, wash the film for about 3 min in water that is within 2 °F of the processing temperature, to remove most of the hypo and then set it in the hypo-eliminator solution for about 2 min before continuing the wash. Before the final wash, you should remove any rem-jet backing that may still be adhering to the film. It can usually be removed by slipping the wet film between two wet fingers, but be careful and be sure your hands are clean. If any of the coating remains stubbornly on the film, you should run a cotton swab that has been soaked in developer over the backing. This will usually remove all of the backing in a single stroke! The final wash should be about 20 min and it should be carried out at a temperature of about 75 °F to insure removal of all hypo. When the wash period is complete, the film should be soaked in a wetting solution (mixed at one-half the recommended strength) for about 5 min. Before hanging the film up to dry, you should inspect it to be sure that the water is 'sheeting' over the film surface and not 'beading up' into small droplets. Finally, lower one end of the film into the wetting solution and run the film quickly back and forth through the bath while you

* For some film/developer combinations (e.g. Kodak 2415 and D-19) dropping the spooled film directly into the developer and beginning agitation as soon as possible will result in more even development.

hold both ends. Then, hang up the film by one end in a clean, dry, dust-free area and hang a clip on the other end to keep the film from curling as it dries. Do not blow dry the film or heat dry it unless you have a drying cabinet that was designed for that purpose. Allow the film to dry overnight before trying to make prints or inspecting it as it is bloated with water at this point and can be ruined easily. Patience is very important at this point!

When processing color films, many of the above processes can serve as guidelines. Some of the major differences between the processing of B&W films and color films include: there are usually more steps when processing color films; the temperature needs to be carefully monitored and controlled; and the agitation used throughout the actual processing stages needs to be consistent and uniform. Our experience has generally shown that the substitute color-processing kits that show up on the market from time to time may not yield the same results as the kits provided by the film manufacturer. We generally find that they may work reasonably well on 'normal' scenes, but can yield very poor results when applied to astrophotographs. Whenever such new products appear, and are given enthusiastic reviews in various photographic magazines, they should be carefully tested for speed and color balance with only one or two astrophotographs on the roll before risking an entire evening's work (sensitometer tests are preferable)!

Another color-processing alternative which is sometimes used for processing astrophotographs is to process color-transparency film as a color negative. The current color-negative process is known as C-41. When color-transparency film is processed as a negative, one ends up with a color negative which does not possess the usual orange base. In addition, the resulting color negative is quite contrasty and the grain is coarse. For many astronomical subjects, the high contrast is not objectionable and longer focal lengths can overcome the increased grain. Orien Ernest's work provides an excellent example of the application of this technique. In this case, the increased contrast from the C-41 process is acting to overcome the lower contrast that is characteristic of cold camera work (this same contrast loss is characteristic of hydrogen-hypered films).

We generally end up sending our color work to a custom color laboratory due to lack of time. Even when this is done, the astrophotographer needs to exercise some care and control. The local, quick-processing service is generally not up to the standards that are needed for this work. After finding dust, hairs, fingerprints, and spots all over your film, you can get pretty upset about a wasted weekend's work! By taking the film to a very reputable custom color laboratory, you are helping yourself and the cost increase over quick-processor prices is generally minimal! Another precaution to take when exposing a roll of film for astronomical use is always to expose some 'normal' scenes (even exposed blank frames will help) at the beginning and end of the roll to prevent the slides from being mounted so as to destroy your photos (even requesting that the film not be mounted is not always insurance against this misfortune). The procedure of including a strobe-exposed color-test chart on each roll can also provide useful information in case a slight color shift should affect the roll of film.

You now have most of the information that you need to pick films and developers and to develop your own film. Kodak supplies vast amounts of information on all of its products and this material can be very helpful in making choices. Kodak is also very good at responding to inquiries about their products. If you cannot get the information that you need, run your own tests to find out what you need to know! This is a good policy to adopt anyway. If we had believed all of the 'good advice' that every 'expert' astrophotographer had told us over the years, we would not be taking photographs of the quality that distinguishes our work!

Basic printing steps*

Once the film is completely dry you are ready to move into the darkroom for the really serious work. Your skill as an 'astrophotographer' is not judged by your negatives, it is the prints that everyone sees and enjoys. If the negative is not printed well, the best of photographs can appear to be of poor quality. If you do not print well, there is no point to making B&W negatives! If you are only interested in records for a nova or comet patrol, there is no real need for making prints until a discovery is made.

The 'darkroom' (or bathroom or closet) should be set up for printing now and the chemicals should be mixed and brought to temperature (70–5 °F is a reasonable range). The developer(s) and stop bath should be mixed as recommended, but we still prefer to use a half-strength fixing bath. The trays should be set up in sequence running from developer through fixer and a large tray of fresh water should be available to move the prints into after they have been fixed. Now, select a negative for printing.

Before placing the negative in the enlarger, all dust must be removed from it. An electrostatic 'gun' or fine-quality camel-hair brush can be used to help loosen up any lint or dust on the negative and then a blast of compressed gas can be used to be sure that the surface is really dust-free. If any dust remains, the process should be repeated until the negative is really clean. If grease spots have appeared on the negative (or any other unusual surface blemishes) it should be rewashed and dried again before attempting to print. Negative care is vital to the production of excellent prints and they must be carefully handled and stored in order to assure their use over periods of years or even decades. The lightweight plastic archival storage sleeves provide a simple and inexpensive means of keeping negatives in a

* We will assume in this section that you already have an enlarger and we will not enter into the debate over condenser- *v* diffusion-type enlargers. Our preference is for a diffusion enlarger, but each can be used to advantage. The diffusion enlarger does have a marked advantage when it comes to printing details in the highlight regions of the print. We will also assume (or urge) that you use only the best quality coated enlarging lenses on the enlarger (in 1980, Rodenstock, Vivitar VHE, Fujinon, Schneider and El Nikkor lenses lead in this field).

clean environment but one still needs to handle the negatives carefully in order to insure that the negatives are not destroyed by scratches and/or greasy fingerprints. There is virtually nothing as unaesthetic as a print made from a scratched-up negative that has a few fingerprints on it.

Now place the negative in the enlarger, adjust for the enlargement you want and focus carefully using a grain focuser. You should be certain, before focusing, that the enlarging lens is wide open to allow through the maximum amount of light. This will help to insure that the focus is as sharp as is possible by reducing the range over which the images will appear to be in focus (i.e. this reduces the depth of focus). We prefer 5 × 7 inch prints and enlargements from 6–12 × . The 5 × 7 inch print is a means of conserving paper and silver while still producing prints that are easily handled and/or mailed. An enlargement of 12 × is about the greatest that even the best negatives can stand. Beyond that point star images become quite large and start to take on a fuzzy appearance that is aesthetically unacceptable to us.

The starting point for making an astronomical print is to try a print on grade 4 paper. After carefully focusing, wait about 5 min with only the safelights on to allow your eyes to adapt to the low-light level. Then turn on the enlarger and slowly stop down the lens. At some point, you will notice that the finer details begin to disappear from view. At this point, stop and turn off the enlarger. Now, place a sheet of enlarging paper in the printing easel and use a piece of cardboard or railroad board to produce a print having a range of exposure from 5 s to about 60 s with 5 s steps (see Fig. 8.1). After completing the exposure, remove the print from the easel and set any timers that you need to allow for processing.

The print should be placed face down in the developer as quickly as you can and begin rocking the tray back and forth gently while turning the print over every 10 or 15 s. The print should remain in the developer for at least 90 s and not more than 150 s, at which time it is removed and placed in the stopbath. Do not cut the print development time short for any reason! If the print starts looking over-developed after 20–30 s, discard it and make another test strip using a shorter exposure. An over-exposed, under-developed print does not have the full tonal range that a properly exposed and developed print is capable of rendering. The print should be agitated during the entire 20 s period in the stopbath. After this, place the print in the fixing bath and agitate it continuously for the first minute in the fixer (after the first minute, you can turn on the lights in the darkroom) and continue to agitate it every 10–30 s for the next 2 min before removing it from the fixer. Since the test print is not to be saved, it can go directly into the wash for 2–3 min before squeegeeing or blotting it dry and inspecting it.

After removing most of the water from the print, set it in a well-lit area and examine the boundary lines between adjacent exposure areas. As the sky region becomes darker, you will see that these lines become less and less distinct until they virtually disappear. When they become difficult to see, you are at the correct exposure zone. If you were lucky, or if you planned the

Fig. 8.1. A test strip for correct printing exposure is shown of the region surrounding Rho Ophiuchii. The correct exposure is in the region where the sky just reaches an almost black tone. Less exposure would result in a featureless grainy grey sky while more exposure only causes a loss of detail. Photograph by Alan McClure.

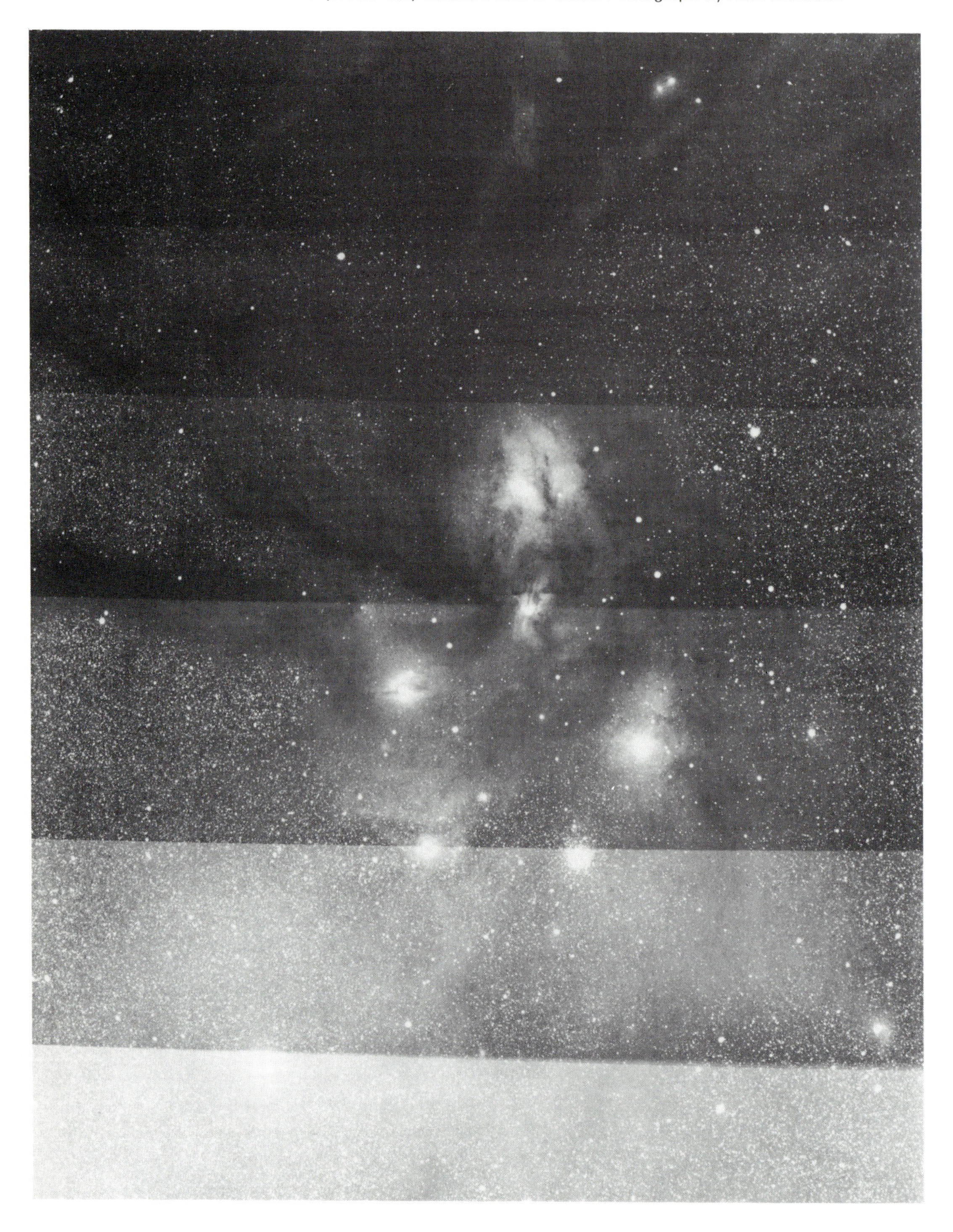

exposure carefully, you may be able to judge contrast from this print and avoid having to make a second test print for that purpose. In order to judge contrast, you will need to examine a correctly exposed area containing a bright, detailed, non-stellar area (e.g. part of a nebula or galaxy in the field). If you want to be sure of determining both parameters with a single print, you will need to make each of the exposures separately and move the easel over slightly between each of the test exposures to insure that the bright area appears in each of the test strips. A printing frame can be made with windows that will allow the test strip (see Fig. 8.2) to be made with very little extra effort. As the price of photographic paper continues to climb, such devices will become more and more attractive.

In any case, you will need a print in which the bright, detailed region has received the optimum exposure. If the area appears to be a dull grey, the paper used is too low a contrast grade, make another test print using the next grade up and examine that print. If, instead, the area appears to be a stark white containing no visible details, the contrast grade is too high and you will also need to try again. When examining the print for a light-toned detailed region, be certain that that region of the negative also shows details. If everything appears just the way you like it, you can proceed on to make a 'final' print. There are some objects that are commonly photographed which just cannot be printed 'straight' (i.e. without using any local manipulation that would cause a change in the apparent reflected density of the print). You can use dodging or burning to let less or more light through to the print in the areas that create problems. It is our general policy to prepare prints with black or very dark skies, to choose a paper grade that shows the fainter regions of a nebula or galaxy quite well and to use other manipulative methods to control the brighter regions of those objects that exhibit wide ranges of surface brightness.

When making prints, it is necessary to think in terms of the aesthetic quality of the image and to keep in mind what you want to see in the print. Some will argue that dodging, burning, masking and other manipulative techniques produce prints that are unnatural and that are inaccurate. This argument ignores the fact that the response curves of the film and the paper have already badly distorted the object's appearance and it also conveniently forgets about the matter of defining what is natural. The object of making a photograph and of producing a print is to show the appearance of some object in more detail than can be seen with the human eye. If the print does not portray all the details that are available, the print is failing in its task. In amateur and professional work alike, the matter of aesthetics enters into the process of preparing photographs for an audience.

Once a 'final print' is made it should be processed as above, up to the fixing step. Since this is a print that is to be kept and possibly circulated, it will need to be placed in each fixing bath for about 4 min. Then, it is placed in a running wash bath for 2 or 3 min and then set in a tray containing a hypo-eliminator solution. The print is agitated continuously in the hypo eliminator for 2 min and then placed into the 70–5 °F wash bath for another 20 minutes. The flow of water in the wash bath should be adjusted until it is flowing at about 1–2 litres/min. Alternatively, the water in the wash bath can be completely changed every few minutes and the print will have to be agitated by some other means during the entire wash period. After this wash, virtually all of the fixing solution will have been removed from the emulsion and the paper base and the print can be considered to be nearly as permanent as the photographic process will allow.

Once the print is removed from the wash it can be dried in one of several ways which will result in either a glossy or a semi-glossy print surface. The glossy print will produce a print with an apparently greater tonal range, but it will also produce a print that is difficult to view because of the highly reflective surfacc. We do not recommend the use of blotter rolls or books for drying prints because of the danger of contaminating the print with hypo if any prints are ever dried that have not been processed to archival standards. Another drying method is to set the prints face-up on nylon screen drying racks. These can be regular aluminum-framed window screens, or they can be constructed using wooden frames. If the prints are dried in this manner they will have a semi-glossy appearance. It will also be necessary to flatten the prints after they have dried. This can be

Fig. 8.2. This trio of prints shows the range of contrast that can be obtained from graded papers. (*a*) A print on grade 1 paper. Note that the fainter details are largely lost and even the brightest areas are greyed out. (*b*) A print on grade 6 paper. In this case, the fainter details are shown quite well but the highlight regions are quite overexposed. (*c*) A compromise print on grade 4 paper representing the best result that can be obtained from such a contrasty negative without resorting to dodging, burning, or masking.

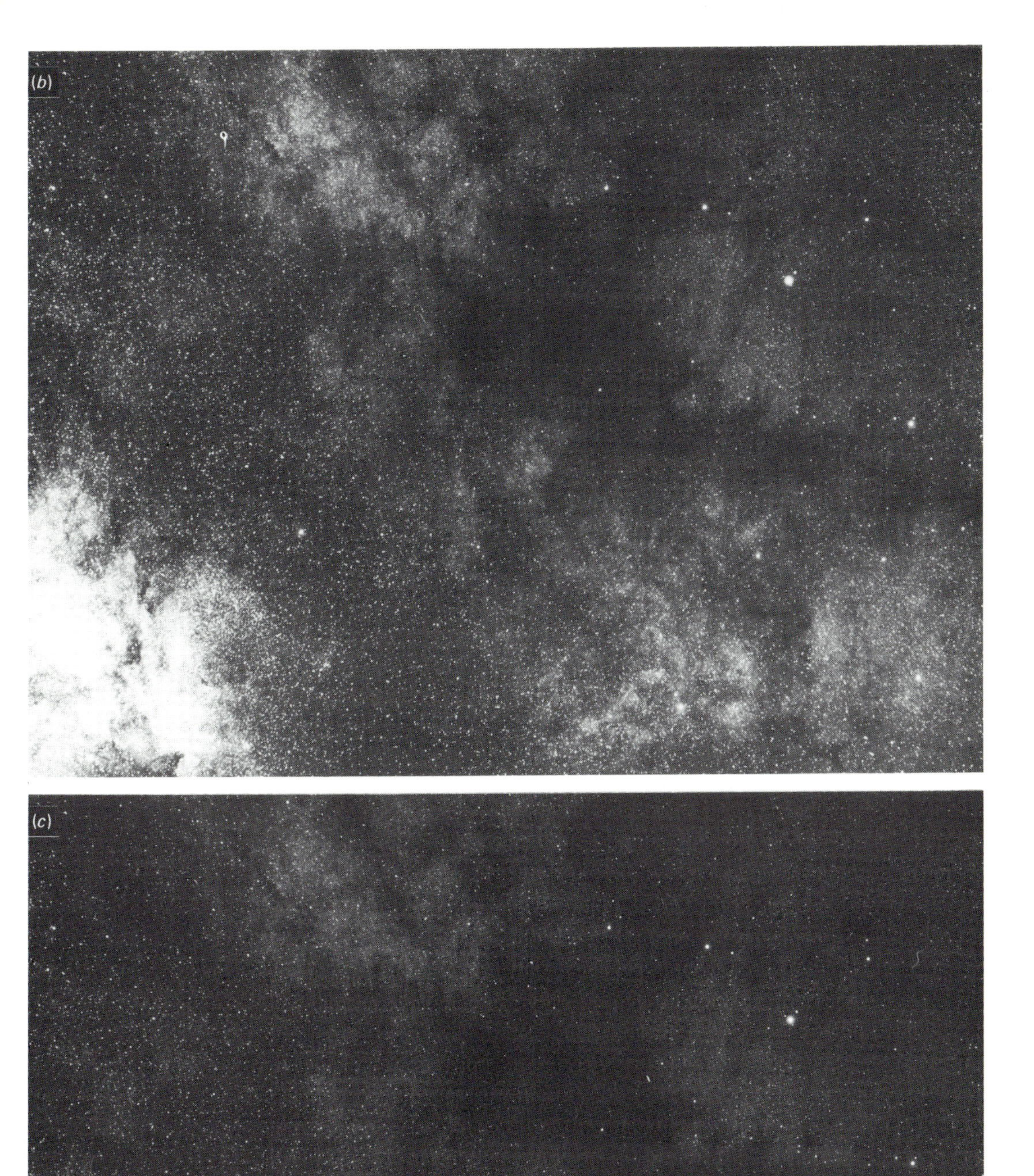
(b)
(c)

accomplished by placing them between cardboard in a warm print-mounting press and then allowing them to cool, or they can be pressed beneath several large books or other weights for a few days.

The other drying method is to use a print dryer. These consist of a highly polished metal sheet surrounding a heating element and can be of either a constantly rolling drum or a stationary design. The print can be placed face-up or face-down on this metal plate and then a canvas sheet holds the print against the hot metal surface for several minutes until the print is dry. The rolling drum model is the easiest to use and can be set to drop the prints automatically just after they are totally dried. If a semi-glossy print is desired, the print does not need any additional treatment before it is placed in the dryer with the emulsion facing away from the metal plate. If a glossy surface is desired the print will need to be treated with a special solution designed for this purpose before being placed in the dryer with the emulsion side against the metal plate. This solution is known as Pakosol or Hi-Gloss and it is essentially a glycerine solution. Producing a glossy print requires some special care, as the dryer must be extremely clean and must be kept in a very narrow temperature range in order to produce a uniformly glossy surface. When using a dryer, it is imperative to be certain that all prints to be dried are processed to archival standards. If prints are placed in the dryer that have not been thoroughly washed, the canvas belt will be contaminated and it will, in turn, contaminate all succeeding prints that are placed on that region of the canvas.

For color printing, the tube-processing method is to be preferred over tray-processing owing to the uniform agitation this method provides and for the convenience of being able to operate with darkroom lights on during the processing stages. Our preference for printing color transparencies is to use the glossy Cibachrome material, while the Type 74 process is to be preferred when printing color negatives. In both cases, the superior color reproduction and the archival qualities of the material make the choice an easy one. Type 74 does not have the archival properties, nor does it have the intense color saturation and sharpness that is characteristic of glossy Cibachrome material (the current Cibachrome 'pearl surface' material is an RC material and is not archival).

There are two varieties of tube processors available at this time and each has its merits. The first and most common is the tube processor which works on a roller base, while the second variety uses a submersible water pump and an impeller belt that wraps around the tube. With this second variety, the tube floats in a water bath and the water pump produces a stream of water which strikes the impeller belt and causes the drum to turn. The roller base processor is smaller and its spin rate is well regulated by the motor base and is quite satisfactory for most applications. The flotation model is best for processes where high temperatures are required or where the temperature must be controlled very carefully. With these flotation tube processors, the weight and size of the tube determine the spin rate and thus the agitation during processing. It is therefore important that the amount of chemicals used with each print be carefully measured in order to insure constant agitation from print to print.

The final choice which needs to be considered when planning to do color printing is whether to use color printing filters (CP filters) or whether to use an actual color enlarger head with dichroic filters. The CP filters are considerably less expensive and work well when placed between the light source and the color negative. However, these filters fade and scratch easily and must be replaced from time to time. In addition, it is desirable to keep the number of filters down to three or less in order to minimize light absorption and keep exposures as short as possible. The dichroic color head puts out a lot of light, and the amount of filtration can simply be set as needed. The only disadvantage of the dichroic color head is its cost. However, for small film formats, up to 6 × 6 cm or 6 × 7 cm, new color enlargers are now available at very reasonable costs. If a color head cannot be purchased, there is absolutely no reason not to use CP filters: the results of printing with CP filters and dichroic filters are indistinguishable.

When doing color prints, the best procedure is to run a series of test strips on a single sheet of paper to fix the proper exposure and then run another series of test strips to establish the color balance. With some practice, color balance gradually becomes relatively simple and, with experience, can be accomplished with one or two test prints. The nice feature of all this is that the color balance does not need to be adjusted very often and, with Cibachrome, the color balance is extremely steady even from one roll to another of the same film. There are numerous manuals on color printing with new ones arriving with each new process. The interested reader is advised to digest a few of these printing manuals before starting an actual printing session. The basic points to keep in mind are that color printing is more expensive than B&W printing, but it is not as difficult as it first appears.

Testing safelights

When one buys safelights for the darkroom, it is normal to assume that the lights are designed for use with some photographic materials and that they really are safe for that application. For most common situations, this assumption is reasonable to a point. However, it is best to know exactly how safe your safelights really are. You may need to move them further away from the enlarging easel, or you may need to turn them off while exposing prints and/or while processing prints.

Virtually all sources that discuss this procedure use a method that does not give a correct indication of the potential effects of the darkroom safelights. The usual process goes like this:

> Remove a sheet of printing paper with all lights out in the darkroom and set it in a location where it is normally used and set up to do a test exposure using the safelights as the only light source. Once you have

all of the paper covered except for the first strip, turn on the safelights and start counting out the various exposures. A reasonable sequence of exposures for such a test would be 30 s, 1 min, 2 min, 4 min, 8 min, 16 min and 32 min. Now process this test strip as you would normally process your prints. The first step on this print that shows any density above the paper base will give you an indication of how long you can safely work with the safelights on.

This would be fine if one was always dealing only with unexposed material in the darkroom but that is not the case, and stray light can do the worst damage to a print by adding a little extra light to those areas that would normally remain base white after they were processed. The result of this contamination is to turn the highlight areas of the print a dull grey with little or no detail. In order to properly test for possible safelight contamination, the following steps must be carried out before carrying out the procedure outlined above.

With all safelights out in the darkroom, no negative in the enlarger, and the enlarger stopped down to its minimum aperture, place a sheet of photographic paper in the printing easel and make a test print. A series of exposures for this test would be 0.5 s, 1 s, 2 s, 4 s, 8 s, 16 s, and 32 s. Then take this sheet of paper, with all safelights still out, and process it as you would normally. After the print has been processed and fixed, examine it to determine the maximum enlarger exposure that can be given without producing any density on the printing paper. Once this is accomplished (and it may take several tries to narrow it down), you are ready to test the safelight properly. Turn out all the safelights, remove another sheet of printing paper and place it in the easel under the enlarger. Now, give the paper the threshold exposure that you have just determined. After that exposure is completed, you can proceed with the safelight test described above. When finished, you will really know how much exposure to your safelights is really safe!

Print-contrast-control methods

Burning, dodging, pre-exposure

We have mentioned the fact that there are a number of objects that simply have too great a dynamic light range to be printed successfully on any paper and show all of the details that one would like to see in the print. Two common examples are M42 (the Orion nebula) and M31 (the Andromeda Galaxy). As one photographs with longer and longer focal lengths, more objects can be added to the list. Some objects that one will find difficult to print when a larger scope is used are M51 (the Whirlpool Galaxy), M64 (the Blackeye Galaxy), Eta Carinae nebula, M82, M22, M13, NGC 253, NGC 7009, and NGC 5128. There are several simple methods that one can employ to aid in preparing better prints of these objects and we will discuss a more elaborate method in Chapter 9 (on advanced darkroom techniques). The simple techniques that can be tried include: burning, dodging, local development, and pre-exposure.

Burning and dodging are basically the same, as one is allowing more light to fall on some parts of the print than on others. The basic difference between one term and the other is the type of tool that is being used. When using a large mask with a hole cut in it to let light pass through onto a portion of the print, one is said to be burning in the image. However, when using a small mask suspended above the print to stop some light from striking one or more areas of the print, one is dodging the print. Some prints require that both burning and dodging be employed in order to obtain an attractive image. The mechanics of both are the same, only the terminology draws a distinct line between the two procedures.

To prepare to burn-in a small area of a print that is otherwise too light when the rest of the print looks fine, first obtain a sheet of heavy opaque construction paper that is larger than the print that is to be burned. Then, under the enlarger, draw the general shape of the area that is to be dodged while holding the paper at the same height that it will be held while dodging (usually 3–4 inches above the print). Next, cut out a hole that matches the shape drawn but slightly smaller. Finally, with a sharp Exacto knife, serrate the edge of the hole that was just cut. Now set up to print and make an exposure on the paper for the best print for overall appearance but do not remove the print for processing. Now, assuming that you have a pretty reasonable idea of the exposure needed for the bright area that you want to burn-in, give an additional exposure to the area through the burning mask and then process the print normally.

The trick to all of this comes in performing the burning of the print. Simply holding the mask still during the exposure, results in a pretty clear image of the mask that is very noticeable on the final image. Moving the mask around too little, results in a less sharp, but still unsightly dark smudge on the final print. If the mask is moved around too much or held too high above the print, some of the nebula that would have been printed correctly will be overexposed by the burning process. Many other combinations of effects can be listed that can produce unsatisfactory results in this process, so the reader should not get the impression that this is simple to use. The act of serrating the edge of the mask helps to feather out the burning effect, but it is not a cure-all. Excellent results can be obtained using burning and dodging techniques and a master printer can obtain reasonably consistent results, as the prints of Ansel Adams prove, but one should be prepared to waste a fair amount of photographic paper on any but the simplest burning and/or dodging projects.

When preparing to dodge a print, the same process is used (and the same problems apply), but one is working with a small solid cutout instead of a card with a hole in it. In this case, the cutout must be suspended above the print and moved around in order to soften the edge of the dodging exposure. The cutout can be taped to a piece of wire that can be used to move it around during the exposure. We prefer to use a wire that has been bent into an irregular shape in order to avoid any chance of obtaining an extra ghost image on the final print. Again, any but the simplest dodging can produce unsightly side effects. In using either of these techniques, one should be prepared to waste a few sheets of paper making practice prints before obtaining the desired result. If not done carefully, the results are considerably inferior to a straight print that has an over-exposed portion. As the old saying goes: 'if you are going to do it, do it right'!

Dodging and burning are referred to by photographers as local-image control. Another means of local-image control that is similar to burning involves local alteration of the development cycle. For this, one can use a cotton swab of an appropriate size and a small container of undiluted print developer at a temperature of 90–100 °F. Just after the image becomes clearly visible in the developer, the print is pulled, slapped onto a sheet of glass and then quickly squeegeed clean of developer (use hand for speed here). Then, the cotton swab is dipped into the warm developer, the excess developer is squeezed out (but not too much) and the swab is applied to the region of the print that would otherwise need to be burned-in, being quickly rubbed around for 5–10 s (one must be careful that excess hot developer does not run across the rest of the print and ruin it). After this, the print is flushed with a water wash and then returned to the developer. This process may be repeated about three times during development, but more applications will be useless and one must be certain not to rub the surface of the print too hard with the cotton swab.

All of the above methods are, effectively, methods of varying local exposure and do not alter the overall contrast of the print. One method that can be used to alter overall contrast, or even local contrast, is known as pre-exposure. This technique can allow one effectively to change the contrast level of graded papers and create papers of intermediate grades. Another method that allows one to have this same ability is described later in this chapter and is known as 'two-solution printing'. Pre-exposure is somewhat more difficult to use, and normally can be applied in a very small number of astronomical situations, but it has one advantage over two-solution printing that makes it useful from time to time: it can be used to alter the contrast of just one area of the print for local contrast reduction.

Pre-exposure is just that: the film or paper is given an exposure before it is used. The level of the pre-exposure is kept to a minimum and, for printing, it is kept as unnoticeable as is possible. The following steps will produce a print that will demonstrate the effects of the pre-exposure technique. Using the test-strip method, start at a very low exposure level and determine the minimum level of exposure necessary to produce any image density on the paper. Then, a second test strip is prepared, starting at this exposure level and increasing by factors of two at each step. Next, using a negative of a large, contrasty object (like M31) set up for a print enlarging the object to a size that will cover much of the paper area. Finally, repeat the equivalent of the step wedge series of exposures just determined and then make a good exposure of the astronomical object on the same paper. The resulting print will show less and less contrast as one moves across the print and will also show the effects of overdoing the pre-exposure level. As one moves from one side of the print to the other, the white areas will start to become grey and the image will become 'mushy' looking. Usually, only the first couple of exposure steps from the step wedge created in this exercise will produce acceptable pre-exposure prints.

As we said before, the pre-exposure technique can be used to reduce contrast in just one (or several) small areas of the print. The method for doing this is quite simple and the example above can be used again to show the effect of this variation. The correct pre-exposure level (as determined by an experiment like the one above) is delivered to the area desired using a mask like that which would be used for a burning exposure and then the final exposure is made using a contrasty astronomical negative (such as one of M31) and the print is processed normally. M31 is an excellent application for this demonstration, and this local-contrast reduction will produce an interesting comparison with the overall contrast-reduction print that was produced in the previous exercise. In general, we have not found many applications for this technique in the printing process. However, it has potential and the advanced worker should be aware of it as a possible tool. Later, we will see a useful application of pre-exposure at the on-telescope end of the photographic procedure.

Improving print quality: two-solution print processing

In our earlier discussion of basic printing, we suggested that Kodak's Dektol be adopted as the darkroom standard print developer. Dektol is a good all-around print developer but it does have its faults, and once well acquainted with the basics of printing it will be beneficial to learn of techniques that will allow further improvement of print quality. Dektol is normally used in a 2 : 1 dilution of the stock solution. Dektol is a very active developer that expends most of its energy in the shadow, or darker, areas of the print while only producing minor changes in the highlight, or brighter, regions. One of the reasons for recommending Dektol as a standard for beginners is that this developer can produce cold black skies given an appropriate exposure and proper paper grade. Most 'astroprints' made by novice printers show a fairly white nebula surrounded by a light grey sky indicating an incorrect choice of paper grade. Once you know how to select a paper that will give you a very good print, you will probably begin to encounter situations when you wish you could have a grade 4.75 paper and you will start burning-in areas of prints to overcome the limitations of graded papers (see Fig. 8.3).

The fact is that it is possible to change the effective contrast range of graded papers! It is possible to develop the highlights a little more without losing details in the fainter areas of the object and you can gain much detail in the highlights without resorting to local manipulations or having to resort to highlight masking. One can achieve a range of contrasts from a single paper grade by varying the type of developer that is used to process the print. Note: you cannot vary the contrast grade by varying the development time without seriously sacrificing print quality! Dektol is considered to be a contrasty developer while Selectol Soft is termed a low-contrast developer. However, there is a difference in the way these two developers act that can be used to the advantage of the photographer (and has been for many years by expert photographic printers in the art field). Selectol Soft is a developer that expends much of its energy in the development of the highlight areas of the print. By using the complementary effects of these two developers it is possible to make dramatic changes in the appearance of the photographic print!

The technique used to take advantage of these complementary properties is to process the print in the two developers sequentially. The usual proceedure is to process the print in Selectol Soft first to develop the highlight details and then to place the print in the Dektol to add density to the deep shadows (or sky in this case). By varying the times that the print remains in each of the developers, one can obtain a range of contrast grades. The essential rule to keep in mind is to avoid reducing the developing times below the point where the print is completely processed. If this is done, you will end up with a print that has a 'muddy' appearance. Some examples of combination development times are seen below.

Selectol Soft (min)	Dektol (min)	Contrast
6	0	Low
3	0.5	
2.5	0.5	
2	0.5	
1	1	
0.5	1.5	
0	2	High

One of the first comments that we hear from people who like to take shortcuts and who feel that this requires too much time and effort is, 'Why don't you just mix the two developers together to change the contrast, or use some other developer like Kodak's Selectol?' The simplest answer to this question is, 'Because it doesn't work that way'! A more detailed explanation includes the fact that mixing the developers together eliminates the complementary effects that we would like to take advantage of. There are some developer recipes that are designed to work like this when they are mixed together, but most of the best printmakers in the world today have abandoned such exotic recipes in favor of the process described above.

We cannot present anywhere near enough information in this single introductory chapter to fill the knowledge gap about darkroom work that plagues many astrophotographers. We have probably swamped some with too much information while taunting others with too little on some of the tidbits that we have had to gloss over in this section. We present a number of highly specialized, but very useful, techniques in the following chapter but know that this will also produce problems for many of our readers. We recommend that the reader interested in improving print quality go through Ansel Adams' book, *The Print*, as well as a number of other works on fine printing. The craft of making a high-quality print is not something that can be mastered in a few hours or days or even weeks.The serious astrophotographer can make enormous improvements in print quality by learning some of the skills of the art photographer. We would also like to note that one cannot learn all that is needed about this complex topic by reading books and articles, but it doesn't hurt either!

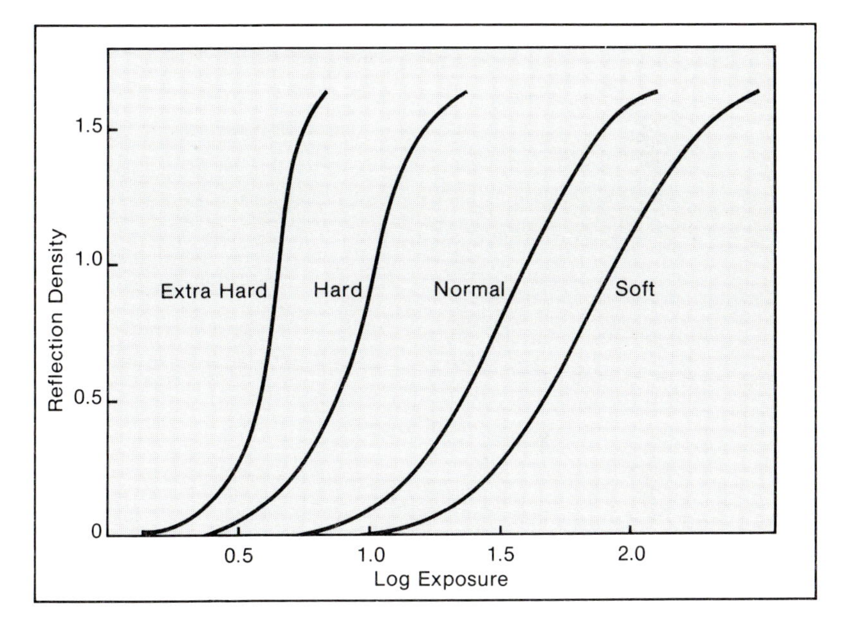

Fig. 8.3. The H-D curves are shown for a range of paper grades. A 'multi-grade' paper would be similar to a 'normal' or grade 2 paper. The small amount of contrast variation available in a 'multi-grade' paper would not vary it to equal the 'soft' paper or the 'hard' paper shown in this plot.

9

Advanced darkroom techniques*

Slide duplication

It is often desirable to produce duplicates of color slides; copies may be needed for submission to a magazine or other publications, copies may be wanted on a more archival color film (like Kodachrome) in order to preserve the color image, or copies may be wanted for friends. Another application of the slide-copying process is to produce positive slides from black-and-white negatives for projection at talks or meetings. Today, equipment for producing high-quality slide duplicates is readily available and also reasonably priced. In addition, a wide selection of equipment is available from which to choose, ranging from simple to very elaborate. Even the simplest arrangement will allow one to manipulate both the color and contrast of the copy, provided that the user is willing to spend a little time experimenting while making the duplicates. While spending more on better lenses can result in a substantial quality improvement (when properly used), the basic concern in choosing more elaborate equipment over a simpler setup is a matter of how much convenience one wishes to pay for.

In general, it is not possible to predict exposure information for any given system and the camera meter cannot be used effectively on astronomical subjects. The exposure required is dependent on so many variables that each system needs to be tested carefully to determine a range of exposures that will generally be found to be of use for preparing well-exposed duplicates. Note that we are stating that a series of tests will provide a range of useful exposures. Because of the great degree of variability that is found in the background density of astronomical slides, no precise value can be determined that will suit all of the possible slides that one would like to duplicate! The result of all this is that a series of exposures needs to be made, bracketing some nominal exposure value, for all slides being duplicated; once the results of this test run are known, a number of excellent duplicates can be made of each slide. Slide duplicating is a process that generally requires several test runs and care in order to produce high-quality results. If one is not willing to invest the time required to perform this work, it is probably best to take the original slides to a high-quality custom laboratory and have the duplicates made professionally.

* The sections on 'Integration Printing' and on 'Masking' use material originally published, in slightly different forms, in *Astronomy* magazine in 1973.

Many custom laboratories which provide a high-quality duplicating service offer the option of turning the 35 mm original into a larger-size duplicate. When planning to use the duplicates for publication submission this option should be seriously considered. The reproductions that can be made from a 4 × 5 inches transparency are markedly superior to any that can be obtained from a 35 mm slide, even when the reproduction is made by the best-quality lithographers.

The basic equipment requirements for duplicating slides call for a bellows attachment with a slide holder. In addition, one will need to obtain either a macro lens or an adapter that will allow a normal lens to be reversed to increase the sharpness of the image at close fucusing distances (the macro lens is the better solution). Beyond this, one will need a 3200 K or 3400 K light source, a small set of color correction (CC) filters and a few color temperature conversion filters.

Once one has the basic equipment, slide duplicating can be used to correct a number of minor problems, including small color-balance shifts and increasing the contrast and density of faint objects. By choosing an appropriate duplicating film, one can achieve a wide range of results. If a contrast increase is desired, a film like Kodachrome 40 is an excellent choice, while Ektachrome Professional Duplicating film is a good choice when a contrast increase is not desirable. The duplicating film should be chosen so as to complement the attributes of the original film. Thus a film, like Ektachrome, that records blues with better fidelity than reds, should be duplicated on a film with opposite characteristics so as to not magnify the color-rendition error. We have generally preferred the results that are obtained by using Kodachrome 40 and have chosen to control the contrast of the duplicate by using the effects of preflash to vary the contrast of the duplicate.

Preflash or pre-exposure in slide duplication allows the use of a film like Kodachrome which produces very saturated colors but also normally produces high-contrast slide duplicates. The pre-exposure allows exact control of how much contrast will increase or decrease in the duplicate slide. The duplicating film is exposed to a low level of light from a neutral source (3400 K for Kodachrome 40) prior to making the actual exposure of the original slide. The amount of contrast decrease compared with a duplicate made with no pre-exposure will depend upon the amount of the pre-exposure. The more pre-exposure, the greater the contrast decrease. The amount of exposure required for the pre-exposure is slight and a variation of a half a stop is quite noticeable. A detailed description of the preflash technique is given in Chapter 10, for interested readers.

Another advantage of the duplicating process is the fact that small corrections in color can be made by using CC filters. Weak colors can be enhanced and the sky color can be brought back to a more neutral color. One must be cautious when trying to correct two colors when duplicating slides, as it is impossible to enhance complementary colors simultaneously.

The addition of red means that cyan will be removed and this is an absolute fact. One must also be careful to avoid overcorrecting by adding too much color. This can result, in its worst manifestation, in colored star images and a slide with all colors shifted noticeably more than the original. To judge the effects of a CC filter pack, look through the camera viewfinder and quickly pull the filter pack in and out of the light beam repeatedly in order to observe the change. Then, it is a good practice to use half of the filter pack that appears to give good results when viewed in this manner. Thus, if adding a CC40Red looks like the proper filter pack, the addition of a CC20Red to the filter pack will probably produce the best duplicate.

The obvious drawback of this process is that it requires extensive knowledge of the effects that can be obtained; this can only be acquired through a considerable amount of experimentation. Like all of the techniques discussed in this chapter, casual applications can produce markedly inferior results. If you want to obtain consistently high-quality results, the price will be an investment of time. In the case of slide duplicating, a high-quality custom laboratory can generally produce far better results than a casual amateur. From the results that we have seen published over the past 10 years, our advice would be to pay a professional to do the slide duplicating job correctly.

One of the most common problems encountered when making duplicate slides is that of producing enlarged star images that show little or no color on the copy slide. While a slight enlargement of star images and some loss of detail is inevitable in the copying process, the effects should barely be noticeable to a well-trained eye. By taking certain precautions when making the duplicates, a duplicate can be made which will show a minimal loss of detail and which can even be considered to be an improvement over the original in many ways.

When setting up to make slide duplicates, one should be very careful about obtaining exact focus. The camera should be focused at maximum aperture, and then stopped down about two stops before making the exposure. Many make the mistake of stopping down the camera to f/11 or f/16 thinking that this will insure a sharp image. In fact, stopping down to such small apertures when photographing small details introduces diffraction effects which actually impairs the sharpness of the image and produces enlarged star images. The optimum aperture of most lenses is frequently found to be between 2 to 3 stops below maximum aperture. The other factor that causes enlarged star images is over-exposure of the duplicate. If one has a dark slide showing an under-exposed nebula and uses the duplicating process to 'intensify' that nebula, large stars will accompany the brightening of the nebulae along with increased noise from the sky background. The amateur who tries to use this technique as a cure for under-exposed original slides will find that his duplicates are markedly inferior to those made from properly exposed originals.

Although there are some benefits to be realized in making duplicates, there are numerous disadvantages. One major drawback is the tendency for highlight blocking, where the brighter areas of a nebula or galaxy merge together and form an image devoid of color and detail. Virtually all duplicates, even when properly exposed, will lose a portion of the highlight end of the density range. This occurs even when special duplicating films are used for the process. To someone who is familiar with the reproduction processes (see pp. 204–8), this loss seems all the more damaging since the highlight details of the 'original' are the most critical areas in the process of reproducing for publication.

Negative intensification

This topic is one which is discussed repeatedly in the literature, dozens of articles on this subject have appeared continually since at least 1917. The general idea is to take details which may be just barely visible on the negative and to make those details more easily visible by some means. The image may be enhanced by any of a number of methods. The original negative can be treated chemically to intensify the existing image or the original image may be copied onto another higher-contrast material to create an image where the details are more easily seen. However, there is no way to bring out details which are not already present on the original developed negative and there are definite losses which accompany any enhancement process. Almost all of the articles which have appeared over the last 50 years have basically reported exactly the same results, some of those articles have revealed some interesting twists which can be used to advantage.

The reader should be aware of the disadvantages which accompany the various enhancement processes, and use these techniques with the realization that the gains achieved by these methods are not without a cost. By knowing the drawbacks, these processes can be used to advantage and not pushed beyond their potential range of usefulness. In general, the use of intensification processes beyond moderation is always to be avoided and these processes should only be considered as a last resort when other alternative solutions have failed to achieve the desired results. The costs generally associated with the various processes can be summarized as follows:

1 the noise level of the original image is intensified along with the original image; this effect is most obviously seen as an increase in the graininess of the image (thus the enhanced image will stand less enlargement than the original); the fog level of the original negative is also to be considered as a source of noise;
2 the gradation of the brightness of stars in the field will generally be lost unless the intensification process is used in moderation (chemical intensification processes are less severe in eliminating these subtleties than are copying methods);
3 copying techniqũes (basically similar to slide duplication) will pro-

duce a slight enlargement of the star images – extreme care must be exercised when focusing the instrument used for the original image to minimize the impact of this effect;

4 if a photograph is intensified which shows both bright and faint nebulae, the brighter regions will tend to lose details and become completely oversaturated as the fainter regions are intensified (in cases where this situation is encountered, a chemical intensification process which has a greater effect on the less dense regions of the negative should be used before attempting any of the copying processes);
5 unless one has reached the limiting magnitude of the instrument, a longer exposure will always provide a preferable result;
6 if the photograph has reached the limiting magnitude of the instrument, integration printing can be used to increase the limiting magnitude while avoiding the above problems (integrating exposures that have been intensified slightly can produce very dramatic results).

Chemical intensifiers

These processes generally involve the use of some agent which will combine with, or cling to, the silver present in the developed image and thus increase the optical density of the image. Some of the intensifier formulas produce an image which is not very permanent and which will need to be copied in order to produce an archival image. Other agents can provide a very marked increase of contrast without increasing the density of the shadow regions, such formulas (known as superproportional intensifiers) are of little use for astronomical purposes. Most of the intensifier formulas work in a manner which increases the contrast of the negative; only a few of the formulas produce a greater effect in the less-dense regions of the negative and all of these involve the use of mercury which produces a non-archival image.

We will discuss the preparation and use of one of the intensifiers which have a greater effect upon the less dense regions of the negative. The formula is that of a compound known as Kodak Intensifier 1; other formulas for different intensifier agents can be found in many of the published photographic laboratory handbooks that are currently available.

The negative must be thoroughly washed before starting this process and it is best to mix the intensifier solution shortly before starting. The solution is made up of

Potassium bromide (22.5 g)
Mercuric chloride (22.5 g)
Water to make 1 litre.

The negative is immersed in this solution until the image is completely bleached to a white color and then it is thoroughly washed. To redevelop the image, it can be placed in one of the following solutions, depending on the degree of enhancement which is desired.

1 a solution of 5–10 % sodium sulfite;
2 any standard developer (D-72 is normally recommended);
3 a '10 %' ammonia solution made by diluting 1 part concentrated ammonia (28 %) with 9 parts water.

The final result will show more and more density as one proceeds from using method 1 through method 3 for redeveloping the bleached image. The redevelopment step must be carried to completion and the only control available over the density of the final result is by selecting the appropriate redevelopment technique. Once again, this method produces an image which is not archival and which will need to be copied in order to produce a permanent image.

Copying techniques

These enhancement techniques produce an overall contrast increase over the entire image, and the user has no means of controlling the areas which will be affected by the copying process (the more dense regions of the negative will always be affected the most by any copying process). There are a number of options available for the use of this process, and the user can choose any of the various methods. In each case, the negative is copied to produce a second negative which is then used to make the final prints.

The original can be copied onto a direct positive material to produce the new negative in a single step, or the negative can be copied twice onto another negative material to produce the new negative. The user also has the option of making the copy negative by contacting the original to make a copy negative which is the same size as the original or by projecting the original onto the copy material to make a copy negative that is larger than the original. With small negatives (e.g. 35 mm), it is preferable to make the copy negative larger than the original and to then print from the larger copy negative. In addition, the user has the choice of a number of materials which can be used for the copying steps. It is strongly recommended that one avoid choosing a material which must be processed in a litho developer, as the contrast level produced by these developers will normally not produce acceptable results.

The easiest and most direct method of producing a higher contrast copy negative is to project the original with an enlarger onto a direct positive material like Kodak Professional B/W Duplicating film (4168). This film, which is avialable in 4 × 5, 5 × 7 and 8 × 10 sheets, is blue sensitive with a speed equivalent to a moderate-speed enlarging paper and this can be handled in a similar fashion to that used in making enlarged photographic prints. Exposure is determined by using the test-strip method while contrast is controlled by varying the time of development (recommended developer is Kodak DK-50). The important thing to remember is that shadow density (or sky density) is largely controlled by exposure while highlight density is controlled by development time (longer development increases highlight density and thus contrast). The manipulation of these two variables make it possible to produce duplicate negatives with any number of contrast ranges that can enhance faint details that would otherwise be unprintable. The primary advantages of enlarged copy negatives are: 1 they contribute less of their own grain structure to the final image: and 2 because they are made the same way as prints, the image can be manipulated mechanically by dodging, burning or masking. The major drawback for some practioners is that the enlarged copy negative may be too large to be printed with a small format enlarger. The solution in this case is either to produce copy negatives for contact printing only or to make them the same size as the original.

Same-size copy negatives are made by contact exposure. Again, as with enlarged copy negatives, Kodak Professional B/W Duplicating film (4168) is used with exposure and processing being determined through experimentation. Although a vacuum frame is preferred for making contact exposures, spring-loaded frames will work well and in an emergency a sheet of glass acting as a weight will suffice. Using an enlarger as the light source the exposure is made through a sandwich consisting of the following components listed from top to bottom: clear glass (as part of the vacuum frame etc.), original negative (emulsion down), duplicating film (emulsion up), and black backing paper (to prevent unwanted reflections). Extreme care must be exercised to ensure even contact during the exposure and to avoid the formation of Newton's rings between the top glass and the original negative. If Newton's rings are a persistent problem a very light application of lithographers offset powder (available at graphic arts stores) to the side of the glass that contacts the original negative is a good preventative. The most effective way to apply the powder is to puff it into the air using a squeeze bottle and then pass the glass through the resulting cloud. The same principles of contrast control apply for contact duplicates as for enlarged duplicates in that manipulation of exposure and development will result in a wide range of enhancement possibilities. For a variation on this technique and for more details in general the reader is referred to the excellent article by D.F. Malin, 'Photographic enhancement of direct astronomical images'.

Mosaic construction

A mosaic is an assemblage of overlapping photographs whose edges have been matched to give in this montage the appearance of a single image. Astronomical mosaics present some particularly difficult problems in their construction which may explain why relatively few are encountered. Unlike most other photographic subjects the night sky does not provide the wide range of varying textures and tones that help hide the join lines of a mosaic. Instead, the photographs must be joined along stretches of featureless sky (avoiding star images) and perhaps across star clouds, smoothly textured nebulae or galactic arms and still be invisible to the eye. Although it seems to be an impossible task, with practice, patience, and the techniques outlined here, quite satisfactory results can be achieved.

The first consideration in constructing a mosaic is the amount of overlap required for each photograph. The requirement here is dictated by the characteristics of the optical system being used. Test photographs should be made on the night sky to determine the degree of vignetting in the field and to detect other edge-of-field aberrations. Once these are established, the photographic fields can be adjusted so that overlap will be just enough to eliminate these edge effects when the photographs are trimmed and joined.

When making the exposures and processing the negatives and prints, every effort must be made to control all of the variables that can alter the appearance of each of the photographs. All exposures should be of equal duration and made under sky conditions as nearly equal as is possible. With very long exposures, it is advantageous to spread the work over a number of nights so that all the exposures can be taken at or near the meridian and thus avoid the effect of atmospheric extinction. Other important factors include seeing, transparency, and sky brightness; all of which can make any single photograph in a mosaic readily distinguishable from the rest and thus ruin the unity of the whole. After the exposures have been made it is imperative the processing of all negatives and prints be identical. Not only should the chemistry be identical but times and temperatures must be held quite constant, as a slight deviation can render one photograph (or several) quite different from the other members of the mosaic. Finally, when making the prints, it is important that all chemistry (especially the developer) is maintained in a fresh state throughout the printing session. As the developer becomes exhausted it becomes more and more difficult to maintain a constant density range from print to print. Of course, it goes without saying, that the density ranges of each print must be matched as closely as possible. A densitometer is a great aid in this situation, but it is not necessary since very close matching of densities by eye is all that is required. All prints should be made on a matte finish, lightweight or single-weight paper and under no circumstances should resin-coated (RC) papers be used.

Once all of the prints have been made with density ranges and background densities matched as closely as is possible on each print, the mosaic is ready to be assembled. The photographs can be either dry-mounted together or joined wet. The latter method gives by far the best result (when done well) but is also the most difficult and requires considerable skill. With the dry-mounting method, finished dry prints are overlapped and cut with a razor or similar extremely sharp instrument so that the images, when butted together, form a contiguous whole. The best results are obtained when the images are overlapped and registered (using strong backlighting), taped into position, and then cut through from the back with the print emulsions resting on a hard surface such as glass. Although a straight-line cut is the easiest (avoiding star images when necessary), better results are achievable by contouring the cut along changes in tonal value. This applies particularly where the join line cuts directly across a nebula or galaxy. Contour cutting requires considerable skill and care to execute correctly and will necessitate some practice for good results. Once the photographs have been cut they must be glued to a piece of heavy cardboard with a workable spray adhesive (workable adhesives allow glued pieces to be lifted and moved if necessary), making sure that they abut as intended.

With wet mounting, the individual photographs are positioned on heavy cardboard and assembled while still saturated with a solution consisting of a few drops of gum arabic mixed to a thin consistency with water. The solution also contains a small amount of formaldehyde as a preservative and glycerin to prevent brittleness when the mosaic has dried. The joints in wet-mounted mosaics, while abutting like dry-mounted prints, are less conspicuous because of a technique called edge feathering. This involves cutting only through the emulsion of each dry print and tearing the paper underneath to give a feathered edge. Further smoothing of the edge may be obtained by additional working with sandpaper. The trick to all this is in the way the paper is peeled away from the scored emulsions: The top overlapping print must be peeled back toward the center of the print while the bottom print is peeled towards the print's edge. What results is that the prints fit together in a wedge-like manner. This allows the photograph segments, when wet-mounted, to be stretched and flattened to create a perfectly matched, relatively inconspicuous join line.

Because of the way that wet-mounted mosaics are cut, great care must be taken to insure a reasonably close match from one print to the next. When cutting, the prints cannot be overlapped and cut at the same time as with the dry mounting method. Thus to insure an accurate match between two prints, a clear acetate 'jig' should be fashioned before cutting starts. To do this, the photographs must first be overlapped in register and taped into position, as is done in the dry mounting method. Then, a piece of clear acetate (~0.004 inches, 0.1 mm thick) is placed over the photographs and key star positions are marked and the cut line is drawn with a fine-pointed drafting pen. Now, the acetate is removed and cut with a razor along the previously drawn cut line. The two halves of the acetate jig can now be used as guides to cut the join lines of each photograph. Extreme care must be exercised in aligning the key star registration points of each side of the jig. In addition, the actual cutting of each photograph using the jig must be done with extreme precision. This method requires much practice to perfect and the necessary skills can be developed by trying the technique out on an already existing photograph. Simply make two identical prints from a single negative, cutting them both in half and then rejoining them as a mosaic.

After the photographs have been mounted by either of these methods they may need some local retouching. Retouching also requires considerable skill and practice in order for the user to become proficient, but one usually has many photographic samples around which can be used for practice before attempting to work on a mosaic that has required several hours' work to assemble. The supplies for retouching are available in any well-stocked photographic supply store and consist of spotting dyes and very high-quality retouching brushes. For black-and-white work, the technique involves mixing the spotting dyes to match the hue and value of the area to be retouched (black-and-white prints are seldom actually neutral grey in hue) and then applying the dye in minute dots until the flawed area around the join line has been rendered invisible. A number of excellent books exist on retouching techniques and the interested reader should obtain one or more of these sources for more information as detailed exploration is not possible in this volume.

After retouching, the mosaic should be sprayed with an aerosol fixative to protect the retouching work and the general surface of the mosaic. At this point, you may choose to rephotograph the mosaic so that the combined images can be recorded as a single negative. The advantage of doing this is that it then becomes possible to make many copies of the mosaic with relatively little effort. The process of properly copying photographic prints is entirely beyond the scope of this book as it is a complex process and an in-depth study could fill volumes. However, let us say that the use of 35 mm cameras should not even be considered if one is seeking high-quality results. The minimum film format for copying work should be considered to be 4 × 5 inch film and high-quality camera lenses are needed as well. Just as with slide duplication, considerable experience and experimenting is needed to find a copying technique that will work with the equipment set that one assembles for the copying task. Again, the interested reader should consult one of the numerous works covering copy-camera use and copying techniques in order to get a better level of understanding of this topic.

Masking

There are a number of astronomical objects which continually cause printing problems for the amateur astrophotographer because some regions of the object are quite faint while other areas are extremely bright. The classic example of this is the Great Nebula in Orion, but other examples include M8, M16, M17, and M31. The reasons for this perplexing problem lie in the fact that the films used in astronomical photography are generally high-contrast emulsions and the printing papers required to produce aesthetically acceptable prints are also of high-contrast. The result of all this is that the printing material cannot handle the density range that is encountered when objects having an extreme inherent brightness range are photographed using typical astronomical emulsions. There are a number of compromise solutions for this problem: use a low-contrast film or printing paper or just print to show the fainter regions and forget about the bright areas. These methods will indeed produce a print, but not a print that shows all the details that are present on the negative and still maintains the aesthetic appearance that is generally associated with astronomical photographs.

A more acceptable solution to this problem can sometimes be obtained by dodging the print during the printing exposure. This can produce very good results, but the inexperienced printer is often discouraged by the fact that dodging is not a skill that comes without much care and practice. If one plans to produce a well-dodged print, one should expect to trash the first half-dozen or more attempts. The amateur who is in a hurry usually gives up after two or three tries and shows prints with light and dark blotches throughout the area that was dodged. Even when one does spend the time necessary to produce a well-dodged print, it is very difficult to produce a series of identical prints when intricate dodging tools and motions are being used. The technique for producing a contrast mask which we will describe here requires a considerable investment in time and tests to determine exposure time and processing times that will produce the results desired, but the time and effort invested is justified by better photographs. Once the basics are mastered, this process can be used in a predictable and repeatable manner. However, be sure to keep notes and records of your experiments for future reference.

Other motives exist for learning and using this process besides the desire to obtain good prints of half-a-dozen spectacular objects. When attempting to print faint nebulosity which appears to be lost in the sky fog level, this process can 'save' negatives that appear to be totally useless. The recent work by David Malin is an example of the application of the masking process to this type of problem, and the printing technique developed by Gum (described later in this chapter) is a logical extension of the masking process to an even more sophisticated level. The masking process has been in use by photographers and printers for decades, but was not generally used by professional astronomers until the 1950s when Bill Miller used some interesting variations of the process to isolate photographically the M87 jet from the rest of that interesting galaxy. We picked it up from Ernest Wallis and developed a technique that we could use regularly to improve prints of these very 'contrasty' objects. We finally published a description of our method in 1973 (in *Astronomy*) for consumption by the amateur community and it has been applied by a number of advanced amateurs since that time.

The 'masking' process involves making the equivalent of a variable density neutral density filter to alter the intensity of the light reaching the printing paper at any given point. It sounds nasty, but it is not as bad as all that! The primary equipment requirements are film and a diffusing material, the other requirements are patience, time, and practice. However, this time investment is only necessary to learn and develop a method which will work with your own printing system and tastes. Once a method has been devised, it can be applied repeatedly to many similar printing problems (the one unfortunate aspect of all this is that some of the testing steps must be repeated any time you change to a new masking film, or the manufacturer develops a new one and eliminates the old one from the market!). There are several methods for producing masks, and we will describe the most commonly used methods prior to describing the one which we have developed to deal with the special problems associated with astrophotography using high-contrast, small-format films.

The easiest way to produce a mask is just to make a diffused contact transparency of the original negative. The image of the original may be softened by introducing some type of diffusing surface between the two emulsion layers or by placing a thin sheet of glass between the two emulsions and then spinning this emulsion-glass 'sandwich' on a turntable below, but not directly below, a light source. As one can imagine, this latter method is best suited for use with larger negatives (i.e. $\geqslant 4 \times 5$ inches).

A second, less direct, method is a variant on the first method above which makes the process of masking smaller negatives much easier. This method involves making an enlarged sharp negative from the original negative (this can be either a one-step or a two-step process, depending upon the choice of internegative material) and then making the mask from this internegative. The internegative and the mask made from it are sandwiched together, in register, and prints are then made by either contact printing or by enlarging (if one wishes to enlarge the masked negatives, a larger-format enlarger is needed). In this method, the negative-duplication process will always result in the loss of some detail and, unless the copy film is chosen carefully, a contrast increase may result.

A third variation that is well suited for small negatives involves making a mask that is the same size as the intended final print. The mask is produced by placing the negative in the enlarger and then placing a diffusing layer on the printing easel above the emulsion that is being used to produce the mask. In this method, the registration of the diffuse mask with the projected image is quite easy. However, there is the disadvantage of having to make a new mask for each size print that is to be made, but this seems a minor annoyance unless one plans on making prints that are larger than 16×20 inches.

The methods above describe the making of a simple contrast-reduction mask which will result in reducing the gross contrast over selected areas of the print while theoretically preserving the local-contrast variations, and leaving other areas of the print relatively unaltered (Fig. 9.1). (Actually, in all regions where the negative is exposed, a uniform large scale contrast reduction occurs – see Fig. 9.2) The problem which we encountered with this involved the preservation of the faintest details on the print. We therefore developed our own variation which is, technically, a version of masking known as 'highlight masking'. In this process, a mask is made which affects only a portion of the print while leaving the rest of the print absolutely untouched. Our own variation on this involved the need to make two separate printing exposures: one without the mask to print the faint outer regions and one with the mask to print the details present in the extremely bright regions. Without this variation, we found it difficult to produce a print which showed all of the details we were interested in.

In order to use this variation, it was necessary to choose a 'masking film' which has a higher contrast index than is generally used. Our choice was Kodak Professional Copy film 4125 which has a Gamma of approximately 0.4 when processed in HC 110 dilution E. When printing through a mask made of this material, all light from all areas of the negative except the most exposed regions will be eliminated. This variation has two major advantages over the 'straight' masking process. First, it allows one to adjust selectively the print density in bright and faint regions to any level and, second, one can alter the contrast in the two regions to suit any desired purpose. Thus, by using this variation, high contrast can be preserved in the fainter regions while greatly reducing the gross contrast in the brighter

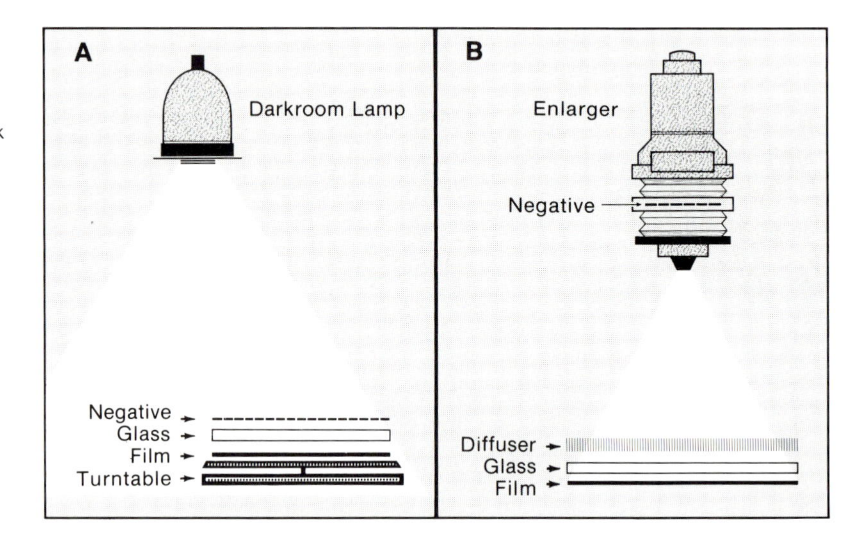

Fig. 9.1. Two methods are shown for making masks. (A) The method best suited for large negatives where the mask will be the size of the negative. When a small negative is used, it is easier to make the mask the size of the final print using the technique shown in (B).

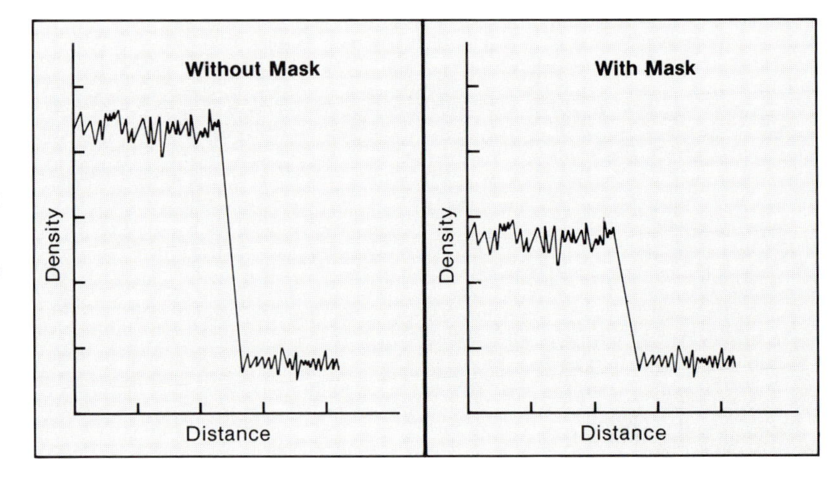

Fig. 9.2. The effects of an ideal mask are shown in this stylized densitometer trace across an unmasked negative and a masked negative. Notice that the gross contrast changes are reduced between light and dark areas while local contrast variations within a bright or dark region are left untouched.

areas. In addition one can preserve the contrast between the dark sky and faint outlying details while piercing into the over-exposed regions that would otherwise be without detail (see Figs 9.3 and 9.4).

This mask is produced using a diffuser consisting of a layer of high-quality tracing velum fixed to a 1/16 inch thick sheet of glass (pressed into contact by another sheet of glass). This is placed above the copy film, an exposure is made and the film is processed and then examined. After a series of tests, an exposure will be found that is close to the optimum value. At this point, the mask should have the appearance of a diffuse transparency with a fairly dark sky. To judge the final exposure level and contrast level, it is now

Fig. 9.3. This negative print shows the appearance of a mask used to make the Orion nebula print shown in Fig. 9.4. Note how the image has been diffused and how light will pass through the less dense areas of the mask (dark in this print) to 'burn in' the brighter regions of the nebula.

necessary to begin making masked prints using the masks that are being produced and to make adjustments based upon the characteristics of the final print.

To produce a masked print, the mask is placed back on the enlarging easel in register with the projected image. If the degree of enlargement is noted, and a plot is made of the positions of the various stars in the projected field, the registration process is greatly simplified. The photographic paper is placed beneath the mask and is taped in place. The first exposure is then made through the mask and then the mask is removed without disturbing the printing paper. Then, an exposure is made to produce a properly exposed sky level. After this print is developed for a standard time of 90–150 s, the exposure made through the mask can be judged and, if not correct, that exposure can be adjusted. Once a dry print is in-hand which appears to be fairly close to having proper exposure values over the entire image, some judgment can be made about the quality of the mask based upon characteristics that are present in the print. Some of the characteristic symptoms which can be used to diagnose problems include those listed below.

1 Dark regions surrounding stars and adjacent to bright wisps of nebulosity are an indication that the mask is too thin, too contrasty or both. The size of the dark zones gives an indication of the sharpness of

Fig. 9.4. Two prints of the Orion nebula region are shown on the same paper grade. The print on the left is a 'straight' print that has not been dodged, burned or masked. The print on the right has been masked using the technique described in this section and the mask shown in Fig. 9.3.

the mask. Other symptoms can be used to give a more definite indication of the exact problem.

2 When the brightest area is printed at the correct density, but the rest of the print is greyed out, or possibly even totally blackened, the mask is too thin.

3 When the brightest region is printed at the proper density and the outer (unmasked) areas appear correct but there is a bright, over-exposed, area surrounding the core of the masked area, the mask is too contrasty.

4 When the brightest region is printed at the proper density and the outer (unmasked) areas appear correct but there is a dull darker zone surrounding the core of the masked area, the contrast of the mask is too low.

When making initial adjustments to the density or contrast of the mask, a timid approach of trimming a few seconds from the exposure time or 10 s from the processing time of the mask will result in wasting a lot of masking film and a lot of time. Each of these times should be trimmed, or increased, by 25% whenever a change is needed until results start to look reasonable. In addition careful records should be kept on the appearance of the print and the changes that were made to the exposure and/or the processing time. Such records will be valuable whenever a new mask must be made at a later date. It is also crucial to record all information pertaining to the making of the masked print (the aperture setting and exposure for each of the exposures). Figs. 9.5 and 9.6 show examples of other masking applications.

Integration printing

Until the recent introduction of fine-grain films for astronomical photography and the development of the hydrogen hypering technique, granularity imposed severe limits upon photographic resolution. These new emulsions do improve upon the information loss caused by granularity and have brought about a dramatic improvement in the aesthetic quality of astronomical photographs since their introduction. Prior to the introduction of these films, like 2415 Kodak Technical Pan and the IIIa-series emulsions, techniques were developed to reduce the effects of granularity when using the coarse-grained emulsions like the 103a-and IIa-emulsions. The technique is known as 'integration printing' and it can yield more than just a reduction of granularity. Integration printing has been known since 1926, but remained little used for quite some time. The technique became more widespread in the 1950s and 1960s and is still in use at this time. Integration printing was introduced to the amateur community in a 1973 article in *Astronomy* by the present authors, and its use in the advanced amateur community has continued to the present.

The success of this process is based upon the fact that the distribution of grains in the photographic emulsion is virtually random. Thus the superposition of two or more emulsion grain patterns will break up the granularity of the image structure, enhancing details of any images which have been recorded on the series of negatives being integrated. Details that have fallen on less sensitive grains, or upon coarser grains, can be 'recovered' using this process because of the random nature of the grain distribution in the photographic emulsion. The technique is known under several alternative names including 'multiple image printing' and 'superposition printing'. The term 'integration printing' was coined by Wm C.

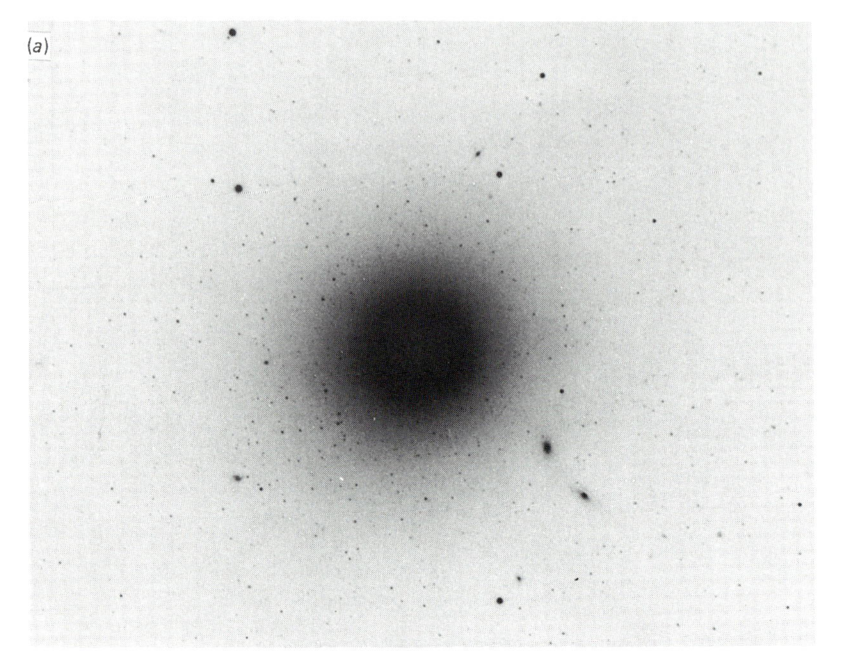

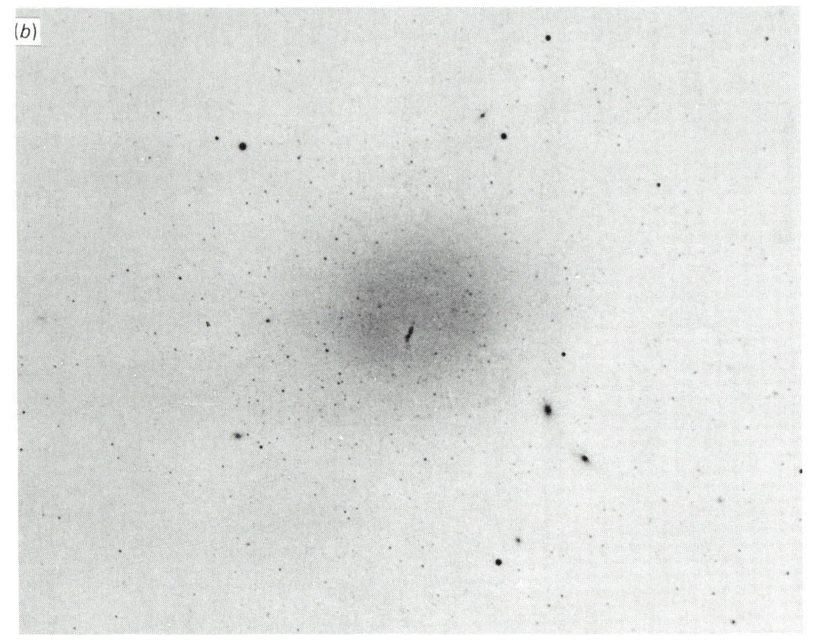

Fig. 9.5. Two prints of M87 are shown, taken with the 200 inch telescope at Mt Palomar, California. (*a*) Shows this object as it is typically seen. (*b*) Has been prepared using a mask that is radially and azimuthally symmetrical in order to remove virtually all symmetrical features of the galaxy and bring out the 'jet' that is located near its core. Photograph and masking by Bill Miller, courtesy Mt Wilson and Las Campanas Observatories and the Carnegie Institution of Washington.

Fig. 9.6. The area surrounding Theta Ophiuchii (also shown in Fig. 8.2.) is shown here after a contrast reduction mask has been applied to reduce the overall contrast between the faint and bright star clouds that are located in this area of the sky. Note the increased visibility of the many small dark nebulae. Photo by the authors.

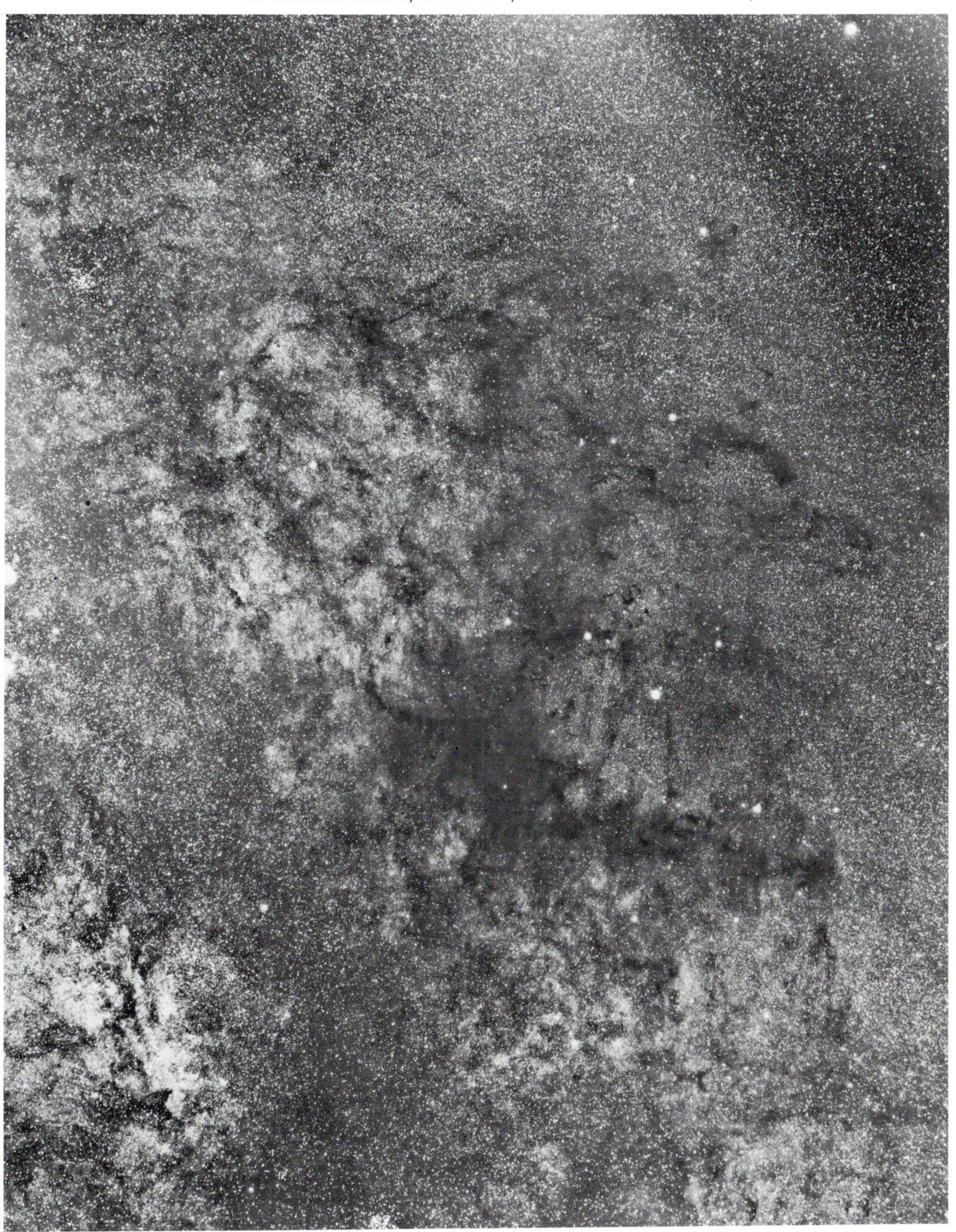

Miller of Hale Observatories, and it is the term which best describes the action of this technique.

The portion of the image which is recovered through the application of this process may be referred to as an information gain; earlier studies have shown that this information gain, or resolution increase, is proportional to the square root of the number of images being integrated (N). Another aspect of the information gain can be measured as an increase of the limiting magnitude on an integrated print when compared with a print made from a single negative. In a 1959 study, a gain of 1.2 magnitudes was measured when 10 negatives were integrated.

Thus, even when fine-grain emulsions are used, integration printing will produce an increase in resolution and an increase in the limiting magnitude. These gains will be accompanied by a decrease in the 'noise' present in the emulsion (e.g. smoothing the grain distribution and eliminating image flaws present in any single negative) and a simultaneous elimination of sporadic 'disturbances' affecting any of the negatives in use (e.g. meteor trails, airplane trails, static discharge marks). Finally, this technique provides additional gains in resolution by recovering some of the information that is lost because of turbulence in the Earth's atmosphere. This latter effect is best seen in the application of integration printing to

Fig. 9.7. Two 12 × enlargements of M51 are shown to illustrate the effects of integration printing: (*a*) was made from a single negative while (*b*) was made by integrating the images from 6 identically exposed negatives. The images used for this figure were taken with a 6-inch f/4.0 Newtonian on 103a-O. Note that the grain is reduced and that fainter objects are shown on the integrated print. Figure 9.8. shows some of the features that can be seen in these prints in order to aid in a comparison.

planetary photography but it is also apparent when applied to the photography of so-called deep-sky objects (see Figs 9.7 & 9.8).

Technique

The basic idea behind integration printing is that a series of images are projected sequentially (not simultaneously!) upon a sheet of photographic material (negative or positive) to produce a single finished image containing information from all of the negatives used. This means that each of the images must be printed **in register** in order to produce the final image. There are a number of methods for registering images for this method, and they range from simple and inexpensive to elaborate and very expensive. We

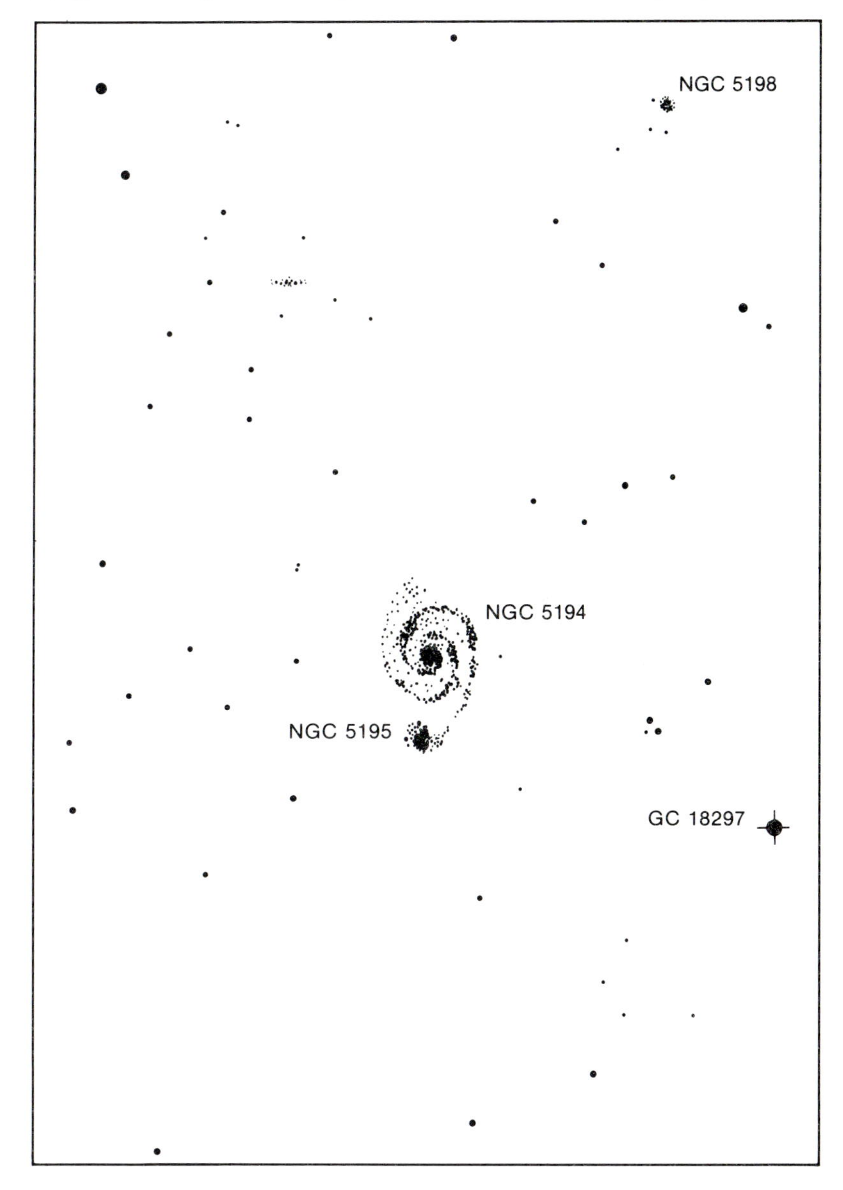

Fig. 9.8. The area shown in Fig. 9.7. is charted in this diagram in order to reveal some of the less obvious features.

have seen a few poorly designed and overly complex registration systems constructed by amateurs over the years and will offer our simple, inexpensive, yet accurate system as an alternative. It is not without its faults, and care must be exercised to insure that the registration is accurate. The most common source of error encountered by the authors in the production of integration prints resulted from problems in changing negatives in the enlarger without affecting the focus or the projected image scale. We will now proceed to detail the procedure which we developed for the application of this process.

Step 1

Our first step was to construct a printing easel with a hinged opaque cover. The upper surface of this cover should be white and should be coated with a material which one can mark on with pencil and then erase without marring the surface. We chose a flat white paint, but also found that a carefully taped-on sheet of ordinary paper would serve well for a single printing session and that this approach would preserve the finished surface of our registration easel indefinitely. This printing easel should be slightly on the heavy side and should be equipped with rubber feet to help prevent any slippage in the course of a printing session.

Step 2

The first negative is placed in the negative carrier, the enlarger is adjusted to provide a 10–12 × enlargement and then focused as usual. The enlarger head is then locked in position to prevent any slippage during the course of the printing session. The position of the printing easel is adjusted for composition, the photographic material is placed in position in the easel and then is securely taped in place.

Step 3

The hinged cover is swung into place over the paper, the enlarger is turned on, and the positions of three to five stars are then marked (in pencil) on the cover surface. In selecting stars to be used for the registration of the projected images, we found that the following guidelines produced fast and very accurate registration.

1. Mark the position of several bright stars in the field to help with the initial rough positioning of the printing easel, but do not use these stars for the actual precise registration.
2. Use star images that are near the limiting magnitude of the plate for the actual registration process. If each of the marked positions are circled on the easel cover, it will be easier to locate the tiny pencil marks during the actual printing session with the lower light levels that are required. Thus, once the field is coarsely positioned using the brighter star-image locations, one will see the registration pencil mark and the actual star image lying within the small circles marked across the field.
3. Use star images that are distributed around the edge of the print field, but which are still unaffected by coma or other edge aberrations.

Step 4

The enlarger lens is stopped down several stops to the printing aperture (which will also increase the depth of focus), the hinged cover is swung out of the way and the first exposure is made after careful consideration of the following variables.

1 The exposure needed to produce a well-exposed print has previously been determined (T).
2 The total exposure time is lengthened by ~25 % (i.e. $T' = 1.25\ T$) to compensate for the fact that N short exposures totaling T s do not produce the same print density as a single exposure of T s duration. This can be partially attributed to the intermittency effect.
3 The exposures required for each of the images is calculated as $t = (T'/N)$, and then the exposure for the first negative is determined by lengthening this value by ~25 % (i.e. $t' = 1.25\ t$). This adjustment is required because of the fact that the response of the photographic material to light is not linear and the first exposure must be lengthened in order to overcome the 'inertia' of the material and to insure that all negatives contribute equally to the final print (see Fig. 9.9). The exposure for each of the successive images are then calculated using the relationship, $t'' = (T' - t')/(N-1)$.

Step 5

The hinged cover is again swung into place, the next negative is placed in the carriage and the star images are registered using the marks previously made on the cover (the enlarger lens can be opened up to full aperture for this step). When changing images in the enlarger, some effort should be made towards an approximate registration so as to minimize the actual shift of the printing easel. This initial precaution will virtually eliminate any possible registration errors which might be introduced by distortions

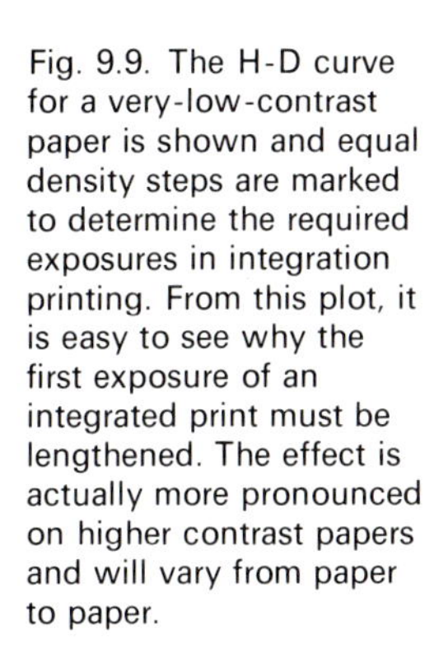

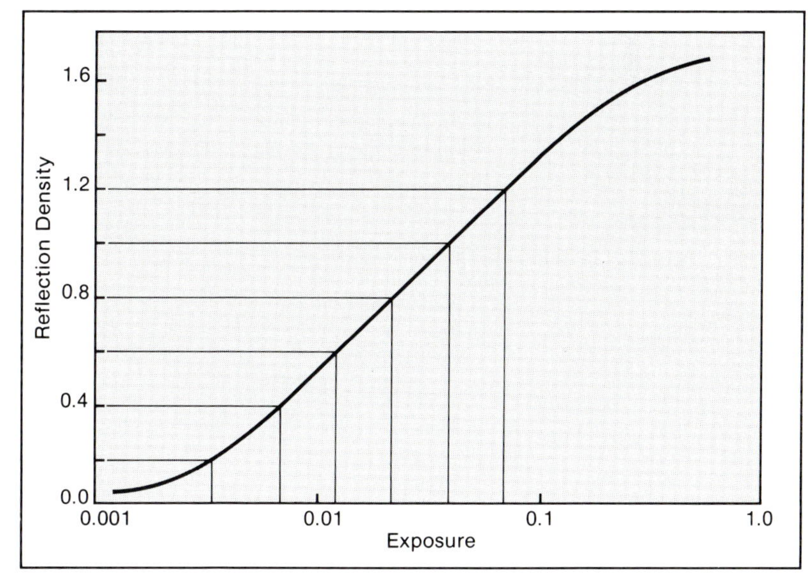

Fig. 9.9. The H-D curve for a very-low-contrast paper is shown and equal density steps are marked to determine the required exposures in integration printing. From this plot, it is easy to see why the first exposure of an integrated print must be lengthened. The effect is actually more pronounced on higher contrast papers and will vary from paper to paper.

present in the enlarging lens. The hinged easel cover is then swung out of the way, the lens is again stopped down, and the image is given an exposure of t'' s.

Step 6

The last step is repeated until all of the images have been printed; then the print is removed and processed in a 'normal' manner.

Conclusions

We have found that this technique is particularly useful in the photography of small objects and objects with extended details which are near (or at) the magnitude threshold of the emulsion. This process allows one to display details that are not available on a single negative. In addition, it provides an increase in resolution and a decrease in graininess. If coarse-grained films (like 103a-emulsions) are being used, the appearance of 10 × enlargements is dramatically improved. In addition to the comments above, the following information will be useful to any person trying this technique.

1 All negatives should be taken under approximately equivalent seeing conditions. An image blurred by poor seeing will degrade contrast and will result in a noticeable loss of detail.
2 All negatives should optimally be of the same density; minor variations can be compensated for by varying the printing exposures.
3 The contrast of the final integrated print will appear to be less than that of a print made from a single negative. Thus, a printing material which will give an acceptable result will probably be of a higher contrast grade than is normally used.
4 While another study found that the integration of 10–15 negatives gave desirable results, we think that excellent gains in information and grain structure can be obtained by integrating as few as 3–6 negatives.
5 If this procedure is being applied to planetary work, it is important to consider the fact that the planet is rotating. All of the exposures used in the integration process should be carefully selected to insure that only the best images are used and all of the images used must have been taken in a timespan which will insure that the planet's rotation does not blur, diminish, or destroy, the expected resolution gains.
6 When using this process on 'deep-sky' objects and taking negatives with a Newtonian over a series of nights or months, the reader should be aware that changes in the orientation of diffraction spikes will occur if the tube is rotated. This can result in some interesting, if not disturbing, results if care is not taken.

The Gum cancellation method

During the process of surveying the Milky Way for faint H-alpha nebulae Colin S. Gum developed a clever method for isolating the nebulosity from the background sky fog and faint background stars. This process does not produce a 'normal' print but, instead, produces a print of the H-alpha nebulosity with most or all of the stars and background subtracted out. It is an interesting method for photographing the full extent of any region of H-alpha nebulosity and can provide spectacular results. At the end of this section we will present an extension of Gum's technique which can be used to produce 'normal' prints with white stars and nebulosity on a dark background, but with virtually no background fog.

The first step of this process involves making a very-well-exposed photograph of the region of the sky being studied in the light of H-alpha. This can be done using a cut-off filter that passes the H-alpha band (like a Kodak Wratten #25 or #29 filter), or by using a medium or narrow bandpass filter which is centered on the H-alpha line at 656.3 nm. If a bandpass filter is chosen, the reader should beware of the problems associated with the use of such filters in the focused beam of light (see Chapter 11). If a fast normal or telephoto lens is used, either filter type can be placed in front of the objective without creating any problems.

The next step requires that a second exposure be taken of the same area of the sky using a filter that passes light close to the H-alpha line but which excludes the light from the 656.3 nm line itself. A Kodak Wratten filter like No. 57, No. 59, or No. 60 will work well for this purpose. Again, this second photograph should be very well exposed and record about the same fog level as the first photograph.

The idea behind Gum's process (not to be confused with Gum printing using the Gum bichromate process) should now be clear. The first exposure records the H-alpha nebula, background stars and some sky fog. The second records the continuum light from the stars, some sky fog, but none of the nebulosity from the H-alpha region. By taking the first negative and superimposing it upon a positive made from the second negative, most, or all of the stars and the fog background will be subtracted out leaving only an image showing the full extent of the H-alpha emission nebula.

The easiest way to produce this result when using 35 mm film is to make enlargements of both negatives on 4 × 5 inch (or larger) film and then produce a film positive of the H-alpha negative. After this is done, both sheets of film can be mounted together, in register, to produce the final product which can be used to make negative (or reversed) prints showing the nebulosity. Normal prints could also be made by producing a copy negative from this 'film sandwich'. If one does not own an enlarger with a glass negative carrier, this latter approach is to be preferred.

As a logical extension of this method, a third photograph of the same region can be taken in the light of H-alpha, or even in white light, which has received an exposure which is short enough to show very little sky fog. Then an enlarged negative of this third negative can be produced and mounted in register with the Gum-cancelled negative. Finally, this resulting sandwich can be used to produce a final print which shows the full extent of the

H-alpha nebula with the surrounding stars on a black sky background. If aesthetics are of value, or if it is desirable to show the location of the nebula against the stellar background, this approach should be of some interest.

On preparing photos for publication

Astronomical photographs pose certain problems for the reproduction processes needed to turn a photographic print into a reproduction that is to be printed by a press. Although astronomical publications vary widely in the quality of their reproductions and seem to frustrate the best efforts at the telescope, there are measures that can significantly improve the odds for better reproduction quality in books and magazines. However, before discussing these measures, it will be of use to gain some understanding of the reproduction process and its limitations.

Virtually all published photographs are halftones and the vast majority of these are reproduced by offset lithography. For both black-and-white and color originals, the halftone and lithographic process impose certain well-defined limitations on the accuracy of reproduction. In virtually all situations the halftone process combined with the lithographic process (or any printing process) cannot accurately reproduce the qualities of the original photograph but it can produce good facsimiles in large quantities, at great speed and at low cost.

Let us start with the fundamentals and first consider how a black-and-white continuous-tone image (like a photograph) is transformed into a printed image, like those found in books and magazines. To print a photograph which can contain a wide range of shades of grey which range from nearly pure white to a very dark black, the image must be converted to a form that can be reproduced with ink that must always be placed on paper at a non-varying density. Printing presses are only capable of laying down solid tones which are defined by the hue and value of the ink in use. Thus a press inked in black cannot print different values of black (which would be seen as different shades of grey) but can only print solid black or nothing at all. How is it possible for such a black/white system to reproduce a photograph that contains an uncountable number of shades of grey?

This paradoxical problem has been solved quite cleverly and involves photographing the original through what is called a halftone screen. This halftone screen breaks up the continuously graded tones of the original photograph and the resulting copy consists of a large number of solid dots of variable size. The dots are arrayed in regular patterns and vary in size in accordance with the density of the corresponding area of the original that was imaged through the halftone screen. Thus, whites or light greys are converted into very small dots and as the shades of the original darken, the dots on the halftone reproduction become progressively larger until their size reaches a point where the dots merge and produce a solid black. The process is, in reality, an optical illusion since no tone gradation actually exists in a halftone. The appearance of tone gradation is a result of the limited resolving power of the human eye. Examination of a halftoned photograph with a magnifier reveals how the illusion of continuous tonal gradation is created. Thus, a halftone reproduction contains no greys, only varying amounts of solid ink on white (the ultimate in pointillism).

Color reproduction for books and magazines is somewhat more complicated but is based on the same halftone principles that are used in black-and-white reproduction. To reproduce a color original, three black-and-white halftones are made through blue, green, and red filters (the additive primaries) and then printed, one at a time and in register, using the subtractive color primary inks: yellow (minus blue), magenta (minus green), and cyan (minus red). In addition, a black halftone is added to compensate for the imperfect nature of the printing pigments. This 'fourth color' (black) helps to increase the density range and deepen the shade of the dark areas of the reproduction. As a further step towards improving upon the quality of the reproduced copy, each separation can be combined with one or more photographic masks that serve to improve both the tonal differentiation and the color balance of the final reproduction. Examination of a printed color reproduction with a magnifier will reveal an array of colored dots corresponding to each of the colors used in the printing process. The technique of making halftones for color reproduction is called color separation, while the printing aspect of the process is referred to as the four-color process.

Although most color separations are done on what is known as a process camera, an increasing number of publishers are turning to electronic scanners. Since these machines combine color separation, masking and screening in one computer-controlled operation they have the advantage of speed and consistency of reproduction. Electronic scanners are also capable of quickly and precisely altering color balance, contrast, and saturation and can also allow contrast to be adjusted in small areas. Although scanner separations are not necessarily superior to the best work done with the camera, their speed and precise controls allow for more consistent good results. As economic pressures push high-quality lithography into the range of luxury and master lithographers become rarer and rarer, the soaring demand for mass-produced reproductions will make the scanner the most common means of producing color separations.

The major problem that arises when photographs are reproduced is the difference in density range between an original photograph and the halftone reproduction. The difference in brightness between the brightest tones (highlights) and the darkest tones (shadows) is known as the density range. With prints, one measures the differences in reflectance values while in transparencies one measures the differences in transmission values. A typical batch of astronomical prints will have a density range of 1.70 to 2.00 with a highlight density near 0.05 (89 % reflectance) and a shadow density of 1.75 to 2.05 (1 % to 0.89 % reflectance). Transparencies usually have slightly greater density ranges than prints, typically showing differences approaching 3.0 (although astronomical transparencies rarely reach such values). By way of comparison, a good black-and-white halftone reproduction will have a range of 1.40 but more typically will be as low as 1.10 to 1.20 with highlights around 0.10 (79 % reflectance) and shadows around 1.20 (6 % reflectance). Four-color reproductions generally have greater ranges since the shadows reproduce darker owing to the superposition of the four ink layers, while the highlight densities are comparable to those of regular halftones. It can be seen from these figures that the halftone reproduction will usually exhibit a compressed density range when compared with an original photograph. The problem facing anyone submitting photographs for reproduction is how to allow for this tonal compression when producing photographic prints destined for publication. In order to gain some understanding of the reasons for the printing adjustments, it will be useful to look at exactly how a halftone is produced.

When photographs are halftoned, the halftone negative is usually given two (and sometimes three) separate exposures in order better to render the details present in the original throughout the entire tonal range. The three exposures are called the main, the flash, and the bump. The main (or detail) exposure produces most of the image detail and is exposed to retain the details of the minimum-highlight densities of the original. The flash (or shadow flash) exposure is necessary whenever the original has a greater density range than the halftone can handle (which is almost always the case). It is a non-image exposure (i.e. shooting just a blank surface) made through the halftone screen, and usually immediately follows the main exposure. The flash exposure serves to add an extra amount of light and mainly affects those dots which did not receive enough exposure during the main exposure. It is the flash exposure that compresses the density range present in the original to a value that can be accommodated by the halftone and the printing process. The bump exposure is used in the rare event that the density range of the original is less than the contrast range that the halftone is capable of rendering. The bump is a very short exposure to the original photograph without a screen that serves to increase the highlight densities and thus expand the density range of the halftone.

Given these realities and limitations of the lithographic process, one can see that the density-range compression that occurs when a continuous tone photograph is converted into a halftone has its primary effects in the shadow regions of the original. Thus, in preparing either black-and-white or color prints for publication, it is most important to maintain good tonal separation in the highlights and mid-tones with the dark shadow areas (the sky background) being printed slightly lighter than they would normally be printed for display purposes. Delicate details that are just slightly lighter than the sky background are much more likely to be retained in the halftone reproduction if the sky is printed with a density no darker than about 1.7. Since a print for publication does not utilize the full tonal range of the printing paper, it will look somewhat greyer and lower in contrast than one carefully printed to D_{max} for exhibition. The best way to monitor the density range of prints is through the use of a reflection densitometer. However, the cost of these instruments generally places them out of the range of the individual. A less expensive alternative is to use the Kodak Reflection Density Guide (Kodak Publication Q-16). Very accurate readings are possible with black-and-white prints and acceptable results can be obtained when using color prints.

On rare occasions, publishers will reproduce black-and-white photographs by the duotone process. With duotones, two halftones are made from the original: one is exposed for highlights and highlight separation while the other is exposed for maximum shadow detail. The two halftones are then printed in register on separate passes through the press. Although two different colors may be employed in printing the two halftones that will make up the final duotone, quite often both halftones will be printed using the same ink. The advantage of the duotone lies in its ability to nearly match the density range present in full range (D_{min} to D_{max}) photographic prints and its increased resolution when compared with conventional halftone reproductions. The primary reason behind the comparative rarity of the high-quality duotone is its increased cost of production.

Further considerations for producing prints for reproduction include the type of paper used and the print size. For both black-and-white and color prints, the best reproductions are obtained from smooth surfaced, but unglossed papers. It is imperative that textured surface papers be avoided at all costs as the resulting halftones will be of inferior quality. Warm brownish toned black-and-white prints are more difficult to reproduce than cool-tone (i.e. bluish) print papers and unexposed areas of any print should exhibit clean pure white. In color reproduction from prints, there are indications that the best results are obtained from Cibachrome prints made from slides. Print size is a problem area as many magazine publishers tend to reproduce photographs much larger than the photographer intended and often beyond normal asethetic limits. One of our solutions to this matter is to insist that a photograph should not be enlarged beyond a specific limit but one can also help by providing the publisher with larger prints. Images should be enlarged as much as grain and the subject will allow without image deterioration and the final print should be 5 × 7 or 8 × 10 inches (with the latter being preferred when fine details are present). Although the risks of a poor reproduction are higher when an original is enlarged in the halftone process, there will always be some loss of detail in the mid-tones and shadow regions of the halftone. The ideal situation is where the photographer knows the reproduction size in advance and can produce prints at the same scale as the final planned reproduction. However this situation seldom occurs and the final reproduction scale is generally unknown to the photographer submitting the prints.

In submitting color slides to publishers the photographer has the option of submitting the original or submitting a duplicate. When an original is submitted, the photographer has no control other than to specify the maximum reproduction scale to the publisher. However, when submitting high-quality duplicates, many more controls become available to the photographer. Aside from the obvious ability to enlarge and crop an original while duplicating, this step also allows one to manipulate the contrast and color balance of the work that will be submitted to the publisher. Again, the 'bigger is better' principle applies as publishers would prefer to see 4 × 5 inch transparencies over 35 mm duplicates and the reproduction quality is generally markedly superior when larger transparencies are used.

By getting to know some of the details of the lithographic process, it is possible to discover its weak points and to prepare works for submission that will minimize the effects of those areas of the process. One of the best guides that the photographer can follow is never to submit an inferior photograph or print for publication. Photograph quality is never improved by the reproduction process. In judging whether a photograph is suited for reproduction, its technical and aesthetic merits need to be weighed in an objective manner without the complications that are introduced by personal considerations. If a photograph is of scientific value, many of the aesthetic considerations can be waived but this is not a satisfactory excuse for sloppy negative care and sloppy darkroom practice. The works submitted for publication should always be of the best quality that the photographer is capable of providing as the printed version of the photograph is the only means that most readers will have of judging the skills and capabilities of the photographer.

10

Hypersensitizing techniques

In the ongoing quest for greater and greater sensitivity in the photographic emulsion, photographers have dabbled like ancient wizards and alchemists: adding a little of this to that and a dash of that to this and then adding the resulting concoction to the emulsion or soaking the emulsion in the mixture. As photographic science and photochemistry evolved, the trial-and-error approach used by the early photographers has given way to a more directed approach in photographic research. The results of some of this research have brought many new emulsions to the photographer, the speeds have been greatly increased, films have been introduced with spectral sensitivities designed for special research problems and Polaroid has brought the realm of 'instant photography' into the world. The astronomy market for film is quite small, yet much research has been devoted to the special problems facing this particular application and a number of specialized emulsions have been introduced into this field. Eastman Kodak has been a solid supporter of the efforts of astronomers and has introduced several emulsion lines of particular interest: the Ia-, IIa-, 103a-, and IIIa-emulsions have all been developed with astronomical applications in mind and it is hoped that Kodak's support for this field will continue to make new films available to the astronomy world.

While new emulsions have been made available which help to overcome some of the problems that are inherent in astronomical photography, the problem of maximizing or optimizing films for particular astronomical applications has fallen largely to the astronomical community. This is largely due to the fact that the enhanced sensitivity of treated emulsions tends to be transitory and the emulsions cannot be stored for any length of time without experiencing a sensitivity loss. Thus, films are normally treated to maximize sensitivity shortly before exposure. These sensitizing techniques are grouped under the heading of hypersensitizing (hypering) although the term is not technically accurate for several of the processes that are lumped into this category.

Of the various hypering techniques that have been applied over the years, only a few have become popular in the amateur community and this popularity has only occurred over the past two decades. One of the probable reasons for this is the need for careful control during the hypering procedure, the poor results when the procedures are not properly applied, the general lack of interest in exacting techniques and the unwillingness to spend time on such projects which exists in much of the amateur community. The information necessary to learn the basic processes exists in a largely non-technical form in magazines and journals that are readily available and most of the processes can be applied with little risk of danger to the photographic experimenter. This field is one where the amateur could potentially contribute to the efforts of professionals by investing the time necessary to investigate the potentials of various processes. For the amateur planning on using any of these techniques it is fair to warn that care and consistency is required in all of these applications if one is to obtain optimum and repeatable results.

It is of importance to realize the difficulty that is involved in measuring the speeds of hypered emulsions and to be aware of the extreme difficulty of comparing the speed of a hypered emulsion with the speed of an unhypered emulsion. If the emulsion is old the effects of hypering will be greater than if a new emulsion were used because of the normal loss of sensitivity that occurs as emulsions age and the fact that emulsions are basically restored to a near new state by many of the processes that will be described. It is therefore important for experimenters to exercise the utmost care and adhere to established professional standards when measuring the speeds and gains of hypered emulsions. It is also wise for the reader to be somewhat cautious about accepting the 'speed' figures reported by amateurs because of generally poor experimental procedures employed. It is our hope that the information contained in this volume will help to correct this situation as there is much that amateurs could contribute to this area.

Cold camera techniques

This technique has the distinction of being the first to really catch on in the amateur community. In 1964, 1965, and 1966, articles by A.A. Hoag and Evered Kreimer appeared in *Sky and Telescope* showing the enormous potentials of cooled-emulsion photography (earlier articles by A.A. Hoag had appeared in 1960 and 1961 in the *Publications of the Astronomical Society of the Pacific*). Over the next decade, a number of other amateurs built cold cameras of various designs and some of the results also appeared in print. Evered Kreimer and Orien Ernest were responsible for producing the most notable results published in this first decade of the cold camera. Most of the cold cameras built and used in this period made use of dry ice or liquid nitrogen to cool the film and employed a vacuum chamber with an optical window to prevent frost from forming on the surface of the film. The vacuum cold cameras designed in this period were heavy, fairly complicated to build, and generally better suited for use on larger instruments. However, during this period, the results produced by the cold camera users served as outstanding examples of the high-quality work that determined amateurs were capable of turning out.

To most, A.A. Hoag, Evered Kreimer and 1964 are all associated with the birth of the cold camera. The actual birth or discovery of the potentials of cooling the photographic emulsion were established in articles published by E.S. King in the 1912 *Annals of the Harvard College Observatory*. King found by observation and careful measurement that plates exposed on cold nights reached fainter limiting magnitudes than those exposed for equal times on warmer nights. Once this observation was made, King attempted to investigate the effects of temperature on sensitivity by exposing plates in colder and colder temperatures and recording the results. He noted that the speed of the emulsions increased as the atmospheric temperature was lowered, and suggested further experimenting as technology advanced. These results then went unnoticed by most until the early 1960s.

The cold camera remained as a novelty device and was used by few until Celestron International introduced a commercial cold camera model in 1974 employing the design of Bill Williams that was published in the December 1973 issue of *Scientific American*. After the introduction of the commercial Celestron–Williams cold camera, the popularity of this device increased dramatically. The new cold camera model was smaller and lighter which made it easier to use on small instruments, and it incorporated a thick plexiglass or glass plug to keep frost from the film surface instead of the vacuum chamber used by earlier versions. This was the feature that finally made the cold camera attractive to amateurs.

The new cold camera model still required dry ice to cool the film, and the plugs could not be used for two successive exposures without a long warm-up period (unless one used a strip of heating tape to keep the plug warm), but the opportunity of using a 'cold camera' and the convenience of not needing a vacuum apparently made this model attractive to many amateurs. In addition, the color photographs made by Leo Henzl and used by Celestron in their advertisements kept the potentials of this device in constant view of amateur eyes month after month. The Celestron–William cold camera was well designed and had the potential capability of obtaining photographs that equalled the results obtained using the vacuum models. With a few minor modifications, it was easily possible to achieve spectacular results. These modifications involved expanding the volume of the ice chamber, the addition of insulation to the ice chamber and the addition of a strip of heating tape to prevent frosting of the front surface of the glass plug. These modifications are required only if exposures longer than about 40 min are planned and are only minor in nature.

One minor problem with this new cold camera was that the Celestron models were designed to use small pieces of 120 or 135 format film. It was rapidly discovered that it was difficult to handle the small lengths of cut film without leaving fingerprints or scratching the film. In addition, it was difficult to process the film using conventional equipment or methods. The processing technique that we found to be the easiest to use and which yielded the most consistent results employed a film-processing drum and thin-sheet-metal plates. Once exposed, several of the film strips were taped securely to each of the sheet-metal plates with the emulsion side up and the plates were then placed in the processing drum as if they were sheets of photographic film or paper. The strips were then processed in the drum using the standard chemistry normally employed for processing the film in use. Because the emulsion base is virtually impermeable to liquids and because all chemicals responsible for the development of the emulsion (color or black-and-white) reach the emulsion from the front surface, the fact that the back of the film remains pressed against the metal plate is of absolutely no consequence. It is important either to coat the metallic plate with a fairly inert material (i.e. paint it) or to use stainless steel for the metal plates. This will insure that the processing chemicals are not contaminated by metal salts dissolved by some of the highly corrosive photographic processing agents.

It is important to note that cooling does not affect all films in exactly the same way. With black-and-white films, the differences appear as variations in sensitivity but when color films are used one sees the sensitivity differences as color shifts. In a series of experiments conducted in 1974, we tested six different color emulsions in a cold camera and found that only one of the six could be used without any corrective filtration and only two others could be reasonably corrected to yield usable results. It is quite fortunate that Hoag and Kreimer used Kodak emulsions in their experiments with color films, for Kodak's Ektachrome emulsions were two of the three that produced reasonable results (the third emulsion was manufactured by Fuji). Thus, when selecting a color film to use in the cold camera, it is wise to start with a film that others are using and then to test other emulsions for suitable color balance. With black-and-white films, it is also wise to start by using an emulsion that is known to perform well in the cold camera and to test others as time allows. One thing to note with respect to the standard black-and-white film in use by most cold camera workers is that Kodak's Tri-X has very little sensitivity to H-alpha emissions; Ilford's HP-5 has almost twice the response to H-alpha light! It is quite important to inspect the spectral sensitivity charts that are supplied by most manufacturers.

Cooling the photographic emulsion actually helps to arrest the process known as reciprocity failure. Thus, cooled emulsions are able to retain much of the speed that they have when used for short exposures making the film appear to be much faster than it would be if uncooled (Fig. 10.1). In addition to this apparent speed increase, cooling causes a slight decrease in the contrast of the emulsion. When using color emulsions on many astronomical subjects, such a contrast decrease helps to produce spectacular results by preserving the details in the highlight areas. When a low-contrast black-and-white film, like Tri-X, is used on a low-contrast nebula it can sometimes create difficulties when producing prints.

When using a cold camera, regardless of the design, it is important to keep the film surface and the surfaces of the optical window or glass plug free of frost. Frost in any of these locations will result in blurring of the image. With either of the two popular cold camera designs, the use of a strip of heating tape or resistors to add a small amount of heat to the chamber in front of the film will help to eliminate any chance of frosting any of these surfaces. However, it is important not to apply too much heat, particularly if a glass or plexiglass plug is used, as the warmth could spread to the film surface and dry ice chamber causing the dry ice to sublimate at an increased rate.

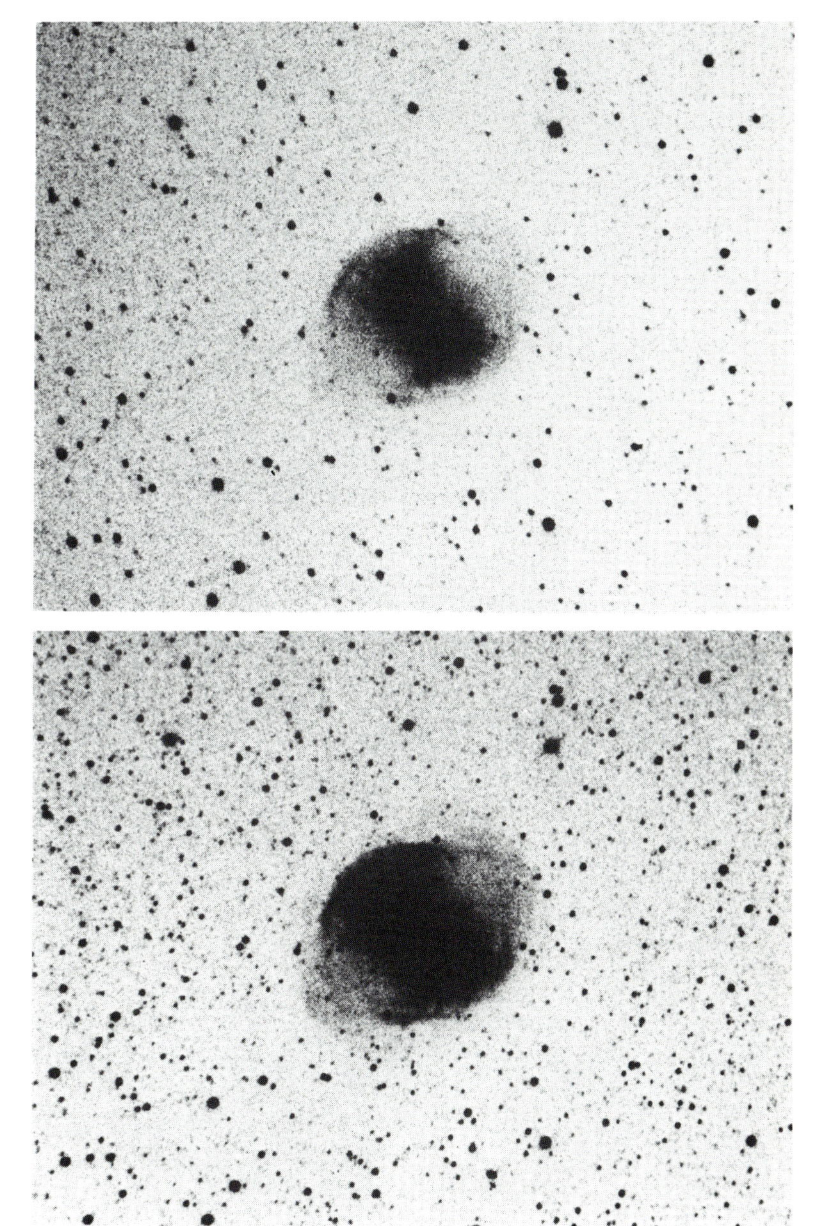

Fig. 10.1. This comparison set of 20 min exposures of M27 shows the gain obtained by cooling Kodak VR1000 in a cold camera. Photographs by Jack Newton using a 16 inch f/5.0 reflector.

In addition to the cold camera designs discussed above there are variations that can be explored. The standard vacuum chamber does not need to be pumped continually if the seals are adequate, and it can also be replaced by a chamber consisting of two optical windows at the end of an evacuated cylinder. Another variation which has attractive possibilities is simply to replace the vacuum with an atmosphere of dry gas. Nitrogen is the usual and preferred choice, but any dry gas will serve the function required here. Thus, one could load the film into the camera and flush the chamber with gas prior to, or immediately after, the dry ice is placed in the ice chamber. The cans of compressed gas used for cleaning photographic negatives offer an inexpensive gas source. If one plans on making a camera that uses a vacuum, a small refrigerator pump will serve the purpose and the small hand-operated vacuum pumps that are currently available offer an inexpensive alternative for use in the field. It should be noted that tests by the authors have shown that the film speeds of several films were found to be higher in a vacuum-type cold camera than in the glass- or plexiglass-plug-type cold cameras.

Hydrogen hypersensitizing

In 1974, Babcock and other Eastman Kodak researchers reported the results of tests of a new hypersensitizing (hypering) process in the *Astronomical Journal.* The technique involved the use of hydrogen gas to increase film speeds during astronomical photography; the results showed marked improvements over other techniques that were currently being used by professionals. This report stressed the importance of removing virtually all moisture and oxygen from the film prior to introducing the hydrogen if serious fogging was to be avoided. To do this, the Kodak researchers used extremely high vacuums (0.000 001 mm Hg) and emphasized that these extreme vacuums were required to purge all moisture from the film. This evacuation procedure created some problems for the researchers as it was necessary to rehydrate the emulsion slowly before processing it.

The possibilities opened up by this technique were exciting and we were definitely interested in trying it out. When it came to such precision vacuum pumps, our lack of cash became the limiting factor. We began to examine the premise behind the high vacuum and look for alternatives. The high vacuum was able to dehydrate the film quickly, but was it really necessary to dehydrate so thoroughly or so quickly? We looked at the possibility of using a low vacuum followed by several purges of the film's container with dry nitrogen gas prior to immersing the film in the pure hydrogen environment. We started off with Tri-X as our test subject and tried a few short hydrogen soaks from 2 to 8 h without any sign of fog. It worked, but were we actually gaining speed. From the literature, we knew that a fog level of 0.1–0.2 above base usually produced the maximum speed gain, so we increased the hypering time until we reached those levels. After

16 h in pure hydrogen at 68 °F and 1–2 p.s.i., we had a 0.2 fog level. When we tried it at the telescope, we discovered that the Tri-X speed surpassed that of Tri-X when used in cold cameras! When we tested the film in the sensitometer, we found that we had produced a gain of 3.8 for a 1 h exposure(Figs 10.2 and 10.3). Our reading, research, and intuition had paid off.

The main reason that this technique was so exciting to us was the fact that it used hydrogen gas. Hydrogen has the property of being able to penetrate membranes that usually stop other gasses. Take a balloon filled with hydrogen (or helium) and tie it down next to a balloon filled with air and see how much faster the hydrogen balloon deflates. It occurred to us that we might be able to hypersensitize a color film using this technique! No other method of chemical hypering will work on color films because of the manner in which the three emulsions are stacked and separated by barrier membranes. When other chemical hypering methods are used on color films (e.g. water or ammonia hypering), the upper layer, which is the blue-sensitive layer, gains significantly more speed than the lower layers, resulting in a very blue image when exposed.

After the success with Tri-X, we tried the process on one of Kodak's Ektachrome emulsions. We again found a tremendous gain with a moderate shift to the blue. The results were promising, and we presented the results at the 1975 Astrophotography Conference held in Southern California. At the same time, we were in the process of finishing off a three-part article for *Sky and Telescope* and they requested the inclusion of the work with the hydrogen process. So, the June 1977 issue of *Sky and Telescope* carried the first published color photographs ever taken with hydrogen-treated film. Since then, the blue-color-shift problem has been solved by Alex Smith who replaced the pure hydrogen gas with a forming gas mixture (92 % nitrogen, 8 % hydrogen) to allow for a longer hydrogen soak and more uniform hypering of all three emulsion layers. In addition Lumicon has introduced commercially manufactured hypering equipment using forming gas for the amateur market. The process is in use by a large number of amateurs at this time and has largely replaced the cold camera in the amateur community. With professionals, this technique is very widely used making available several emulsions with very desirable characteristics that would otherwise be too slow for astronomical applications.

When first faced with the thought of using hydrogen gas a vision of the Hindenburg comes to many people's minds. Hydrogen can be explosive and very dangerous if some simple precautions are not observed. The greatest danger is from igniting the gas as it comes from the tank and producing a blowtorch effect. When hydrogen gas is ignited in this manner it produces a flame that is invisible in normal daylight, making it a particularly dangerous source. Hydrogen will not explode unless the hydrogen–oxygen mix reaches a level where there is more than 4 % hydrogen. Thus, it would be necessary to release the entire contents of a

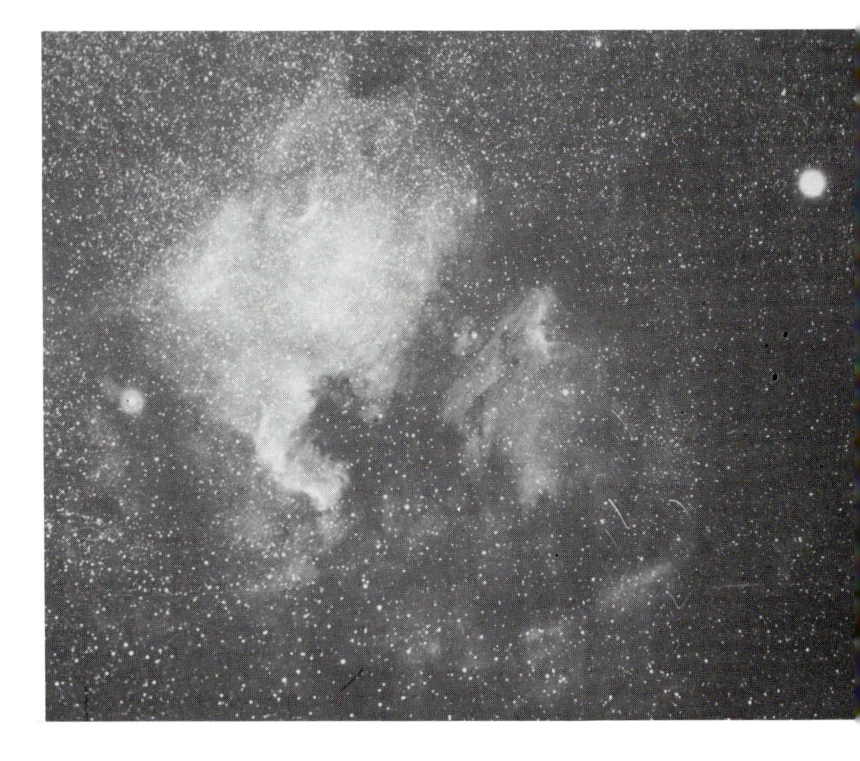

Fig. 10.2. The dramatic effects of hydrogen hypering on Kodak 2415 Technical Pan are shown on this pair of photographs of the North American nebula. A better comparison could have been made by equalizing the sky levels, but the illustration does provide a qualitative 'feel' for the speed gain created by hydrogen hypering. Exposures were 15 min at f/1.65 using a No. 92 filter. Photographs by Pat Walker.

40 ft^3 tank into a $10 \times 10 \times 10$ ft room and then add a spark in order to produce an explosion. The one time during this hypering process where there is danger of an explosion is when the hydrogen is removed from the film chamber. If an explosion-proof vacuum pump can be purchased, the film chamber can be evacuated mechanically. If no such pump is available, the chamber should be purged several times with nitrogen in order to remove the hydrogen after hypering is complete. The escaping hydrogen gas should always be vented outside the laboratory.

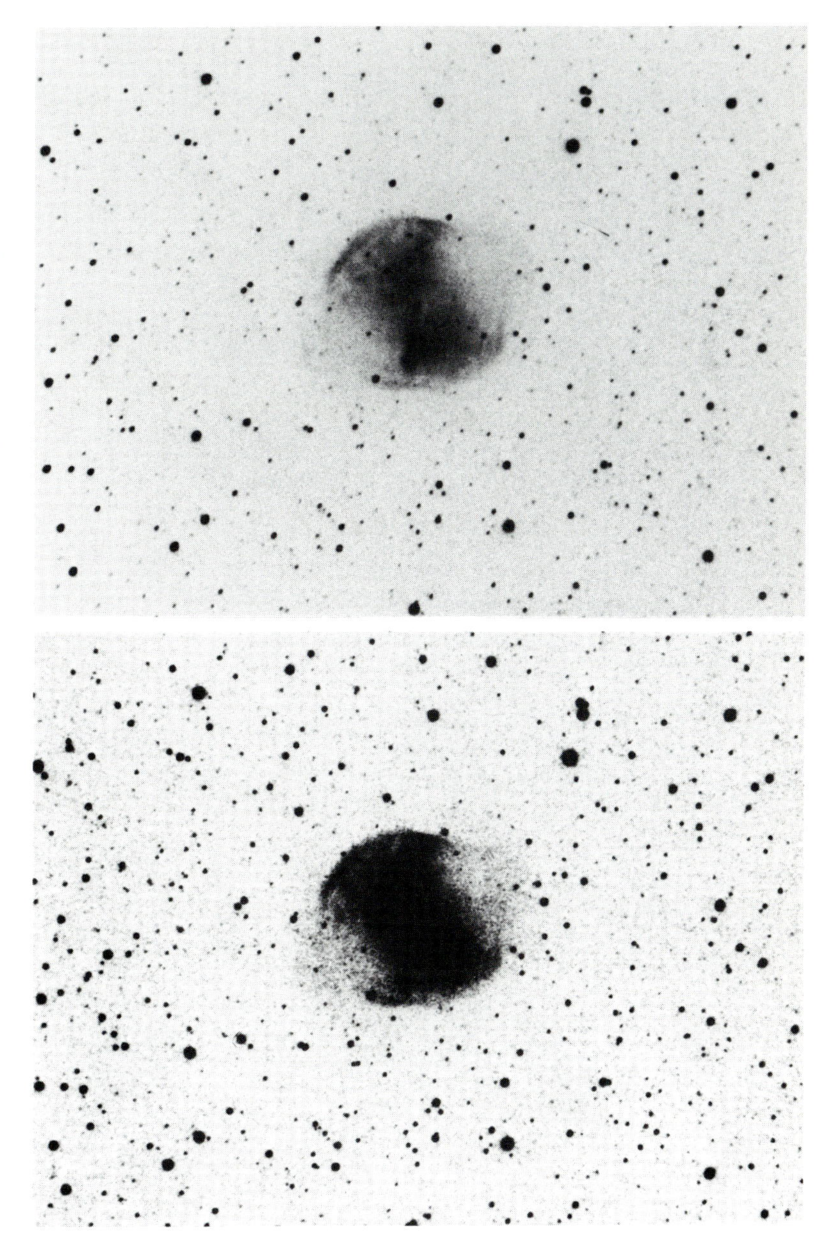

Fig. 10.3. The speed gain obtained by hydrogen hypering High Speed Ektachrome is shown in this pair of photographs of M27. The negative prints were made by projecting the color slide onto Kodak Panalure paper in order to produce a panchromatic rendering of this object. Exposures were 60 min at f/10, using an 8 inch Schmidt Cassegrain system. High-Speed Ektachrome (ISO 160) was hypered for 16 h at 68°F (20°C) in pure hydrogen gas at 3 p.s.i. Photographs by the authors.

There are a number of points that should be kept in mind when using hydrogen or forming gas to hypersensitize films. The procedure that we have adopted in our work with this process is outlined below.

1 When using roll films, the film is removed from the manufacturer's cassette and spooled onto a stainless steel processing reel. The primary supplier of hydrogen hypering equipment (Lumicon) sells one hypering chamber that is designed for use with film left in the cassette. Our tests have shown that the film is unevenly hypered and that fog is also non-uniform when the film is treated while in the manufacturer's cassette. All films placed inside any hypering chamber should be arranged in a manner that insures free circulation over the emulsion surface!
2 The film is placed in the hypering chamber which must be light-tight and air-tight. This chamber needs to have inlet and outlet valves to allow easy flushing with the hypering gasses. This chamber should be capable of withstanding 25–30 p.s.i. and also be capable of maintaining a 29 inch vacuum for at least 1 hour.
3 Execute either step 3 or 4, depending on the equipment available.'* If a vacuum pump is available, the chamber should be evacuated to produce a vacuum in the range of 26–29 inches. After the vacuum is achieved, nitrogen can be introduced to the chamber. By repeating this evacuation–gassing cycle several times one is sure to obtain a reasonably dry environment inside the chamber. The film should be left in this environment for at least 24 h at $\sim 68+$°F to insure desiccation. A desiccation period of 48–72 h is preferable as it insures the removal of virtually all moisture and thus helps to minimize the chemical fogging created by the hydrogen soak that follows. It has been found to be very beneficial actually to 'bake' the film in the nitrogen gas since the effects of nitrogen baking and hydrogen hypering are additive. During this period, the nitrogen gas should be changed several times to help with the desiccation process.
4 If a vacuum pump is not available, the chamber should be 'flushed' 4–6 times with nitrogen gas in order to provide a dry 'oxygen-free' environment. Each 'flush' is performed by first pressurizing the tank to $\sim$28 p.s.i. with nitrogen and then releasing the gas from the tank via the outlet valve. If five 'flushes' are carried out, the remaining

* The desiccation process that is described in items 3 and 4 is an essential and critical step in the hydrogen hypering process and should not be taken lightly. It is essential that virtually all moisture and oxygen be removed from the film prior to the introduction of the hydrogen gas as the presence of either with hydrogen will bring about the rapid formation of unwanted emulsion fog. The desiccation period should also be carried out at temperatures near 20°C before any heat is applied to the chamber in order to avoid another possible source of chemical fogging. Any additional fog created during the nitrogen or hydrogen soaking periods means that the hypering process will need to be terminated early and that the film will therefore not be hypered to its optimum speed.

atmospheric gas content is reduced to 0.4 % and the remaining oxygen content inside the chamber is reduced to 0.08 %. As with the process outlined in step 3, the film should be left in this tank for 48–72 h and the atmosphere inside the tank should be changed 3 or 4 times during this desiccation stage. The use of a vacuum pump, as outlined in step 3, helps to reduce the amount of nitrogen required at this stage of the process.

5 At the end of the desiccation time, the vacuum pump should be used to remove all nitrogen from the hypering tank prior to introducing the hydrogen or forming gas. **Do not use the vacuum pump to evacuate the chamber after the hydrogen is introduced to the chamber** unless it is an explosion-proof model! If a vacuum pump is not available, the tank should be 'flushed' 4–6 times with hydrogen to insure a pure environment. All hydrogen gas 'flushed' from the tank should be vented outside through a length of plastic tubing, as an added safety precaution.

6 After introducing the hydrogen, the film is left in the tank for the amount of time required fully to hyper that particular film. During this period, the hydrogen environment should be changed several times. At the end of the hypering time, the tank should again be filled with nitrogen in order to stop the fogging process and to preserve the sensitivity of the film until it is time for its use at the telescope.

7 With roll films, it is necessary to respool the film before it can be used in a camera. With film that has been so thoroughly desiccated, there is considerable chance of causing static discharges when the film is advanced or rewound. We have found that such discharges can be prevented by grounding the camera or cassette whenever the film is advanced or rewound.

When a hydrogen-hypered film is removed from the inert nitrogen atmosphere and exposed to atmospheric oxygen and moisture, it will quite rapidly begin to lose its enhanced sensitivity. Some films may retain a large portion of their increased speed for as long as one week while others will show a marked decline in sensitivity over the course of a single day. The 103a- series spectroscopic emulsions exhibit a very rapid loss of sensitivity once exposed to the atmosphere. We found, and later studies confirm, that storage in a nitrogen environment will greatly slow the sensitivity-decay process; cold storage in a nitrogen environment will virtually arrest the decay process.

When processing films that have been desiccated as these have, it is wise to presoak the film in a solution of water mixed with a few drops of a wetting agent. The film should be placed in this solution for about 1 min before being drained and placed in the developer. It is important to realize that the use of a presoak makes it necessary to lengthen the development time by 1 min or more, in order to compensate for the time required for the developer to replace the water that has already been absorbed by the film.

With films that have been sensitized by any process which causes fogging to occur, it is especially important to avoid using developers that rely upon chemical fogging compounds to produce apparent increases in speed. Some of the developers that are currently available which rely on chemical fogging include Diafine, Accufine, Microphen and FG-7. When such developers are used on films that have been chemically fogged or exposed to a uniform fogging light source (e.g. pre-exposed, or exposed to substantial sky fog levels) the results will be quite disappointing. If you are uncertain of the action of a developer, *do not use it without testing* its action in a controlled situation. When any developer or developer additive professes to cause an increase in film speed, one's suspicion should be raised that the developer employs a fogging agent.

Constructing a hypersensitizing chamber

The commercial hydrogen hypering chambers manufactured by Lumicon are very well made, but are somewhat highly priced for many amateurs. Also, if a club wishes to hyper batches of film for its members, or if plates or film larger than 4 × 5 inches are to be hypered, these chambers are too small. When we started using this process, no commercial units were available and we looked very carefully at the problem of constructing a practical chamber from readily available materials for a very reasonable cost. Over the years that we have been using the hydrogen hypering process, we have constructed several chambers of differing sizes using the following general design and have been totally satisfied with the results. The basic requirement for a hydrogen hypering chamber is to construct a container which meets the requirements listed below. The chamber

1 must be light-tight and gas-tight;
2 must be made so that it can be opened and resealed easily to allow film to be placed in the chamber and then removed;
3 must have entry and exit valves to admit and release the gasses;
4 needs to be equipped with pressure and vacuum gauges to monitor the pressure inside the tank;
5 should be wrapped with some kind of heating element and equipped with a means of varying and monitoring the temperature inside the tank (in addition, it should be well insulated to minimize temperature fluctuations).

The chambers that we have constructed have been made from an aluminum cylinder sealed off at the top and bottom with aluminum plates. The cylinder wall needs to be thick enough to allow for the bolts that will be needed to seal the chamber for the hypering process. Thick walls also help to minimize any thermal fluctuations by increasing the thermal inertia of the tank.

The chamber can be heated by wrapping several feet of plumbing heating wire around the tank and then using a small Variac (variable voltage transformer) to control the input voltage. The thermometer used to monitor the temperature inside the tank can be mounted with its probe inside the tank, or simply mounted in the outer wall of the tank. A temperature control device can be used to control exactly the temperature inside the chamber, but the temperature can be controlled to within a few degrees by simply wrapping the tank with a half-inch layer of ensulite to minimize the loss of heat from the hypering chamber. By plotting the internal temperature of the insulated chamber against the voltage applied to the heating coils, one can plot a relationship between voltage and temperature that can be used to set the tank temperature to almost any desired point. If the tank is well insulated, the relationship will be a linear one and the range of temperature variation (compared with outside temperature) will be slight. The graph so produced will also show the amount of temperature fluctuation that can be expected at any given voltage (see Fig. 10.4).

Water and ammonia hypersensitizing

Many emulsions will show a sensitivity gain when exposed after being immersed in a distilled water bath and vigorously agitated for several minutes. After being treated in this manner, the film is usually dried quickly in a stream of cool, filtered air. Many older sources recommend the use of pure alcohol baths for rapidly drying water-hypered and ammonia-hypered emulsions. More recent work has shown that the use of an alcohol bath actually causes a sensitivity loss and that far greater gains are obtained when the film is quickly air dried. This process is capable of yielding sensitivity increases of about 2 or so, but it removes the excess bromides from the emulsion that are left in by the manufacturers as a preservative.

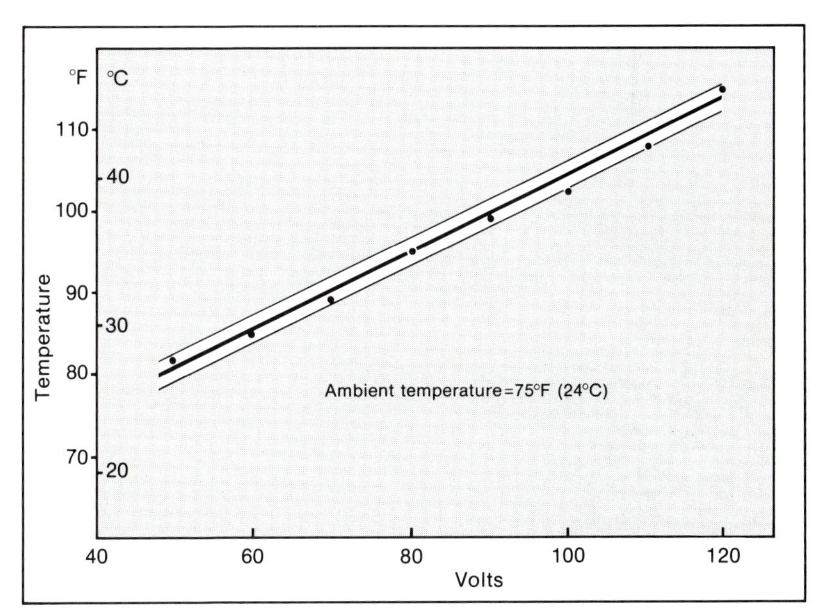

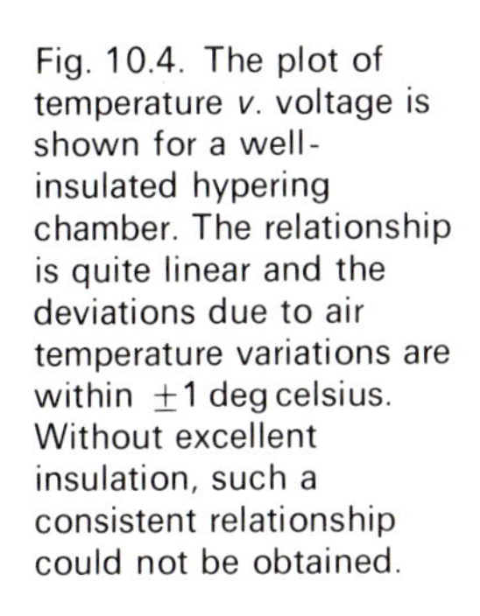
Fig. 10.4. The plot of temperature *v.* voltage is shown for a well-insulated hypering chamber. The relationship is quite linear and the deviations due to air temperature variations are within ±1 deg celsius. Without excellent insulation, such a consistent relationship could not be obtained.

Because this preservative is removed from the film, it is necessary to expose the film shortly after it is treated. If one is interested in using the film at its maximum sensitivity, the film can be stored at dry-ice temperatures for quite some time without any sensitivity loss. It is however very important to be certain that the film has been dried totally before exposure, as a damp emulsion can be as little as one-third the speed if dry (Wm.C. Miller). This technique is more easily executed than the ammonia-hypering method described below and the results are also much more reliable.

Water hypering, like ammonia hypering, shows a greater effect when used on emulsions which are exposed to red and infrared wavelengths. In fact, the greatest gains seen with water-hypered emulsions occur when this technique is applied to infrared emulsions. For instance, a water-hypered I-N emulsion will show a gain of about 4 for exposures of 1 h to wavelengths around 800 nm.

If the water bath is exchanged for a 6 % solution of reagent grade ammonium hydroxide at around 6 °C, and the film is immersed for a period of 4–6 min, greater speed gains may be produced, especially when red- and infrared-sensitive emulsions are being used. When used on Kodak's IV-N Spectroscopic emulsion (infrared-sensitive) speed gains of up to 40 may be expected when measured at a diffuse density of 0.10 above base + fog. However, this technique is very sensitive to many factors and extreme care must be exercised to obtain uniform sensitizing and avoid irregular chemical fogging. Since the ammonia solution is rapidly exhausted during the hypering process, it is advisable to use a fresh ammonia solution for each plate or roll that is being treated. In order to insure uniformity from plate to plate, all of the ammonia bath should be mixed at one time so as to insure that the concentration of each trayful of solution is identical.

When using this technique, vigorous agitation is essential during the ammonia-immersion period. When we briefly experimented with this process, we were advised by Bill Miller to use a weak 1 % acetic acid solution after the ammonia bath in order to obtain the best results. This mild acid bath neutralizes the ammonia solution and allows the emulsion to be handled without producing further non-uniformities. After immersion in this mild stopbath, the film should be slipped into a water bath containing a generous quantity of photo-flo so as to allow the water to drain uniformly from the film. Much of the non-uniform fogging than can occur with this process has been associated with water spotting during the drying period. To help avoid this problem while the film is drying, it is important to wipe the moisture off the back side of the film or plate since the cooling that is caused by these droplets will affect the overall appearance. In general, one can expect to obtain sensitivity gains in the range of 4–10, but larger gains will be obtained in conjunction with near-infrared- and infrared-sensitive emulsions. The plates or films sensitized using this process should be exposed immediately, unless stored at very low temperatures. If the film is stored at dry-ice temperatures, it can be kept for as long as a month, while temperatures of $-90\,°C$ will allow the emulsions to be stored for several months.

Nitrogen baking

One of the findings that led to the discovery of the effects of hydrogen gas on the photographic emulsion was the fact that the removal of oxygen and water from the emulsion prior to exposure produced a sensitivity increase. Another feature of this treatment is that emulsions can be baked for prolonged periods in a nitrogen atmosphere and only show a small fraction of the fog that would have occurred if the film had been baked in a normal atmosphere. Emulsions are usually baked in a nitrogen environment at temperatures in the range from 50–65 °C. The optimum baking time must be determined for each emulsion by gradually increasing the baking time on a number of film samples until the processed samples show a fog density of around 0.2 above base. Nitrogen baking can be used on color emulsions if the film is thoroughly desiccated prior to treatment and if the baking temperature is kept fairly low. As the baking temperature is decreased, the baking time must be increased, a shift from 65 °C to 50 °C will require about triple the time to achieve the same results. The baking temperature must always be kept below 80 °C as any emulsion will be damaged at this temperature.

The chamber used to bake film must be light-tight and air-tight, just like the hydrogen-hypering chamber described above. In fact, the hydrogen-hypering chamber is ideal for nitrogen baking, provided that the chamber is wrapped with heating wire or tape and then insulated with an outer layer of ensulite. In addition, a small variable-voltage source (like a Variac) is needed to regulate the temperature. See the section on hydrogen hypering (pp. 214–21) for additional details on the construction of a heated hypering tank.

Pre-exposure

Pre-exposure has been used for decades in 'normal' photographic applications to produce one of the following results:

1 to provide the threshold energy or exposure to the film to permit the recording of lower light levels (i.e. to increase the apparent film speed);
2 to reduce contrast by providing a larger pre-exposure than used to provide a threshold exposure.

Pre-exposure is just that, an exposure that is given to the film before it is exposed to the actual subject. The pre-exposure is made using an evenly illuminated target which has no surface details while the lens is either removed or the image is defocused in order to prevent any distortion or contamination of the details that are present in the actual subject. This technique is frequently used by landscape photographers to record scenes which are excessively contrasty. Pre-exposure, like other sensitizing methods that rely upon fogging, produces gains in the lower range of the exposure scale while leaving the upper range relatively unaffected. Both of the goals described above can be achieved using either black-and-white or color films. However, in color photography, the color of the shadow areas will be affected by the color of the light used in the pre-exposure.

In astronomical photography, this technique has been applied almost exclusively to black-and-white work and it has had relatively little success. A study in the late 1960s showed that gains of nearly a magnitude were possible when the exposures were over 90 min in duration. The conclusions that have been drawn with regard to the use of pre-exposure show that the technique is most useful when photographing extended objects that are at the very limit of detection. In such applications, a pre-exposure density of 0.40–0.45 has been found to be very beneficial to the performance of the photographic emulsion. Some explanation for the poor performance of pre-exposure on relatively short exposures may be offered by the possibility that skyglow and light from nearby cities may already be acting as a quasi pre-exposure mechanism and thus masking the effects of pre-exposure when used outside the laboratory. We had tried this technique on spectroscopic emulsions and found very little in the way of speed gains, but we began using it on color films in 1978 and found the results to be very encouraging.

Using a pre-exposure on Ektachrome 200 color transparency film which produced a fog level of only 0.03 **below** the normal unexposed film density (D_{max}), we obtained a gain of about 1.4 in the lowest light levels recorded. The actual sensitivity curves for one emulsion sample can be seen in Fig. 10.5 for the pre-exposed film and for the untreated film. The pre-exposure produced a secondary effect which we felt to be most desirable in color work: the sky background consistently recorded as a dark neutral grey. Without the pre-exposure, the sky background is usually recorded as some color that is totally dependent upon which of the three emulsion layers happens to have the highest sensitivity. Thus, one can usually

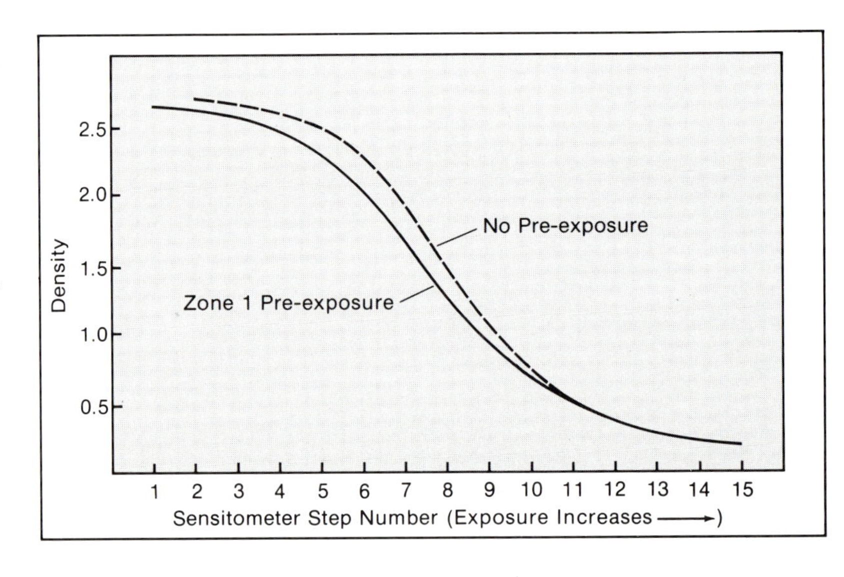

Fig. 10.5. The H-D curve for Ektachrome 200 color transparency film is shown without any pre-exposure (dashed line) and with a pre-exposure equivalent to 1 stop greater than an exposure that would produce no density change (solid curve). Note that the pre-exposed sample shows a 1.4 × speed gain in the shadow regions.

identify the film used for a particular photograph (and sometimes the processing method used) by the color of the sky in the photograph.

In order to pre-expose color films, it is necessary to use a light source that will produce a neutral color. If daylight color films are used, the color temperature of the pre-exposure device must be adjusted to a 5500 K color temperature. We had decided to use a 12 V light source that could be operated using a car battery in the field as its power source, so we measured the color temperature of the light using a color meter. We found that the color temperature of this particular bulb was very close to 2350 K and, using a color temperature conversion chart, found that a No. 78 Wratten filter would convert this source to the required color. It is important to keep in mind that the color temperature of a light will vary as the voltage changes and as the bulb ages. It is therefore important to be sure that the batteries and the bulb used in any portable light source are in good condition.

In order to insure the even distribution of light on the film plane, the light must be evenly diffused and the surface of the light source should be placed quite close to the camera lens which is set at its infinity focus. The device that we designed and built was constructed to allow the film to be pre-exposed in a 35 mm camera just prior to exposure on the telescope.* An alternative approach would be to build a device for exposing plates in the darkroom away from the telescope as described by W.C. Miller. The light in a device such as ours can be diffused using opal glass or layers of high-quality frosted acetate which will not alter the color of the light source. The device shown in Fig. 10.6 uses two diffusing layers separated by a length of tubing to provide the even illumination required for this process.

The proper pre-exposure must be determined by experiment for each camera used since most camera meters will produce different readings, and camera shutter speeds can vary considerably. We will define the basic pre-exposure level as the exposure that is necessary to produce a barely perceptible fog on the color transparency film. We adopted the term 'Zone I' for this exposure. When an unexposed strip of transparency is compared with the region that had received the Zone I exposure, the exposed area will

* The camera must have provision for double exposure

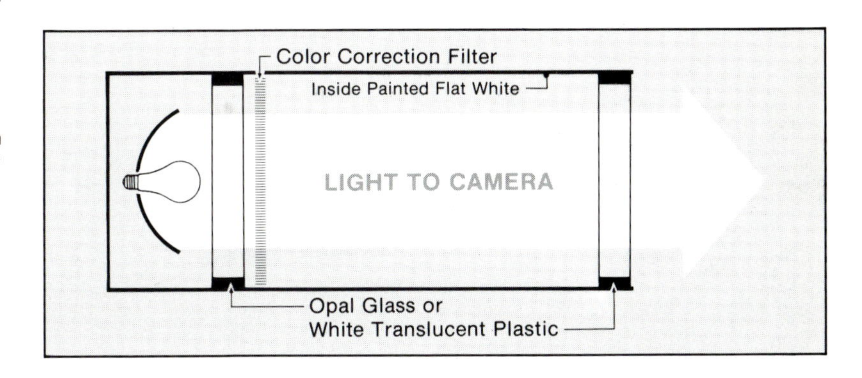

Fig. 10.6. A simple preflash device is shown. The camera is placed against the far right side with the lens set at infinity for the pre-exposure. The color correction filter is chosen so as to correct the color temperature of the light source to 5500 K. The overall size of the device in use is 10 in L × 4 in W × 4 in H and it is operated from 12 volts direct current (VDC) for field use.

appear just slightly lighter. When actually measured on the densitometer, this 'Zone I exposure' was found to correspond to a density of 0.03 below D_{max} (the density of unexposed color transparency film). We also adopted the scheme founded and defined by Ansel Adams in his works on the Zone System. An exposure of one stop less than Zone I is denoted as Zone 0, and an exposure that is one stop more than Zone I is termed Zone II. These terms were adopted as a matter of convenience and simplification. The actual densities produced by a Zone II exposure will vary depending on the contrast of the material in use, and the actual density level of other workers' Zone I exposures will also vary unless a densitometer is used for calibration purposes. In astronomical applications, we found that pre-exposures of Zone 0, Zone I and Zone II could all be used in specific situations to increase the effective speed of color transparency films (we did no testing of color negative films). Examples of these gains on a sample of Ektachrome 200 film can be seen in Fig. 10.7.

Our initial sky tests were run using a 135 mm f.l. lens with pre-exposure levels of Zone 0 and Zone I. The initial results showed that the Zone 0 preflash produced excellent results with this system without adding unnecessary fog. When we tried the Zone 0 pre-exposure on a 6 inch f/6 system, we were greatly surprised to find a total lack of sky fog on the film. After a couple of more tests, we found that exposures through systems of longer focal lengths and slower f/ratios can easily accept Zone I and Zone II pre-exposure levels. After using this process in a number of systems, we were able to compile the following guidelines.

1 When using wide-angle lenses (e.g. 24 mm to 35 mm) at f/ratios of f/4 to f/5.6, a Zone 0 pre-exposure is adequate.
2 'Normal' lenses in the 50 mm range at f/2.8 to f/4 require no pre-exposure, but a Zone 0 pre-exposure may be desirable to produce a neutral sky color.
3 Moderate telephoto lenses used at f/4 can normally accept a Zone I pre-exposure if the main exposure is not unduly excessive.
4 When telescopes with focal lengths from around 24 inches to over 80 inches are used having f/ratios from f/5 to f/11, it will be found that pre-exposures of Zone I or Zone II can be used depending on the brightness and contrast of the object. In general, the slower f/ratios can accept the greater pre-exposure levels. Objects which are photographed with the greater pre-exposure levels can always be copied to increase the contrast and to restore the sky to a darker level.
5 Pre-exposure can also be used to reduce the contrast of objects which are normally beyond the brightness range that can be accommodated on color emulsions. The film strip which we had given a Zone 0 pre-exposure showed a contrast decrease from a Gamma of 2.06 to a Gamma of 1.13.

Additive hypersensitizing methods

We have tried to give the reader some background on a number of useful hypering techniques. However, the literature is vast and yet the information bank on this topic is far from complete. New information is appearing constantly in the literature and one needs to keep track of the changes whenever dealing with a moderately new method like hydrogen hypering.Other problems arise from the fact that professionals do not normally use a number of emulsions that are commonly used by amateurs.

Fig. 10.7. The effects of a 'Zone I' preflash on Ektachrome 200 are shown in this pair of photographs. Both photographs are 1 h exposures at f/2.8 using a 135 mm lens and were taken on the same evening. The top photograph received no pre-exposure while the bottom photo received a 'Zone I' level pre-exposure (see text).

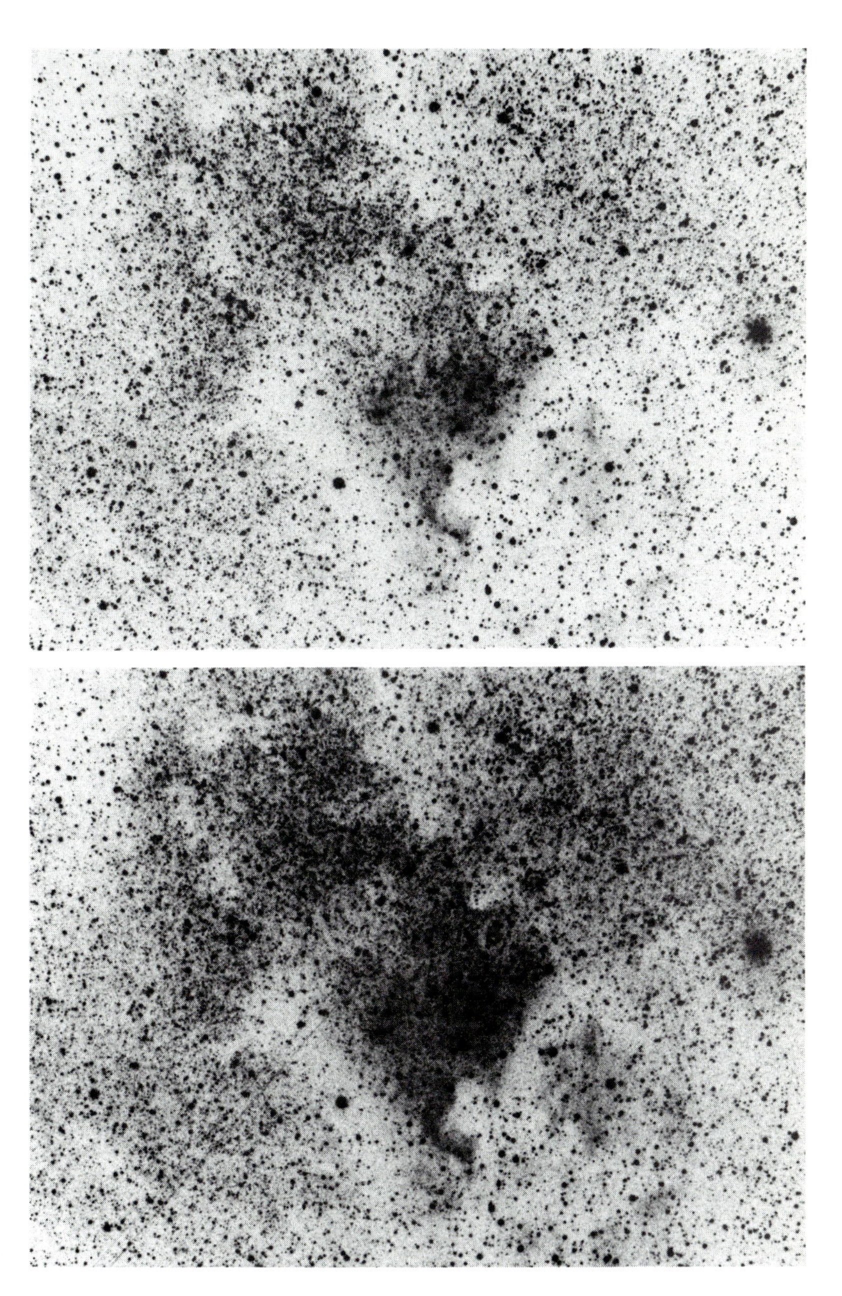

Thus, it is improbable that one will ever find a professional article dealing with the effects of using Kodachrome 64 in a cold camera. This is an area where an amateur, who is willing to perform some careful work using a sensitometer and densitometer, can make significant contributions, at least to the amateur community. However, we must emphasize that this work needs to be carried out in a carefully controlled manner in order to be of value. Estimates of film speed from uncontrolled exposures at the telescope or from photographs of dimly lit objects inside a darkened room are of little quantitative value. Several amateurs in recent years have reported enormous speed gains using hydrogen- and ammonia-hypering methods which have later been refuted by professionals using properly controlled techniques. Such 'false alarms' result in criticism and distrust of amateur work in the professional community.

Some of the known hypersensitizing methods can be used in conjunction with other hypering methods to produce gains that exceed the gain produced by either of the methods when applied individually. Such additive effects generally do not affect all films in an identical manner and may only work well on a small number of emulsions. A few additive effects produce such superior results that they are used as standard procedure by observatories (like the known additive effects of nitrogen baking and hydrogen hypering) but many do not produce enough of an advantage to merit their use. Additionally, if one tabulates all of the known hypering methods and then calculates the number of possible combinations, the number of variations is staggering. When one then considers that the effects of each of these combinations needs to be tested for a number of films, it should be clear that there are enough projects available to keep amateurs and astronomy graduate students busy for many years.

There are some guidelines that can be used when looking for methods that might work together to produce additive effects, and thus save a great deal of time that would otherwise be wasted. Tables 10.1 and 10.2 tabulate a number of hypersensitizing methods in a format that has been adopted by several authors and shows the mechanism(s) responsible for the sensitivity gain and the effects of the method on a number of commonly used films. One thing that should be obvious to all looking at Table 10.2 is that few films have been tested with any but a small set of the hypering methods that are available (blank spaces indicate no data available). In some cases, it would appear that the effects of a hypering method can be generalized to some extent over a given class of film, but it should also be noted that the degree of effectiveness does vary considerably within a given emulsion class. One of the few universals that appears from Table 10.2 is that none of the bathing methods will produce acceptable results when used on color emulsions. If one is seeking to find a combination of effects that will work together to produce higher sensitivity during long exposures one can certainly eliminate any combinations that use one or more methods that is unacceptable for that particular class of film. Also, it would seem useful to

Table 10.1. *Common film hypering techniques and the mechanisms that are responsible for the gain achieved with each of these techniques*

Presumed mechanism	Baking			Soaking		Nitrogen bake and hydrogen soak	Vacuum	Cold	Bathing			Water bath and hydrogen soak	Preflash
	Air	Nitrogen	F.G.[b]	Nitrogen	Hydrogen				Water	Ammonium hydroxide	Silver nitrate		
Further chemical sensitizing	(#)	(#)	#		#	#						#	
Reduction sensitizing	(#)	(#)	#		#	#			#	#	#	#	
Remove oxygen		#	#	#	#	#	#	#					
Remove water	#	#	#	#	#	#	#	#					
Raise pAg^+										#	#	#	
Stabilize exposed centers								#					
Increase neutral Ag stability								#					
Provide threshold exposure	#	#	#	#	#	#				#	#	#	#

[a] The symbol '#' implies an active process.
[b] Forming gas.
After Smith & Hoag.

Table 10.2. *Effects of a number of common hypersensitizing techniques on several spectroscopic emulsions and several other commonly used films*[a]

Film or plate	Baking			Soaking		Nitrogen bake and hydrogen soak	Vacuum	Cold	Bathing			Water bath and hydrogen soak	Preflash
	Air	Nitrogen	FG	Nitrogen	Hydrogen				Water	Ammonium hydroxide	Silver nitrate		
103a-O	s	E	E	E	E	V	E	A	s				s
103a-E		E		s	E								s
098		E		E	V	V	E		E	s			
IIIa-J	s	V	V	E	V	V	m	E	A				m
IIIa-F	s	E	V	E	V	V	m		A	m			
2415 TP		E	V	E	V	V		A		E, A			
Tri-X		E	V	E	V	V		V	E	E			
Fuji 400		m	V	m	V	V w/FG		V	A	A	A	A	m
EK 400		m	V	m	V	V w/FG		V	A	A	A	A	m

[a] Blank spaces indicate that no data are available. s, slight effect; E, effect; A, avoid; m, moderate effect; V, very effective; FG, forming gas.
After Smith & Hoag.

avoid combinations that tend to conflict with each other. Although the combination of water-hypering and hydrogen-soaking would seem to conflict on the matter of removing water from the emulsion, this combination does work. When seeking to find out which methods will work for an emulsion that is not listed in the table, one should look at the 'general emulsion class' to which that film belongs. Thus, 103a-F and 103a-D would fall into the class with the other 103a-type emulsions, while HP-5 and Plus-X will fall into the class with Tri-X. However, a major problem arises when trying to lump color emulsions together into a single class since not all will produce an acceptable color balance when used with either gas-hypering or cold camera methods.

There are very few hypering techniques which employ combinations of two or more methods. The one such method that is in almost universal use today is the combination of nitrogen baking and hydrogen (or forming gas) soaking (or baking). This combination technique was apparently tried or reasoned out independently by several researchers (including the authors) in 1975, just after hydrogen-hypering started to become widespread in the professional community. Cooled-emulsion photography also involves the combined effects of exposing in a vacuum and cooling the film to stabilize the exposed halide centers. Other combinations that have been tried with some good results include hydrogen-hypering plus water bathing, water bathing plus ammonia bathing and pre-flash plus nitrogen baking. Each of these combinations has produced a positive result, but the basic facts are that the gains did not equal an alternative method that was available at the time the research was done. However, one needs to be aware that new methods are only tested on one or two films in regular use and that the results of one series of tests cannot be extended to all of the emulsions that are available (e.g. the effects of ammonia on IV-N emulsions). The shortlist of combinations that are known to have been tested and which have produced some positive results when used on at least one emulsion can be summarized in the following list:

Nitrogen + hydrogen
Nitrogen + forming gas
Vacuum + hydrogen
Vacuum + forming gas
Nitrogen + cold
Nitrogen + hydrogen + cold
Nitrogen + forming gas + cold
Preflash + nitrogen
Ammonia + water
Ammonia + Silver nitrate

11

Deep-sky astrophotography

Wide-angle photography

For many amateurs interested in astrophotography, wide-angle photography represents the first step towards 'seeing' more of the universe that the telescope opened up for the curious mind. For others, this may just be the first step on the road to learning the means of taking exciting and very beautiful photographs. Regardless of the motive, the common 35 mm camera equipped with a 'normal' or 'wide-angle' lens from 24 mm to 58 mm in focal length is the usual tool for the first ventures into this intriguing field. While this is generally an introductory topic, we will discuss some of the problems that frequently confront the beginner and present ideas for improving those wide-angle photographs. The techniques and solutions extend to formats other than the common 35 mm camera, for instance, a 200 mm lens and a 4 × 5 inch sheet of film cover about the same area of sky as a 35 mm camera with a 'normal' lens. The spectacular photographs by Alan McClure in the early 1960s represent the extension of wide-field photography to longer focal length lenses and larger format films.

One of the most common experiences encountered by the beginner is to find a wide-angle slide which seems beautifully guided at the exact center, but which otherwise appears to be an unguided photograph of polar star trails. This effect is commonly known as 'field rotation', and it is caused by poor polar alignment. We have discussed this problem in detail in Chapter 3 along with a method of polar alignment which is capable of yielding very precise results. While this method may require 45–60 min, or more, to achieve near-perfect alignment, this time is easily justified when one

Table 11.1. *Angular field sizes of lenses commonly used with 35 mm format cameras.*

Focal length (mm)	35 mm field size (deg)
24	53 × 74
35	37 × 53
50	26 × 39
135	10 × 15
200	7 × 10

considers the loss of several long exposures or of perhaps an entire night's work.

Another problem which can be very annoying and even frustrating is the appearance of elongated or, possibly, even multiple-star images. This problem may affect every image taken on some nights and only one image from another night. The problem appears to be a guiding error, but there are times when you feel absolutely certain that the guidestar never left the crosshair. The problem is actually being caused by flexure in one or more of the many components which make up the telescope/camera complex. When the mechanics of guiding a 35 mm camera, with a 50 mm lens, using a 60 mm f/15 refractor at 100 × (as the guidescope) are carefully investigated, it becomes quite clear that it should be virtually impossible to produce a *badly guided* photograph! If the guidestar wanders halfway across the field of the guiding eyepiece, the error will not be resolved!

If this difficulty is encountered, there are several possible causes of the problem that are commonly encountered in the design and construction of small portable instruments and in the methods of attaching the camera to the main instrument. One of the weakest links in this mechanical chain arises from the fact that many amateurs attach a camera to a fiberglass telescope tube by means of a 1/4 × 20 bolt and a wing nut. The fiberglass tube is one of the weakest structural components of any amateur telescope and this means of attachment only provides a pivot point for the camera to turn on. If the camera must be attached to the telescope tube, it should be securely attached to a flat, rigid surface. If the tube is circular, a short length of U-channel can be used to provide the flat surface and a good-sized washer inside the tube will provide a much firmer anchor for the camera and its platform.

Another common cause for lengthy star trails on wide-angle photographs results from using a cheap ball-and-socket camera mount to attach the camera to some part of the mount (or sometimes this is used to attach the camera to the telescope tube!). These ball-and-socket mountings are not designed for the stresses which develop as the 1 lb camera swings through the 15 deg arc of a 1 h exposure. In effect, one more weak link has been added to the telescope mount. These camera mounts barely meet the needs of normal photography and they will certainly not fulfill the stringent requirements of astrophotography.

The weak areas which we have discussed above are the usual culprits creating the gross flexure which causes the star trails seen in wide-angle astrophotographs. If the weak areas described above are corrected and trailing still occurs using short f.l. lenses it will be necessary to go over the entire instrument very carefully looking for components which are not securely fixed in place. Areas which are particularly suspect include: guidescope supports, secondary spider and mount, and the primary mirror cell. Since this sort of thorough 'debugging' is not usually required for wide-angle work, the detailed discussion of the design considerations for photographic telescopes is to be found in Chapter 2. In general, this sort of detailed instrumental analysis should be considered a prerequisite for long-exposure astrophotography through the telescope whenever a guidescope is to be used.

A common error encountered in wide-angle astrophotography is the constant use of lenses at maximum aperture. Since nearly all SLR lenses are designed to perform best at 1–2 stops below full aperture, the use of the lens at full aperture often accentuates lens aberrations, produces excessive vignetting, and will result in heavy and uneven levels of sky fog. Common camera optics are designed to meet the needs of the average camera user. These requirements are not nearly as stringent as those for star photography which is the toughest test of any lens. Lens aberrations are not readily apparent in portrait work or in landscape photography. Additionally, in most everyday applications, the photographer is only interested in the central region of the photograph, leaving edge aberrations unnoticed. Furthermore, with most photographic applications, lenses are rarely used even close to full aperture which makes the presence of aberrations and vignetting even less noticeable.

For astronomical use, the simplest test for determining the optimal aperture setting for any particular lens is to make a series of guided exposures on the night sky using a color reversal film (to allow detection of color errors without introducing a second optical system into the test). The exposures should all be somewhat over-exposed (e.g. the sky should not be totally dark) to allow the full effects of lens aberrations and vignetting to become apparent. The series should begin with the lens set at full aperture with each succeeding exposure being stopped down by half-stop increments. At the same time each succeeding exposure should be increased by 50 % to maintain approximate equivalency. If a lens needs to be stopped down more than 2 stops from full aperture to reduce aberrations and vignetting to acceptable levels, it will be of limited use for astronomical work. Most lenses that fail these tests will be perfectly acceptable for 'normal' work but, once in a while, a lens will be found that is defective by the manufacturer's standards and which can hopefully be traded in for a new lens at no charge.

There are many lens aberrations present in even the most sophisticated optics, only a few are serious enough to noticeably affect results in most wide-angle astrophotography. The most noteworthy aberrations include chromatic aberration, coma, and distortion. Of the three aberrations, only distortion is not easily detected with a star test and will require its own separate test to be revealed. However, chromatic aberration and coma are readily apparent when the star test transparency is examined with a 8 × or 10 × magnifier.

Fortunately, chromatic aberration (or more precisely, longitudinal chromatic aberration as discussed here) is well corrected in most SLR lenses. Not until focal lengths become longer than about 150 mm (for 35 mm format) do the characteristic color errors become evident. Typically, this aberration shows itself as a blue halo around the brighter stars throughout the entire field. Stopping the lens down does not generally improve the situation by a significant amount. While most of the long telephotos suffer from this aberration, there is considerable variation in its degree when comparing lenses made by different manufacturers. A few lenses that employ exotic glasses and/or crystal elements have eliminated the problem entirely. Those lenses that do exhibit unacceptable color error can still be used in black-and-white photography when used with yellow or red filters. Likewise, haze or ultraviolet (UV) filters will help eliminate some, or all, of the blue halo for color work.

Coma is an off-axis aberration that increases in magnitude with the distance away from the center of the field and is common to nearly all lenses when used at maximum aperture. In its classic form, the normal diffraction pattern is changed into a fan-shaped configuration with the brightest part of the image at the apex which points toward the center of the field. When combined with lateral chromatic aberration and/or astigmatism, comatic images can take on the familiar moth or seagull forms and may show color fringing. Coma is probably responsible for most of the poor imaging seen at the edges and corners of amateur astrophotographs, being particularly evident because of its asymmetrical shape. While coma is common to most SLR lenses when they are used wide open, it can be virtually eliminated, in many cases, by stopping down the lens 1 or 2 stops. If stopping down does not correct the problem, the lens probably suffers from excessive astigmatism and/or lateral color error.

Distortion does not affect the sharpness of individual star images but rather alters their positions owing to variation in magnification with field angle. In order to test for this aberration, one needs to photograph a target that is made up of a grid pattern of straight lines (a brick wall will serve as a crude test pattern). Extreme care must be taken to make sure that the film plane is absolutely parallel with that of the target. The resulting photograph (negative or transparency) is then inspected with a magnifier. A distortion-free lens will record all the lines as perfectly straight and parallel over the entire field. However, if the lines bow inward or outward, the lens suffers from pincushion or barrel distortion respectively. For most astronomical applications, distortion is not a serious problem. However, if plate reduction or mosaics are planned, the work can be considerably more difficult to accomplish with distortion present.

While vignetting is not strictly a lens aberration, it is discussed here because it is so commonly encountered in wide-angle work. All lenses suffer from vignetting to some degree and while many variables contribute to the problem, two primary factors are responsible for most of the problem. The first of these is created by the basic geometry of cameras and lenses. Illumination at the edge of the image will be less than at the center owing to the greater distance between the lens and the edge of the film when compared with the center (effectively giving the lens a slower f/ratio). The more serious factor is created by the internal structure of the lens barrel and the supports for the many elements that are part of the lens. As one moves off-axis, the full aperture is no longer seen. Instead an elliptical aperture is seen which becomes smaller and more elongated as one moves further from the optical axis. Fortunately, these losses can be reduced or eliminated by stopping down the lens. For most lenses, a 1 or 2 stop reduction will cut the light fall-off in the corners to $\leqslant 50\,\%$ (1 stop), which is acceptable for most applications. If one intends to use the lens for taking photographs for mosaics or for measuring stellar magnitudes, it will be necessary to stop the lens down further to reduce the vignetting effect even more (note: for measuring stellar magnitudes, the light fall-off can be measured and used in an equation to compensate for the off-axis vignetting).

The star test described earlier should give a good indication of the aperture that will reduce vignetting to acceptable levels for most work. However, more demanding applications will require a more precise test. For this test, a series of photographs at half-stop intervals need to be made of a smooth, evenly illuminated target (the north sky on a clear day will serve well for this test). After processing the exposures, which can be either black-and-white or color, each must be measured with a densitometer. The measurements must proceed from corner to corner at regular intervals ($\sim$2 mm spacing will suffice). The results of this test will give the exact distribution of light fall-off for each aperture setting and will indicate the maximum aperture where vignetting is reduced to nearly its minimum value.

The present authors have used 35 mm camera lenses from 24 mm through 750 mm f.l. and have generally found that lens quality cannot be predicted. The fact that a given manufacturer has made several lenses of excellent quality for astrophotography is no indication that another lens by the same manufacturer will perform well. Specific experiences include a 24 mm f/2.8 lens which produces images so small at f/4 that the magazine which commonly printed the photos had to throw the images out of focus in order to reproduce the slides. Another 55 mm f/1.7 lens by the same manufacturer can be used at f/2.8 without any problems, while a 58 mm f/3.5 macro lens by the same lens maker cannot be used for astrophotography unless stopped down to f/4.5 and yet another of their lenses, a 200 mm f/3.5, cannot be used without a contrast filter owing to its poor image quality.

The tests outlined here will show the maximum aperture that can be used while reducing the effects of vignetting and aberrations to acceptable levels. They should also give the photographer a good indication of proper exposures, given specific films and site locations, to produce the best possible astrophotographs. With this knowledge, it should be possible to realize a substantial increase in the quality of one's astronomical photography. The benefits of stopping down a lens are two-fold: not only are aberrations and vignetting suppressed but sky fog is reduced allowing better all-around contrast. With the new super-fast color films that are becoming available, 20 min exposures at f/4 will produce excellent results. Since photographs taken with 50 mm lenses should not need to be guided, given a good-quality clock drive, there should be little reason to sacrifice quality for a shorter exposure. There is more to this field than merely gathering light on film; quality and aesthetics separate mediocrity from excellence.

If you are taking photographs on a regular basis as part of a nova or comet patrol, the above tests and procedures will greatly improve the usefulness of your photographs. The presence of vignetting and aberrations will not only make it less likely that you will find anything outside of the center of the frame, these defects will make it difficult to measure the positions of any object found as well as making it hard to perform accurate measurements of the brightness of any objects located. Since the probabilities of finding any novas or comets are so slight when small-aperture, short-focal-length cameras are used, there is no point in reducing the odds any lower than they already are.

Meteor photography

This is an area of astronomical photography which requires both perseverance and a healthy measure of luck. The requirements in terms of equipment and a knowledge of photography and photographic materials are minimal. Some tips that can help to improve the odds of capturing meteor trails on film include:

1. a fast film should be used to improve the odds of recording a meteor trail due to the meteor's rapid motion across the sky;
2. a fast lens (of excellent optical quality) should be used for this same reason – the probability of capturing meteor trails on film diminishes as $(1/e^{f})$;
3. the probability of capturing meteor trails increases in proportion to the area of the sky covered by the photograph;
4. the frequency of visual meteors increases steadily as the night progresses, reaching a peak at dawn;
5. the average frequency of visual meteors is greater in the summer months, reaching a peak in August (see Fig. 11.1 & Table 11.2);
6. the frequency of visual meteors is, by definition, greater during a meteor shower;
7. if photographing during a meteor shower, it is best to point the camera slightly away from the shower radiant point.

After one applies these tips to the task of capturing meteor trails on film, the rest is simply a matter of luck. The average cumulative exposure required to record a single meteor trail can range from a few minutes to several dozen hours, depending on the methodology and luck of the photographer.

Beyond the mere act of recording meteor trails on film, one can establish two or more stations for photographing meteors and determine the altitudes at which the meteors occur. If this is combined with the technique of using a beam chopper in front of the lens to break the trail at known regular intervals, one can determine the actual speeds at which the photographed meteors move through the Earth's atmosphere. If one is lucky enough to record this information on one of the rare meteors that manage to make it to the Earth's surface, the information obtained from your photographs can be used to narrow down the location of the impact point and may lead to the recovery of the meteor fragment(s).

Photographing aurorae

This is a field which normally affects only astrophotographers living in the more northern or southern latitudes but sometimes, after an intense solar flare, the ionized portions of the Earth's upper atmosphere can produce visible aurorae at latitudes of 30 deg and less. Major observatories often detect the effects of invisible aurorae at these latitudes, but the actual visibility at such latitudes is an uncommon sight. In any case, any astrophotographer faced with a display of these phenomena will be reaching for a camera. These are incredible apparitions, even when color is

Table 11.2. *Major meteor showers*

Name	Date	RA	Dec.	Hourly Rate	Limits
Quadrantids	4 Jan.	232	+50	40	~24 h
Lyrids	21 Apr.	271	+33	8–15	20–2 Apr.
Eta Aquarids	6 May	338	+1		4–13 May
Draconids	29 June	231	+54		<24 h
Delta Aquarids	28 July	342	−16		25 July–4 Aug.
Perseids	12 Aug.	45	+58	50	4–16 Aug.
Giacobinids	10 Oct.	262	+54		<6 h
Orionids	21 Oct.	96	+15	10–20	15–25 Oct.
Leonids	16 Nov.	152	+22		15–20 Nov.
Geminids	13 Dec.	113	+32	60	9–13 Dec.
Ursids	22 Dec.	207	+74	18	~24 h

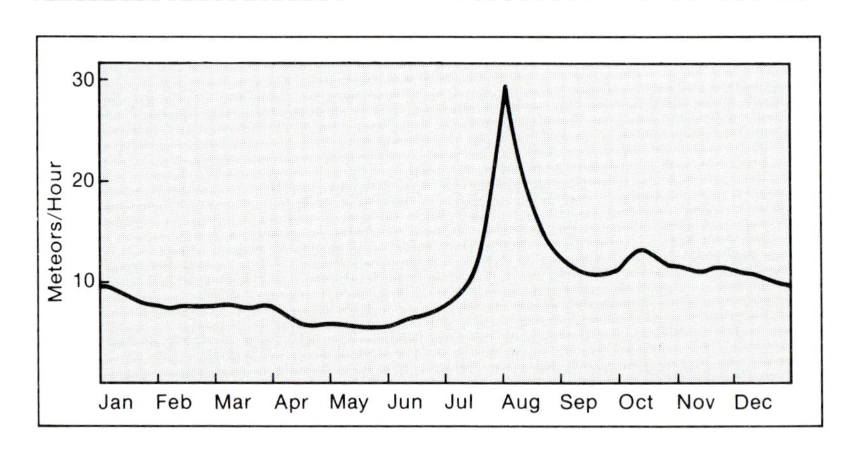

Fig. 11.1. The average nightly meteor rate is plotted for a year. The pronounced peak in August is a result of the annual Perseid meteor shower.

not visible to the naked eye, and photographs can be truly spectacular.

The major problem in photographing these apparitions is the fact that they are not unusually bright and do not remain in any fixed form for any great length of time. Some may dance across the sky in continual motion, while others may remain relatively stationary for a minute or more, but vary in brightness and extent during that period. The problem is to get an exposure which is long enough to show the color and details of the aurora while not allowing those details to become overly blurred by the motion of these dancing specters. While this sounds easy, sometimes it is not possible and the only photographs that can be taken will show a diffuse colorful cloud with a few blurry ridges and rifts within the colorful cloud and a few blurry ridges and rifts outside it.

The basic guidelines for photographing aurorae are to use the fastest color film available and a fast lens at its maximum aperture. In order to help cut exposure down to a minimum, it is also a good idea to plan on having the film pushed 1 stop (there is no film yet available that will do a good job on auroral phenomena without having to resort to some sort of abusive course of action). If the aurora appears to be around 3rd or 4th magnitude and shows no color, exposures of 30–60 s are in order at settings of f/1.4 to f/1.7. If the aurora is brighter than this and color is visible, one can try a series of decreasing exposures based on the visual appearance of the aurora. The process of bracketing exposures over several stops is an excellent idea owing to the difficulty of estimating the actual brightness of one of these dancing clouds. If the aurora is not of a fast-moving variety, a good approach would be to open up the shutter for as long as the aurora appears to remain in a relatively stable configuration (or as long as is needed to obtain a reasonable exposure). If the aurora is one of the faster-moving types, the only alternative is to shoot a large number of frames in the hope of capturing the aurora during one of its less-active phases. It may require a lot of film before you finally get a photograph that even comes close to catching the beauty that these dancing apparitions present to the naked eye.

Nova patrols

For many years, amateurs have sporadically sought to establish systematic photographic programs to search the skies for novae and comets. These survey efforts have generally been short lived and, because of poor strategy, doomed to failure. In considering such a program, a number of aspects of the problem should be considered in order to enhance the possibilities for success.

If small-format cameras are to be used in such a patrol, the limiting magnitude of the camera/film system severely limits the probability of success. For example, with a 55 mm lens at f/2.8, the limiting magnitude will be in the range of 10.0–13.0, depending on the film that is used in the system. Films like 103a- and hypered 2415 Technical Pan will give the fainter limiting magnitude while color films will generally give less satisfactory results. Fainter limiting magnitudes can be reached by using telephoto lenses or larger-format film with longer-focal-length lenses. However, this is not the only problem which needs consideration.

With most normal and wide-angle lenses, vignetting is a serious problem when the lens is used wide open and can amount to more than a magnitude of lost light at the edges of the frame. The problem of vignetting affects most lenses at f/ratios faster than f/4.5 and thus the lens in use should be stopped down to at least f/4.5 in order to insure that the entire field is more uniformly illuminated. If this is not done, only the central portion of the field will reach the fainter magnitude, and only the central portion of the frame will really be of use for a patrol comparison. Many object to the longer exposure times required at slower f/ratios; many should never attempt anything as trying as patrol surveys. If one cannot adhere to the stringent requirements necessary to perform this task well, one should not start it.

When patrol exposures are taken from an urban or suburban environment, the sky fog level will be found to vary from night to night. As a result when patrol frames are being compared or 'blinked' to search for novae or other objects, the sky from such plates will appear to flicker as one switches from one frame to another. This flickering will create a distracting effect which can easily influence the effectiveness of such a program. If the blink-comparator device has a means of varying the intensity of the lights illuminating each of the frames, the apparent sky fog densities can be evened out, eliminating the flickering effect. However, it has been shown that variations of plate fog from sky light does affect the speed of films and thus a well-fogged frame will also have a fainter limiting magnitude than a less-fogged plate. The best solution to this problem is to patrol from dark sites. However, since darkness is at a premium and getting scarcer year by year, the alternative solution is to perform patrol work using yellow, orange or red filters to severely reduce the effects of the light-polluted night sky.

Processing is also of prime importance in such work. If the film processing is not carefully controlled, all of the stars in the field will appear to vary, even though the exposures are identical. If color films are used (hopefully we have presented enough evidence to eliminate this as an option already), the color of the sky will also vary from batch to batch (from manufacturing variations, processing variations, and sky fog effects). All of these variations add up to distractions that may well affect the probabilities of detecting any novae which may actually appear on the film.

The point to consider when looking at these various complications is that these distractions are additive and that the task of identifying novae, comets, or even asteroids will be made considerably more difficult by their presence. In assessing the probabilities which one recent popular survey program would have for the discovery of a nova, we estimated that all of these distractions would amount to a 'loss' of information corresponding to at least 2 magnitudes! From the results (and lack of results) of several years of such a survey, it actually appears that the information loss amounts to at least 3 magnitudes! Thus, instead of looking for novae brighter than magnitude 10–13, the participants were limiting themselves to a search for novae brighter than magnitude 6–9 (since many of the participants of such haphazard patrol efforts may choose to use color film because of the ease of getting this film processed, one is generally looking at the lower end of this range). The odds are terribly slim in such a situation of finding a nova before it is sighted by a careful visual observer, and the odds of catching a nova on the rise (this is actually one of the prime purposes of nova patrols) are even slimmer under such conditions.

Nova patrol efforts are worthy projects, and they need not be doomed to failure if consideration is given to some of the problems that can lessen the probabilities of success. These problems can be greatly reduced with a small amount of effort and do not add significantly to the work involved in such a survey effort. Without such care, a poorly conceived survey is virtually doomed to a small number of discoveries of fairly bright novae, and the odds against catching the nova on the rise are even slighter. To summarize the steps that can be taken to enhance the odds of success (the following list is ordered partly by level of priority):

1 use a black-and-white film, preferably one designed for long exposures or a film that has been hypered for this purpose (a high-contrast film like 103a- or 2415 Technical Pan is to be preferred for this task).
2 stop the lens down to eliminate vignetting, extend exposure times to compensate (~30–45 min exposures) – or, use a special-purpose, fast, unvignetted system designed for this work;
3 use medium focus telephoto lenses (e.g. 135 mm f.l.) to reach fainter limiting magnitudes, or go to a longer lens in conjunction with a larger-format film;
4 use filters to eliminate the background variations caused by fluctuations in the brightness of the night sky;
5 process your own film under controlled conditions, do not rely on laboratories for such work;
6 process each night's work immediately and perform the comparison work without delay.

If one is really interested in increasing the odds of making a nova discovery, consider a project that patrols galaxies for the appearance of supernovae. This could be done with a medium-focal-length telescope to photograph individual galaxies, or simply use a camera and a telephoto lens and repeatedly survey a cluster of galaxies to really improve the odds of making a discovery. With two frames exposed using a 135 mm lens, one can cover the central core of the coma cluster and thus hit several dozen galaxies with just over an hour's effort. This would seem to be one area where efforts should be concentrated since the odds of making a significant discovery are very good.

Finally, we come to the matter of what to use for the task of comparing patrol photographs. Over the past 10 years a number of designs for usable blink comparators have appeared in the amateur literature. Some of these designs have been extremely expensive to implement, requiring two matched precision slide projector units or two matched video cameras, but these systems are best suited for showing comparisons to large audiences. Many less-expensive alternatives are available, and a very simple system can be constructed very inexpensively which will function quite well.

Such a simple device can be made by constructing two small (~3 inch) cubes from $\frac{1}{8}$ inch sheets of translucent white plexiglass. Each of these cubes is fitted with a small 9 V or 12 V light source and then mounted beneath a pair of stereo viewers. The two film frames are placed atop each of the cubes and adjusted until the images line up. Then the voltage is switched from light *A* to light *B* repeatedly to create the blinking effect. The switching can be performed manually with a two-way switch, or it can be accomplished by a circuit which will cause the switching to occur automatically. A useful feature with which such a device should be fitted is a rheostat on each of the light sources so that the apparent brightness of the skies can be equalized prior to blinking the frames. With film formats larger than 6 × 6 cm, this arrangement needs some alteration in order to compensate for the necessary separation of the two images. In this situation, the stereo viewer can be fitted with an array of mirrors to allow the viewing eyepieces to remain in their proper positions.

Narrow-field photography

In this section we will consider the objects and problems associated with astrophotography using focal lengths from about 24 inches and up. We have discussed a number of instrumental design factors in detail in Chapter 2 and we will only give a brief review of those areas which relate to this topic.

Focal ratios from f/5 through f/7 are to be preferred for this work. The lower limit of f/5 produces a usable coma-free field, while f/6 will produce a coma-free field 1 inch in diameter, and an instrument of f/7 provides a coma-free field which is large enough to allow off-axis guiding on reasonable-looking star images (assuming a Newtonian is used). While there are a number of catadioptric systems available with f/ratios of f/10, such slow systems are generally less suited for astrophotography. With these systems, exposures are quite long, even using hypered or cooled film unless a focal reducer is used to reduce the f/ratio. If such a catadioptric system has good optics, and if one is willing to endure exposures of 1–2 h in length for fairly bright objects, they can produce good results as witnessed by the work of Leo Henzl, Ron Potter, Bill Iburg, and others. In each of these cases, the optical-tube assembly was used in conjunction with a well-built mounting equipped with a high-quality clock drive which makes the task of guiding long exposures a great deal easier.

The optical quality of commercial catadioptric systems should be carefully checked using whatever test means are available as some of the mass-produced commercial units are incapable of producing sharply focused star images even under the very best seeing conditions. If the optics of the system are better than $\frac{1}{4}$ wave, yet still produces soft fuzzy star images, the main mirror (which is shifted manually for focusing) is probably shifting during the exposures. The problem of 'mirror flop' is often present in these systems and while an off-axis guider can help avoid trailed star images, one cannot continually adjust the focus during a long exposure. The fact that the mirrors in many of the commercial systems do shift during long exposures makes the use of an off-axis guiding system mandatory.

Our general philosophy is to use instruments for the purpose for which they were specifically designed, and not to attempt to use instruments for a task for which they are unsuitable; there is no such thing as a general-purpose telescope. These lightweight compact catadioptric systems were designed to be inexpensive and portable for the casual observer. Although usable, they are not well suited for the demanding requirements needed for high-resolution astrophotography and one should not expect them to perform as well as a massive, rigid system with superb optical and mechanical components designed specifically for this demanding work.

When considering objects for photography, one needs to realize that negative enlargements greater than about $12\times$ produce unsatisfactory prints, regardless of the film being used. If the film is grainy, like 103a-emulsions, the grain on prints from greater enlargements will be totally unsatisfactory. If films like Kodak 2415 Technical Pan are used, the grain will be quite satisfactory on such prints, but the faint star images will begin to grow in size on greater enlargements and the prints will take on the appearance of an out-of-focus photograph.

Using this $12\times$ enlargement as one criterion, and combining this with the desired image size on the $12\times$ print, one can deduce that the smallest usable images of non-stellar objects will be in the range of 1.5–2 mm on the film, assuming that the guiding and focus are absolutely perfect. With objects like globular clusters, where the image is composed of numerous stellar images, longer-focal-length systems are required to resolve the densely packed stars. When shooting globulars, a focal length of 48 inches can be considered a minimum requirement for shooting some of the larger and less densely packed clusters. Using the image-scale formula in Appendix C, we can calculate that the size of the smallest objects to be photographed (S) with a system of any particular focal length (f) should be:

$$S = 202.5/F(\text{inches}) \text{ arc min (for a 2 mm diameter image).} \qquad \textbf{(11.1)}$$

Another technique that is useful in determining the size of an object in the camera field involves the use of overlays in conjunction with large-scale atlases. The overlay has a rectangle drawn on it that corresponds to the size of the photographic field of the telescope. This overlay can then be used to see the size of an object as it would appear if photographed, or it can be used to find the best possible position and orientation for the photographic field prior to making the actual exposure. One obvious advantage of using this system is that one can check the exact positioning of the telescope when trying to photograph objects that are below the limit of visibility.

Determining exposures for various objects is a rather vague art which one develops after lots of experience. Assuming that an f/6 system is being used with a film with a speed similar to that of 103a-F, we can provide a few general guidelines which should be useful. Another technique for obtaining exposure information is to write down exposure data from photographs published in various magazines and journals. This can be a valuable habit to get into; not only can one gather exposure information, but one can also discover new objects to photograph. The objects included in the catalogue in Appendix A include an interesting variety that are seldom photographed by amateurs, while leaving out a great number of the objects that are widely known to all.

Getting back to the topic of exposure guidelines, we will discuss a number of general 'rules of thumb' based on the following assumptions: (*a*) the f/ratio of the system is f/6, and (*b*) the speed of the B&W film in use is close to that of 103a-F (when processed in MWP-2). When exposures for color films are mentioned, we are assuming that a fast unhypered color film is being used (remember that hydrogen-hypered color film can reach the speed of 103a-F). Given these assumptions, the reader can make exposure adjustments for the particular system in use and obtain exposure values which should yield usable photographs. One of the most basic 'rules of thumb' is that you can photograph all of the details that you can see in about 15 minutes with a fast B&W film or in about 30 min with fast unhypered color film. The exposures given in Table 11.3 for various object types are for the average objects in that grouping, and are not universal exposures for all the objects of that type.

In the case of most globular clusters and some planetary nebulae and diffuse nebulae, the exposures required to obtain sufficient faint details can cause the details in the brighter regions to become severely over-exposed when high-contrast films are used. In these situations, the use of a low-contrast film or the use of a special developer can provide a dramatic improvement of the results. The use of a film like hypered HP-5 or Tri-X on globular clusters produces superb results. However, in the case of H-alpha nebulae, the 656.3 nm sensitivity of HP-5 is greatly reduced and the sensitivity of Tri-X is even lower. In this situation, the use of a special long-scale developer like POTA on a fine-grained film like 2415 Technical Pan should be considered.

The basic requirements for excellent-quality photographs with moderate-to-long focal-length systems include the topics below which are covered extensively in the chapters listed:

1 excellent seeing (Chapter 12);
2 excellent focus (Chapter 3);
3 excellent guiding (Chapter 3).

Comet photography

Among the most spectacular astrophotographs seen in the journals and magazines are those of comets. Faint comets that are within the reach of amateur instruments are usually present during 2–4 months of every year. However, those comets that we all dream of, the 'great comets', may only appear once every two or three decades. The last great comet to be seen was Comet West in 1976, before that, Comet Bennett made a very nice show. Halley's Comet returned in 1986 for one of its less memorable passages through the inner solar system, but it still was an interesting object for amateurs and professionals alike.

The anatomy of a comet has been subdivided into several distinct structures by astronomers: the nucleus, coma, and tail. In a small faint comet, the nucleus and coma may be the only structures present. However, as comets approach the sun, a tail structure usually develops which generally points away from the sun. The tail is generally the most

Table 11.3. *General exposure guidelines for deep-sky photography*[a]

	Exposure times (min)	
Object Class	Bright	Faint
Open clusters		
Black-and-white[b]	20	
Color[c]	40	
Globular clusters		
Black-and-white	20	
Color	40	
Galaxies		
Black-and-white	30–40	40–90
Color	60–90	90–120
Diffuse nebulae		
Black-and-white	20	60
Color	40–60	90–120[b]
Planetary nebulae		
Black-and-white	10–20	40–60
Color	15–30	40–90[b]

[a] Assuming f/6 system.
[b] 103a-F-type emulsion.
[c] Unhypered 'fast' color film (reduce by ~2 for hypered color).

spectacular feature and may be very short or may spread brilliantly across 90 deg or more of the night sky. This structure is often perturbed by the irregularities of the solar wind and its shape may change rapidly in just a few hours. The tail which is usually seen is composed of 'dust' that is being shed from the 'dirty snowball' that is the nucleus, this is known as the type I tail. Another structure which may develop is a second tail composed primarily of gasses which are being sublimated from the nucleus. This is referred to as a type II tail. Finally, a third variation in the tail may occasionally develop which is known as an anti-tail. When present, this is usually seen as a small spike pointing opposite the main tail. The comet's nucleus may, or may not, be visible depending on the brightness and extent of the coma. The coma is a diffuse cloud surrounding the nucleus and it may show some very interesting structures from time to time. Each of these structures can be photographed and studied using amateur size instruments.

Since the comet's tail is so very profoundly affected by the variations in the solar wind, the changes that occur in that region are very interesting to study. Amateurs may also provide valuable information in such time studies, as time on large instruments is generally not available for the entire period that a comet can be studied from the Earth. If the comet is large and bright, photographs with lenses ranging from wide-angle through moderate telephoto can be made using high-speed emulsions to study changes in the comet's tail, while relatively short exposures taken through telescopes with focal lengths of 24 inches and up can reveal short-term changes in the comet's coma. Color photographs of these spectacular voyagers can be most interesting.

The brightness range that is encountered in a comet is extreme and exposures must be chosen to suit the type of photograph desired. The nucleus, coma and inner tail regions are quite bright compared with the rest of the tail. Exposures showing the outer reaches of the tail will result in a grossly over-exposed coma, while photographs showing details in the coma will very seldom show any trace of the comet's tail. Another factor that must be considered when a type II tail is present is the fact that a type II tail is blue in color while the remainder of the comet is generally neutral.

When photographing large comets, the angular span of the comet must be estimated in order to select a lens that will allow the entirety of the tail to be contained on the film. The angular field of various lenses can be calculated (see Appendix C), and Table 11.1 summarizes the field size of a number of common-focal-length lenses. The general principles of wide-angle photography apply to this task (these are discussed at length, see pp. 232–7). A fast lens is generally desirable for this work, but the lens should be stopped down enough to eliminate aberrations at the edges of the field. Depending on the lens quality, f/ratios from f/2.8 to f/5.6 may be required. A problem that is frequently faced by the comet photographer is that of starting or continuing exposures in skies that are normally considered too bright for astronomical work. Experience alone will provide insight into this problem but one should generally avoid having shutters open unless major details are visible to the unaided eye or through the main instrument in the case of fainter comets.

The problem that faces anyone attempting to photograph a comet is the fact that the comet is continuously in motion relative to the background stars. When photographing with 'normal' lenses (focal length less than 200 mm), it is almost always acceptable to guide on a star and to ignore the motion of the comet during exposures of 10–20 min. In some cases, the comet's motion can be so rapid that even such short exposures with 50 mm lenses will show the motion of the comet and it becomes necessary to guide on the comet. Also, as longer focal lengths are used, the comet's movement becomes a serious matter and it is virtually always necessary to guide on the comet itself. It is best to reduce the guiding magnification to the minimum that is practical, which is around twice the focal length (in inches) of the camera being used (i.e. if the camera has a 36 inch f.l., the minimum guiding magnification (M) would be $M=2F=72$ power). If the comet's nucleus is visible, the guidescope should be centered on it and this will then be the 'guidestar'. In many cases, no well-defined nucleus is visible and it is necessary to use one's imagination to invent one! Actually, the head of the comet (nucleus + coma) is centered in the guidescope and one keeps the crosshair centered on this large diffuse blob. When this approach is necessary, it is important not to 'overguide' and to make as few corrections as is possible. It is also important to avoid staring at the image in the guidescope for prolonged periods as this tends to dull the visual definition and makes it extremely difficult to judge the tracking quality correctly.

One problem which may arise when photographing a fast-moving comet is unique to comet photography: you may be unable to keep up with the comet's motion using many conventional variable-frequency drive controls. With comets which have very close approaches to the earth (particularly those which become circumpolar) it may be necessary to run the drive at speeds of 0.5–1.5 times the normal siderial rate (i.e. 30–90 Hz) which is not within the range of many of the common drive control units. These periods of extremely rapid motion are generally short, and will usually last for less than a day or two. It is possible to avoid this problem by not photographing during times when the comet's motion is so extreme but this may also be the most spectacular period for photography. An alternative, and much more desirable, solution is to construct or purchase a drive control which is equipped with the option of using such extreme drive rates. Another alternative is to equip the drive with a universal gear and a special clutch which will allow one to override the drive manually by the twist of a knob.

If you designed a telescope that uses an off-axis guiding device instead of a separate guidescope, you may have just realized one of the biggest problems that faces users of these fine accessories: you cannot photograph a comet through an instrument which has an off-axis guider and no guidescope (at least not if you want to have the nucleus in the photograph)! If you have chosen to use an off-axis guider as a means of avoiding the various problems associated with differential flexure, you will be quite disappointed when the next 'big' comet comes by. Even if you have a separate guidescope on your instrument, you will have to face the problems of eliminating flexure unless you just count on having many of your comet photos ruined. If you use an off-axis guider, you still need a separate guide telescope and you need to eliminate as much flexure from the entire system as is possible in order to use it for this sort of special occasion.

One aspect of comet and asteroid photography that is distinctive of these fields (at least in the amateur realm) is the topic of plate reduction. This involves the measurement of the positions of stars and the comet (or asteroid) in order to determine the object's exact position in the sky at the time of the photograph. These measurements are then sent to a central clearing station which correlates all such data and uses them to determine the orbit of the object. Exposures designed for plate reduction need to be fairly short owing to the motion of the comet or asteroid and the measurements from the film or plate need to be done with great care. In this situation, measurements from plates are superior to measurements from film owing to the added dimensional stability of the glass plate. However, measurements of very good quality can be made from film and this should not discourage anyone from entering this field. One needs to use a pocket calculator (slide rules and trigonometric tables also work, but calculators and computers are nicer) or a computer, and one also needs to have access to a 'measuring engine' to measure the plates. A measuring engine can be built if one has access to a lathe and mill and is a very good machinist or, alternatively, one can be located at a local college and used there with the approval of the astronomy department.

The process of measuring the plate is not exactly simple, but is not excessively difficult either. First, one needs to measure the coordinates of three or more stars on the plate whose exact positions can be looked up in a source like the *Smithsonian Astrophysical Observatory Star Catalogue* (SAO). Then, the star positions need to be corrected for proper motion (also listed in the SAO Catalogue) and a system of equations is set up and solved for the constants needed for determining the right ascension (RA) and declination (Dec.) of the body with the unknown position. Finally, these constants are 'plugged into' a set of equations with the x and y coordinates of the body, as measured from the plate, and the equations are solved for the actual RA and Dec. of the body. A summary of the simplest systems appears in the September 1982 issue of *Sky and Telescope* which also contains an article on the construction of a measuring engine.

Photography with filters

Since the mid-1970s, the use of filters in astrophotography by amateurs has increased markedly. The gross increase in light pollution from major cities is responsible for much of the increased popularity of filters. The introduction of special interference filters which block much of the light from the various types of street lights has also been responsible for this popularity surge. In addition, the need to use red cut-off filters with the commercial Schmidt cameras has introduced a large number of amateurs to the use of filters for astrophotography. The increase of light pollution throughout the USA will apparently continue until the population discovers that having lights on all night does not buy security and safety. As this plague grows, the need for filters will also increase. In the near future we will probably see the introduction of more special-purpose filters including some very efficient bandpass filters to the amateur market.

There are a number of reasons for using filters in astronomical photography and we shall present a brief discussion of some of the fundamentals for using filters for differing purposes. The motives for using filters can be listed as follows:

1 increasing visibility to H-alpha (Hα) nebular features;
2 light-pollution reduction;
3 eliminating UV for color photography;
4 decreasing visibility of lens aberrations;
5 color correction (discussed in Chapter 7);
6 tri-color astrophotography (discussed in depth in Chapter 7).

Nebular photography

Photographing faint H-alpha (Hα) emission nebulae is made considerably more difficult by the growing presence of light pollution as most of these nebulae are quite faint. In order to reduce the sky brightness and accentuate the fainter nebular filaments, the sky brightness must be reduced while allowing much of the light from the nebula to pass freely through the filter. Fortunately, for amateur and professional astronomers, the street-light manufacturers have miraculously left this region of the spectrum relatively untouched as they have developed brighter and cheaper sources of light pollution. The vast majority of light from these nebulae is radiated in just a few very narrow bands of light of hydrogen with the majority of this light concentrated in the deep-red band at 656.3 nm. The reader should note that the blue-reflection nebulae, such as the Pleiades nebulosity, do not radiate in this manner and that the techniques described below will not be applicable to them.

Thus, deep-red filters are used for photographing Hα nebulae, and a film must be selected which has high sensitivity at the 656.3 nm wavelength. Unfortunately, most commercial films, including Tri-X and Plus-X, have virtually no sensitivity in this region (the fact that hydrogen nebulae can be recorded on these films is due to the presence of hydrogen emission lines in the blue and violet regions of the spectrum and to the fact that there are lines from a number of other elements scattered throughout the visible and UV spectrum, see Fig. 11.2). The filters can be chosen from several sources, they can be made of glass or gelatin, but they should be of the best optical quality.

Most of the high-transmission filters that are readily available for this work are known as cut-off filters. These filters block all of the light below a particular wavelength and then pass most of the light (up to 90 %) from longer-wavelength radiation. Among the common gelatin filters that are available are Kodak's Wratten filters No. 23A, No. 25, and No. 29. Each of these filters leave the 656.3 nm line almost intact (~90 % transmission) while blocking lower-wavelength radiation. The cut-off wavelengths (the transmission at this point drops below 50 %) and the Hα transmission values for these three filters are:

Wratten filter no.	Cut-off wavelength (nm)	Transmission @ H-alpha (%)
23A	582	89.5
25	600	88.3
29	622	90.8

Since the Hα transmission of each of these three filters is within 3 % of any of the other two, there would seem to be little reason to choose one filter over another. In fact, since the transmission of the No. 29 filter has the highest transmission, it would seem to be the best choice of the three. In addition to its slightly higher transmission rate, the No. 29 filter offers another advantage which is not apparent until the light curves from various common street-lights and the natural skyglow curve for the

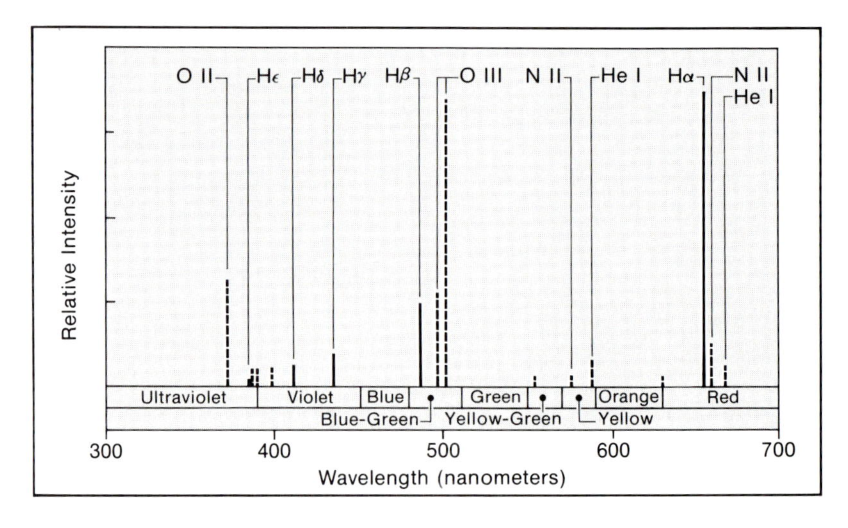

Fig. 11.2. The spectra of a typical emission nebula is shown with spectral lines identified. The regions of the spectra corresponding to commonly used color names are labeled in accordance with accepted color standards. The nebula used for this example is the Trapezium region of M42. Other nebulae spectra vary widely from this example; the lines that generally vary in intensity from nebula to nebula are shown with dashed lines.

atmosphere are examined. The outcome of such a study (see Figs 11.3, 11.4 & 11.5) is the discovery that the No. 29 filter is cutting out more light pollution (primarily from high-pressure sodium lights) than either of the other filters. The net result of choosing a No. 29 filter over either of the others discussed is a photograph showing the same amount of nebulosity but showing less sky fog density on equal exposures.

Many will shy away from using the Wratten No. 29 filter because of the amount of light it knocks out when used on a normal scene. However, if a room is illuminated with red light and then examined with the No. 29 filter, very little light will be blocked by the filter. If one is interested in photographing Hα nebulosity, and these cut-off filters represent the available choices, the No. 29 filter is to be preferred over either of the other filters (the application of the No. 29 filter to galaxy photography, or Tri-color work is a totally different issue and these conclusions are not to be generalized).

The other alternative to the gelatin cut-off filters discussed above is the glass cut-off filter with an interference film coating. Until 1982, such coated cut-off filters were not generally available to amateurs at a reasonable cost. At that time, Lumicon introduced an Hα pass filter with a 90 %

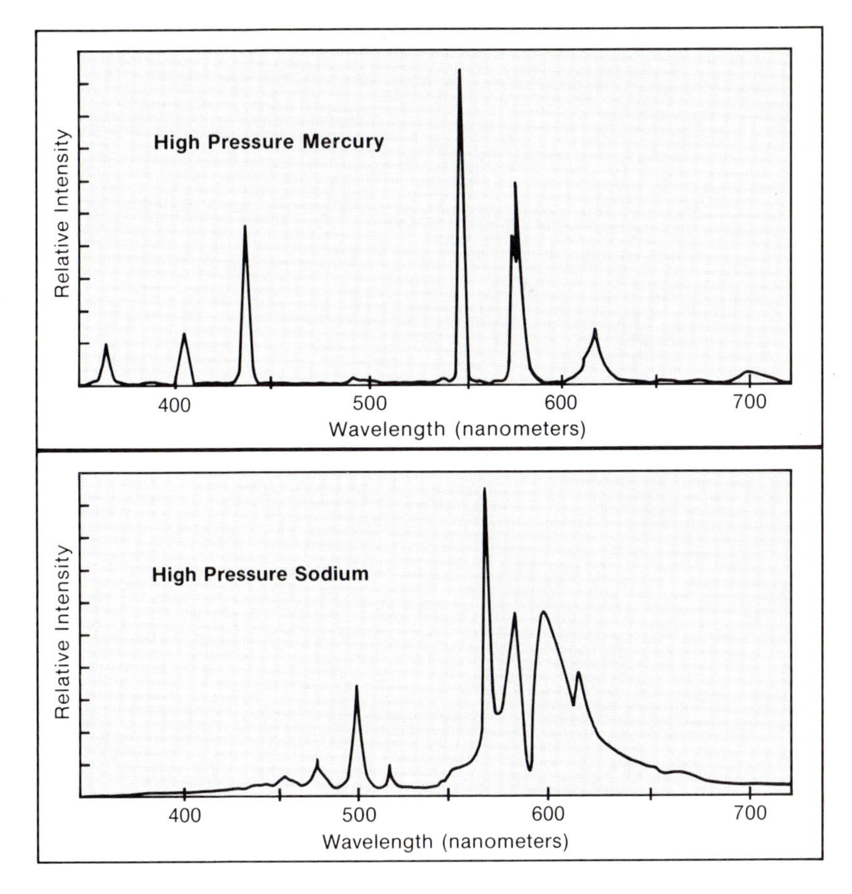

Fig. 11.3. The spectra of high-pressure mercury and sodium lights are shown. The spectra were recorded by a low-resolution scanning spectrometer and thus the singular spectral lines appear as broadened peaks. In the case of the high-pressure sodium light, the discrete spectral lines are superimposed upon a broad continuum in the 550–650 nm range (spectra courtesy of Jack Marling).

transmission factor for a very reasonable price. This filter has a cut-off wavelength of ~640 nm and thus rejects even more light pollution than the Wratten No. 29 filter. Special filters like this can be ordered (or even specially designed) from any number of sources such as Oriel, Corion and Andover.

When using any of the above filters for Hα nebula photography, it will be found that the exposure will need to be increased by at least 50 % in order to record the same nebular density. This is due to the fact that these filters reject the light from a number of other nebular lines which would otherwise contribute to the formation of the image. This 'filter factor' has been reported by Dr J. Marling of Lumicon and agrees well with the results obtained by the present authors. This same filter factor will apply to any of the filters discussed above or any others with a 90 % transmission factor at the 656.3 nm wavelength. One of the major advantages of using these filters for Hα photography is that exposures can be extended many times beyond the normal sky fog limits to record nebulosity down to the true limiting capability of the camera in use.

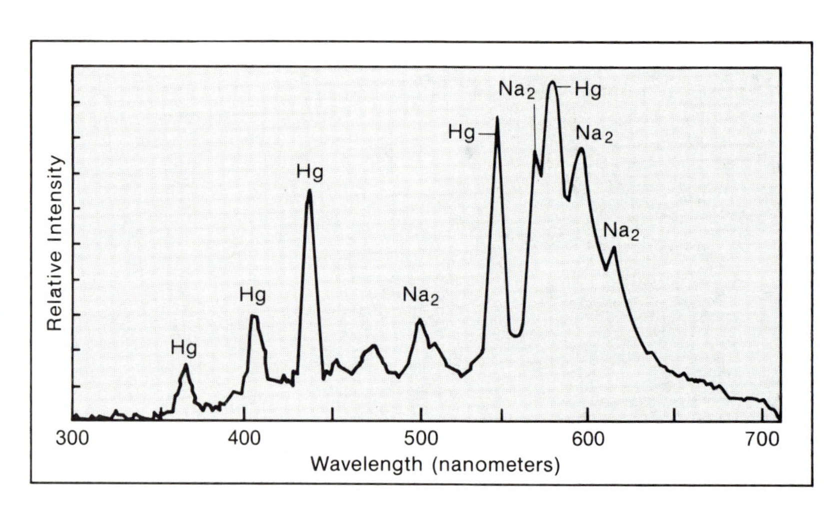

Fig. 11.4. The spectra of the night sky in Livermore, California. The urban 'sky light' is composed largely of light from artificial sources (mercury, sodium and tungsten lights) with an extremely small component caused by natural sky glow sources. The spectra was recorded by a low-resolution scanning spectrometer and thus singular spectral lines appear as broadened peaks. This effect causes the discrete spectral lines to be superimposed upon a broad band of continuum light (spectra courtesy of Jack Marling).

Relative Intensity
Hg
Hg
Hg
Na2
Hg
Na2
Hg
Na2
Na2
300
400
500
600
700
Wavelength (nanometers)

Fig. 11.5. The curves of four commonly used 'red' transmission filters are shown to illustrate how the cut-off wavelength approaches the 656.3 nm hydrogen line as one moves from one filter to another. The transmission of all filters at 656.3 nm is within 4 %.

Relative Transmission
Kodak Wratten Filters:
23A
25
29
92
400
500 525 550 575 600 625 650 675 700
Wavelength (nanometers)

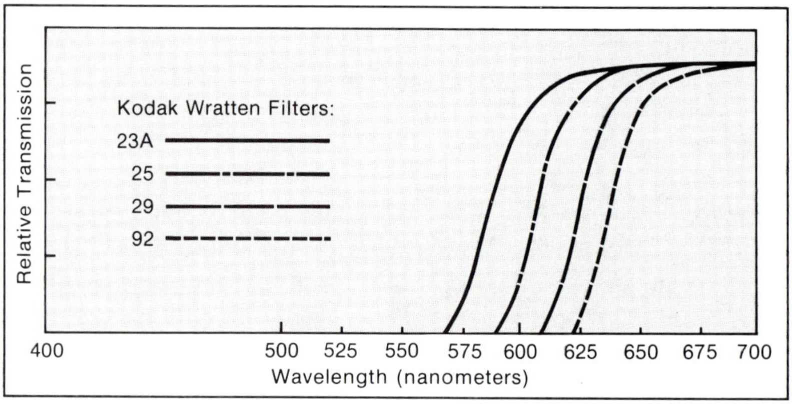

Light-pollution reduction

In 1978 a new telescope accessory appeared on the market which has steadily grown in popularity since its introduction. The 'nebular filter', as it was called, was first conceived and designed by Del Woods. In its earliest form, it was designed strictly for the enhancement of nebular details through the elimination of a large portion of the atmospheric light pollution produced by emission-line sources like the mercury and sodium vapor lights. The nebular filter was, and is, an important addition to the accessories available to the amateur. Since its introduction, a variety of filters have been introduced for differing purposes from Hα and H-beta (Hβ) filters, to those which eliminate much of the light pollution while passing broad bands of the spectrum for the observation and photography of galaxies. With such filters, the limiting exposure times in moderately light-polluted areas can be extended by factors of 3–6, allowing useful work to be carried out in areas that would otherwise be useless for astrophotography. Many of these special purpose filters have been developed and marketed by DayStar and Lumicon.

The nebular filter and its related variations are of a fundamentally different design than the common dye filters like the Wratten No. 29. These filters are constructed by depositing many extremely thin layers of a semi-transparent film. As the light crosses a boundary between adjacent layers, part of it is reflected back. The nature of these interface reflections and the nature of the multilayer coating result in a multitude of reflections by trapped light. As these rays are reflected within the multilayer coating, they set up interference patterns which cause the light of certain wavelengths to be eliminated from that which is transmitted by the coating. For this reason, they are known as interference filters. By judiciously choosing the spacing of the reflecting layers, these filters can be designed to pass light of differing wavelengths. The light passed by the filter can be confined to a very narrow band of the spectrum less (than 1 nm in width), or it can be spread out over a very wide band of the spectrum. The nebular filters which have been designed thus far generally combine some medium-bandpass features with a broad-bandpass feature in order to transmit as much of the spectrum as they can while still eliminating much of the heavily light-polluted region between ~535 nm and ~630 nm (Fig. 11.6).

Interference filters are designed for use in a beam of parallel light with all of the light rays striking the surface of the filter perpendicularly, as a change in the angle of incidence results in a change in the wavelength that is passed by the filter. In the case of a narrow-bandpass filter (<20 nm) in a focused beam of light, the transmission peak can be shifted away from the emission line being studied as one moves away from the optical axis across the photographic plate. With a medium- or broad-bandpass filter, this problem is not as serious and it is feasible to use such filters in the beam of focused light (provided that the f/ratio is not too small). In this situation, the peak transmission wavelength will be shifted to a lower wavelength than desired, and the transmission at the wavelength of interest will be reduced (Fig. 11.7). The use of narrow- or medium-band interference filters in front of a normal or wide-angle lens should be avoided because of the wavelength shifts which will occur away from the center of the field (light from off-axis stars strikes the filter at an angle thus encountering a filter with a shifted bandpass peak). As with all glass filters, when placed in the focused beam of light, the focus will be displaced, and thus focusing must be carried out with the filter in place.

Eliminating UV

When photographing Hα nebulosity in color (using color film or the tri-color process), the actual color of the nebulosity can be adversely affected unless the UV light below 400 nm is blocked from reaching the film. This is due to the presence of an OII emission line at 372.6 nm which is generally present in such nebulae and which can be a third of the intensity of the Hα-emission line. This UV light is recorded on the blue layer of the color film. The addition of this blue color to the red caused by the Hα emission line dilutes the dark red color of the Hα and turns it to a color which ranges from magenta to a pale pink. When photographing in color and seeking accurate color rendition, this line must be removed from the light reaching the film. This can be done with a UV cut-off filter like Kodak's Wratten filters No. 2B or No. 2C.

Decreasing lens aberrations

Many lenses, old and new, are corrected for light from the visible spectrum only. When light outside of the visible passes through the lens, it is not brought to a sharp focus as the rest of the light is. Two factors then cause the UV light to form blue halos surrounding many of the brighter stars in the field. The first is that the optical system itself is responsible for dispersing the UV light more than it does the longer wavelengths, and the

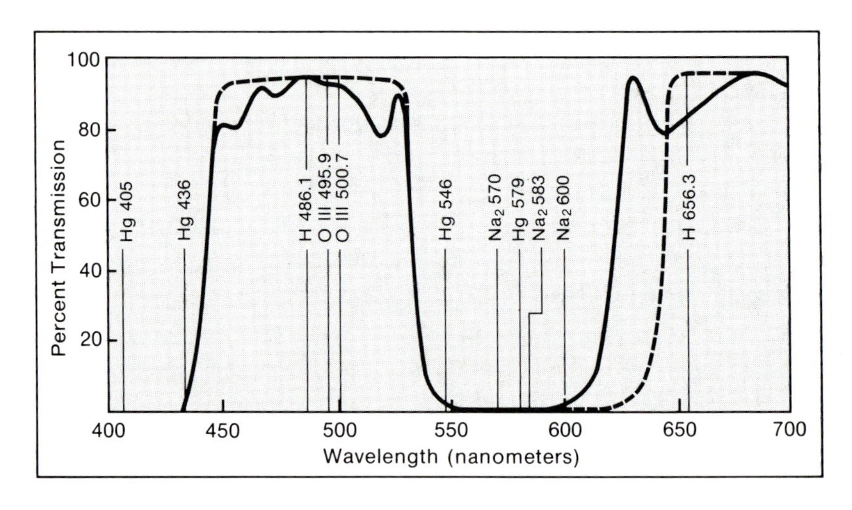

Fig. 11.6. The transmission curve of an 'ideal' nebula filter is shown (dashed line) along with that of an actual nebula filter (solid line). In addition, prominent lines of hydrogen nebulae and light pollution sources are shown.

second is that the photographic emulsion is very sensitive to UV light while being totally insensitive to IR (infrared) radiation. Thus, the addition of a UV cut-off filter serves a two-fold purpose: eliminating UV-emission lines that affect the color rendition of Hα nebulosity and eliminating UV light from stars which would otherwise contribute to the formation of blue halos about many of the brighter stars in the field.

For many of the newer lenses which form halos around stars when used for astrophotography, a UV cut-off filter will create a dramatic improvement in the quality of the stellar images. However, when many of the older lenses (especially aerial-camera lenses) are used for astrophotography, it will be found that the UV cut-off filter does not eliminate the halos around the stars. In some cases, it will be found that a yellow filter like the Kodak Wratten filter No. 12 will be required to eliminate the halos, while other lenses may not be corrected even when a red filter is used. If the use of yellow or red filters does not improve the quality of the images formed by a given lens, the lens should be stopped down and the filter tests run again. With old and new lenses alike, there is always a possibility that the lens will not function well when used for astrophotography. This possibility is a significant consideration when older lenses, including Second World War aerial lenses, are applied to this demanding photographic task.

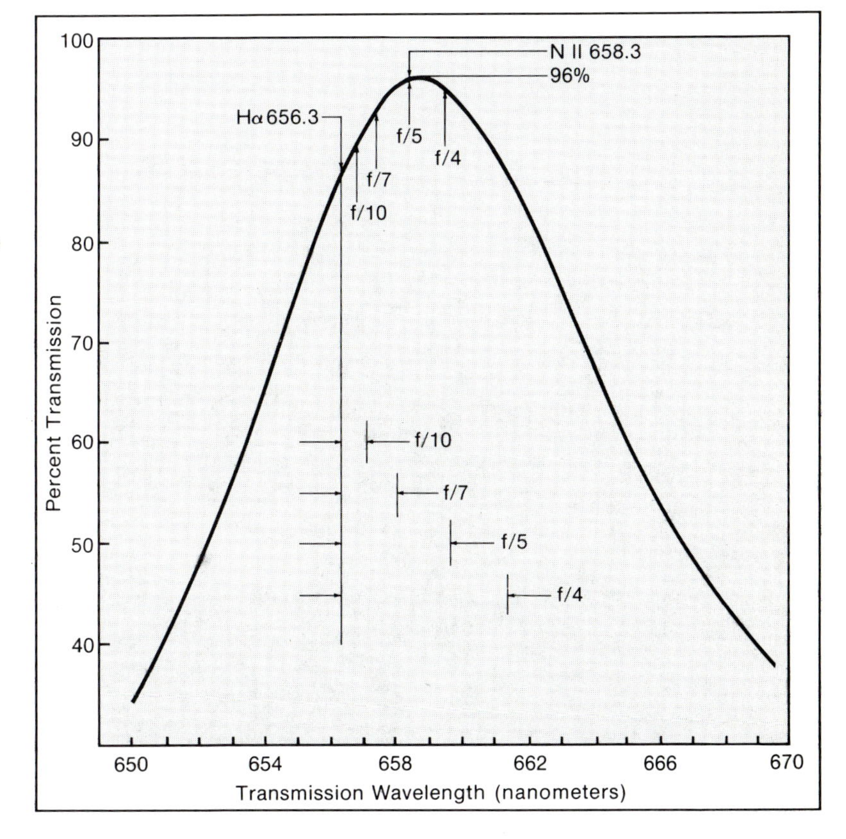

Fig. 11.7. The transmission curve of a narrow-band interference filter is shown and the effects of using different f/ratio systems are shown for both the peak transmission position and the 50 % transmission point. As the f/ratio of the system is decreased, the peak transmission point would shift to wavelengths lower than the 658.3 NII line. At an f/ratio of f/5.0, the peak transmission point would lie directly over the 656.3 nm hydrogen line.

Care in using filters

Using filters in astrophotography introduces a new set of variables which must be recognized in order to avoid discovering some rather annoying problems the hard way. The filters which are the easiest and least expensive to use are the Kodak Wratten filters. These filters are of optical quality and can be used in front of the lens or in the optical path between the lens and the film plane. However, these filters are very delicate and are easily warped by moisture in the air. When these filters become warped, it is false economy to continue to use them. They should immediately be discarded and replaced with new filters. These filters are also easily scratched, but a small number of surface scratches will not harm the image unless the filter is directly in contact with the film. To minimize scratching these filters, they should be handled with great care and always stored in the tissue and foil liners in which they are sold.

When these delicate filters are used with some of the film holders made for the commercial Schmidt cameras that have been introduced, the filter can either be attached to the front of the film holder, or slipped into the film holder with the film. Attaching the filter to the front of the film holder is the best solution as it minimizes the danger of scratching and allows the camera to be used without the need of refocusing (the design of many of these film holders is such that the introduction of a filter in the film holder in front of the film will shift the film surface out from the focal plane by the thickness of the filter; at these short focal ratios, such a focus shift will be noticeable). If the camera is refocused for use with the filters in this position, the Schmidt camera cannot be used afterwards *without* a filter.

If glass filters are used in the beam of converging focused light, an additional factor must be considered. These filters will cause the focus to be shifted away from the lens by a small, but detectable, amount. It is therefore necessary to focus with the filter in position. With the dense red filters, like the No. 25 or No. 29 filters, it will be almost impossible to focus in this configuration unless the knife-edge method is used, owing to the attenuation factors on continuum sources like stars. The obvious solution to this problem is to substitute a clear filter of the same optical thickness for the purpose of focusing. The additional problems encountered when glass interference-type filters are used in the focused beam of light have been discussed previously (pp. 253–54).

When glass filters (or optical windows) are used in the focused beam of light, there is another possible problem which is often not expected. As the light passes through the filter, a small fraction of that light is reflected back to the front surface of the filter and then reflected back through the filter to the focal plane. This doubly reflected light is brought to a focus in front of the normal focal point at a distance equal to twice the filter thickness. If the filter is too thin, this halo of light can be intense enough to record around the brighter star images. For this reason, it is best to select filters which are several millimeters or more thick. Another solution is to use filters which have received anti-reflection coatings on both surfaces.

12

High-resolution photography

Hope for the best,
expect the worst.
Jules Verne

The title of this chapter is probably somewhat deceiving as the areas in the previous chapter on deep-sky astrophotography should be considered as high-resolution work; the sections in this chapter that are devoted to lunar and solar eclipse photography, while being subtopics of lunar and solar photography, are generally less demanding than most other areas in this field. We have placed in this chapter a number of topics that require very strict adherence to the requirement for excellent seeing and, while we have stressed that excellent seeing is needed in order to obtain the best possible results from virtually any instrument, we have reserved the discussion of seeing, including theory and applications, for this chapter alone.

While we have already discussed narrow-field astrophotography in the previous chapter, we do not want any of our readers to have the misconception that excellent seeing is not needed to get maximum results when taking long-exposure deep-sky photographs using a telescope. From the discussion on resolution in Chapter 4, you will remember that there are three parameters which can limit the resolution of the telescope: seeing, aperture, and focal length (or film resolution). When photographing in extremely poor seeing (3–6 arc sec), the seeing will impair performance of telescopes as small as 6 inches aperture and focal lengths as short as 16 inches! However, in 1 arc sec seeing performance is not restricted unless the aperture is greater than 24.8 inches or the focal length extends beyond about 130 inches. Again, we will stress that seeing and focus are the two most critical factors when trying to obtain the best possible resolution from a given system. Learn what the term 'seeing' really means, learn how to measure accurately the seeing, then search for sites that tend to have good to excellent seeing much of the time.

Seeing

The quality of 'seeing' is one of the most important elements in high-resolution photography and is considerably more important than most realize in deep-sky work where resolution would appear to be less critical. We are familiar with the theoretical appearance of a highly magnified perfect stellar diffraction pattern as it should be seen through a telescope. Unfortunately for most, this familiarity is more the result of exposure to published illustrations than from direct telescopic observation. Aside from poor optics, the primary reason for this situation is the condition of the atmosphere which we call seeing. Seeing is a term used to describe the irregular distortions and motions of telescopic images that are caused by random changes in the density of portions of the Earth's atmosphere. The relative importance of these two components of seeing (distortion and motion) are closely tied to telescope aperture and the size or the density of seeing cells in the Earth's atmosphere.

As light from a star passes through a moving atmosphere of non-uniform density, an initially plane wave of light will become distorted. The stellar diffraction pattern when observed telescopically under these conditions becomes blurred, loses contrast and often undergoes rapid changes in its internal structure. Under certain conditions, the image itself, although it may be well defined, will be seen to move about erratically, jumping and bouncing about randomly many times a second. Under very poor conditions these motions can amount to as much as 5 arc sec in any given direction. While image distortion is common to telescopes of all apertures and increases directly with aperture, image motion is observed only in small apertures and decreases with increasing telescope size.

In an effort to understand the relationship between telescope size and associated seeing effects, a brief description of the appearance of the diffraction pattern using different apertures under varying seeing conditions will be useful (see Fig. 12.1). With small telescopes (apertures of 6 inches or less) nearly perfect diffraction patterns can often be observed. As seeing conditions deteriorate, the diffraction rings will begin to pulsate, rotate, lose contrast and can blur into the central disk (the Airy disk) becoming virtually invisible. Under extremely poor conditions, the star image may enlarge and become nebulous without any discernible structure. In addition, an image in a small telescope may exhibit pronounced motion with or without distortion of the diffraction pattern. Indeed, at times, the diffraction pattern may be virtually undistinguishable or may be nearly perfect while moving about wildly in a circle some 8–10 arc sec in diameter! With moderate aperture instruments (on up to about 60 inches) the Airy disk and rings are almost never seen. When seeing is good, the image will appear as an amorphous disk with a small central concentration several tenths of an arc second in diameter. Poor seeing can cause the amorphous seeing disk to expand to 10–20 arc sec and, on rare occasions, it can grow to as much as 1 arc min in diameter. Under these conditions the seeing disk can often be seen to pulsate and, at times, it may break up into a

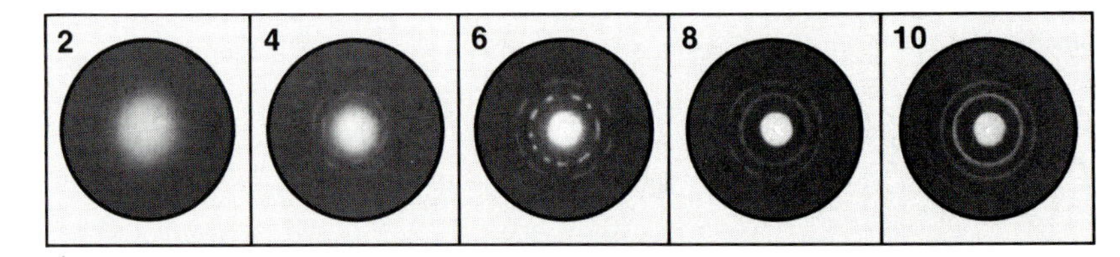

Fig. 12.1. This illustrates the appearance of highly magnified star images under varying seeing conditions. The instrument is a 5-inch refractor used at approximately 300–400 power. The number above each image corresponds to the seeing rating from Table 12.1. Diagram adapted from *Astronomy* (July 1977, p. 76) by permission of the publisher.

number of segments with rapid changes taking place in the internal structure. Using larger-aperture instruments, the diffraction disk displays less internal motion but the pulsating aspect of the diffraction disk becomes increasingly noticeable as the aperture increases.

To account for these aperture-related seeing effects, it is necessary to look at a simplified model of optically effective atmospheric turbulence.* In order for telescopic imaging quality to be unaffected, incoming star light must pass through a homogeneously stratified atmosphere (i.e. where temperature, pressure and density of the air are functions purely of elevation above the ground, at least through the troposphere). This is, in fact, what meteorologists call a 'standard atmosphere', an idealized and artificial situation. Turbulence in the atmosphere disturbs homogeneous stratification; however, for turbulence to be optically effective (i.e. able to modify the incoming wavefront), it must result primarily from variations in atmospheric density. For practical purposes, a simplified model is generally assumed where the atmospheric density variations exist in the form of spherical cells of uniform internal density which differs from that of the surrounding air. Since these variations in density translate into variations of the refractive index of the air, it can readily be seen that the cells will act as lenses that can defocus light and deform the wavefront as it passes through the cell.

Where the telescope aperture is larger than the density cells, only part of the beam of light entering the telescope is defocused. Thus, while part of the light forms an undisturbed image, a defocused image is superimposed producing an integrated image that is enlarged, washed out, and of lower contrast. When several of these cells intrude into the light beam a higher proportion of the light is thrown out of focus (which magnifies the effect further). The reality of the situation is considerably more complex: the cells are actually extremely variable in size, shape and density. When the real cell structure is combined with rapid motions of the seeing cells, it becomes quite clear how the greatly enlarged and pulsating images of poor seeing can be created in large-aperture instruments.

When the telescope aperture is smaller than the density, or seeing, cells, the cells act more like prisms than lenses. As a prism, the cell will tend to deflect the entire beam entering the telescope and thus the image moves about rather than being defocused. Many irregular and rapidly moving cells will produce the relatively sharp and erratically moving images commonly observed with smaller apertures. In a similar fashion, the combined effects of image motion and defocusing will be observed when the mean size of the density cells is equal to the telescope aperture.

This seeing model, in addition to explaining the different effects seen using different aperture instruments, also gives some indication of the size of

* This discussion is largely based on J. Stock's Site survey for the Inter-American Observatory in Chile, *Science* (5/21/1965), p. 1054.

the turbulence structures in the atmosphere and adds some understanding of the mechanism behind the deterioration of seeing with increased zenith angle. Using this model, and assuming that it is a reasonable approximation to the real situation, one can deduce that the largest optically effective density cells are smaller than the apertures of large telescopes ($A > 60$ inches) and likewise the smallest cells must be close to about 1 or 2 inches in diameter in order to account for small-aperture-seeing effects. This model also shows that as one looks through an increasingly longer column of air (i.e. as the zenith angle increases) the randomly moving variable density cells will affect seeing in direct proportion to the square root of the length of the total air mass which the incoming wavefront must traverse.*

Atmospheric turbulence resulting from the density cells described above is caused largely by the mixing of air masses of differing temperatures. The origins of temperature variations in the air can range from the immediate microclimate of the telescope to the large-scale global air movements that give rise to the varying climates of the world. All of these effects on seeing however are confined to the troposphere (the layer of air from sea level to about 50 000 feet) since the air above lacks the density to be optically effective. The most important seeing factor is the local climatic and topographic condition and, more often than not, seeing is affected by conditions in the immediate vicinity of the observing area.

We will begin by looking at sources at or near the telescope that can cause air-temperature fluctuations and bad seeing. Sources of heat around the telescope can include electronic gear or even the body heat of the observer. More often than not, the greatest source is from night time re-radiation of heat collected by equipment exposed to the sun during the day. Even with enclosed instruments, radiation cooling of the dome or slide-off roof can destroy local seeing. To prevent these problems, efforts should be made to keep the heat of warm objects from rising in the optical path of the telescope. Particular care must be taken with open-design telescope 'tubes' and portable instruments should be shielded from the direct rays of the sun to prevent heat build-up in metal parts and other areas that conduct heat well. Telescope shelters or observatories should be finished with aluminum or white paint as both will absorb little in the visible region of the spectrum. Some titanium dioxide paints act as a black body in the infrared and can thus ruin night time seeing with strong long-wave radiation. Because of this, titanium dioxide compounds are well suited for solar telescopes while aluminum is a better choice for night time work.

After consideration of the immediate area of the telescope, the local geography must be examined since surrounding vegetation and topog-

* Some sources relate seeing effects to the secant of the zenith angle. While this works well with angles of 65 deg or less, away from the zenith, at greater angles the secant rule predicts seeing effects that are much greater than actual observations can confirm.

raphy can often play an important role in either maintaining or disrupting the quality of seeing. Just as telescope parts and enclosing structures which are exposed to the direct rays of the sun absorb and later re-radiate heat, the ground surrounding an observing site tends to absorb short-wave solar radiation during the day and re-radiate long-wave radiation at night. In order to minimize the effects of this problem, sites should be selected with enough surrounding vegetation to keep ground heating to a minimum. For similar reasons, locations that are closely surrounded by water bodies will also ameliorate local turbulence set up by ground radiation.* Another important aspect of the local environment is the effect of topography on the movement of air around the observing site.

The mixing of air masses of different temperatures is the primary cause of turbulence and, therefore, of seeing effects. This mixing is accomplished during daylight hours through convection resulting from the re-radiation of heat by the ground. However, at night, after the ground has cooled to air temperature (or below), the movement and mixing of air masses is largely induced by local temperature inversions that are greatly influenced by topography. Within the troposphere (where nearly all seeing effects are generated) the normal condition is indicated by a regular and predictable decrease in air temperature with increased altitude. However, with a temperature inversion, a layer of cold air is formed at ground level inverting the temperature gradient since warm air lies above it. Although inversions are not 'normal', they tend to be very stable since the colder, heavier air at the bottom of the inversion is very resistant to changes in vertical distribution. Despite this stability, local, relatively shallow inversions are a primary source of bad seeing since pronounced turbulence exists at the interface between the top of the inversion and the region where the local air returns to the normal temperature gradient.

Although there are a number of mechanisms that can cause temperature inversions, the most common type we have encountered results from topographically induced air movement and pooling. In the cool dry mountains of Southern California, ground temperatures drop rapidly after sundown. Most of the radiated heat collected during the day escapes quickly into the air since the clear dry air above the ground absorbs very little of the energy passing through it. As a consequence, air temperature will continue to drop during the night. Eventually, the ground temperature will fall below that of the air and will then begin to cool the air in immediate contact. With the formation of this layer of cold air over the ground, a local inversion is established. Air cooled on mountain slopes during this process, being heavier than the surrounding air, will begin to flow down the slope much as water flows down a stream. Rivers and lakes of cold air are then

* Although establishing observation towers that are 10 feet or more above the ground will greatly reduce the effects of ground radiation, such structures are impractical for portable instruments and are difficult and expensive to build owing to the stability that is required for astronomical purposes.

formed in the canyons and valleys between mountain peaks, greatly magnifying the inversion effects. Often these cold air flows are quite noticeable in the vicinity of stream channels and ravines and are a good indication of topographically controlled air movement.

We have often experienced a classic example of this mechanism at work in Lockwood Valley, located at the base of the south slope of 8900 foot Mt Pinos some 60 air miles northwest of Los Angeles, California. At one of several small, private observatories in this valley (that is also located very near a drainage channel) one will often begin a clear evening with fairly good seeing (4–6). However, as the ground cools rapidly, cold air flow coming down the slope of Mt Pinos will begin to pool in the small valley basin where the facility is located. Instruments located on the observatory grounds will almost immediately be engulfed by bad seeing (1–3) as the ambient air temperature drops rapidly. At the same time, seeing in the dome will remain good for an hour or more after this cold air flow is first detected owing to the fact that the dome is elevated some 12 feet above the ground. Seeing tests we have conducted have indicated that a regular local inversion exists at this site (and throughout much of this valley). Using a lightweight 5 inch Schmidt–Cassegrain, it was found that moving to a ridge some 50 feet above the ground and just a short distance away almost always resulted in a marked improvement of the local seeing.

Curiously, this site nearly always experiences one or two months (April–May) of relatively good seeing (6–8) almost every year. This is probably the result of gross seasonal shifts in large-scale air movements over the Southern California region. Every year at this time, the area experiences a massive influx of marine air (often 4000–7000 feet thick) pulled in from the Pacific by the development of a thermal low in the interior deserts. The influx of this moisture-laden air is probably responsible for greatly retarding the formation of local inversions since this moist air can more readily absorb the energy radiated by the ground and thus maintain a relatively higher temperature during the night.

At locations outside of mountainous regions, the formation of local inversions will depend on such factors as the type and amount of ground cover and, also, on the amount of moisture in the air (dense coverage and moist air will retard the formation of inversions). While local inversions formed at locations of relatively flat topography are usually shallow, they will still ruin the seeing once they rise above the telescope aperture. On windy nights, inversions will not form either in mountains or in the flatlands. However, wind virtually always results in bad seeing. While all of the conditions for inversion formation can still exist, especially with regard to cold air formation at the immediate surface, the mixing that accompanies wind precludes any stable stratified air mass from forming. Instead, wind tends to mix cold surface air with the warmer air above and, in the process, creates microthermal eddies that destroy the seeing.

At a few locations, conditions are such that very deep inversions are created that can actually improve seeing. The Los Angeles basin is one such area where the combination of atmospheric, topographic, and oceanic conditions produce inversions which may be several thousand feet deep. The basin can be imagined to be a bowl cut in half with its open side facing the Pacific Ocean, its base lying at sea level and the rim being the surrounding mountains with elevations ranging from 6000 to 10 000 feet. Beyond these mountains (outside of the bowl) lie deserts that, when heated during the spring and summer, produce thermal low pressure areas that tend to draw the cool air off the ocean and into the basin. As this cool air fills the basin, being trapped by the surrounding mountains, the resulting inversion will continue to deepen, sometimes reaching depths of 4000 feet. When the inversion reaches these depths, the seeing can be quite good even though the air can be murky and hazy with both moisture and pollutants.

The explanation for this phenomenon probably lies in the fact that the warm air that is being displaced upward with the deepening inversion is cooled adiabatically, reducing the temperature differences at the interface of the inversion and the air mass above it. This reduction in thermal differences will act to reduce turbulence and thus improve seeing conditions. In contrast, local shallow inversions (100–600 feet or 30–180 m) caused by radiation cooling will not, for the most part, reach sufficient depths to cool significantly, through uplift, the air mass above it. In some mountain regions however, these local inversions can reach significant depths as cold air accumulates and pools during the night. In fact, this provides an excellent explanation why many such locations (including the previously discussed site in Lockwood Valley, CA) will have improved seeing during the early morning hours just prior to sunrise.

At this point, some ideas should be forming about what conditions are more likely to produce good local seeing. First of all, there should be adequate vegetation to prevent the ground from being heated during the day (a surrounding water body serves the same purpose). Narrow valleys and shallow basins should be avoided since these channel and pool cold air during the night. The foot and slopes of mountains should also be avoided since extensive cold air movements downslope will exist here. Good candidates for excellent seeing will hence be found on top of broad-backed, gently sloping mountains of moderate elevation (5000–6000 feet) which have good vegetation coverage. Sharp peaks will, in all likelihood, have poor seeing due to rapidly ascending or descending air currents along their steep slopes. By picking a site above 5000 feet elevation, one gains the added advantage of being above most of the haze and convective turbulence so characteristic of the lower atmosphere. For those who live too far away from mountain locations, gentle rises are the best compromise and the vicinities of streams, river beds and shallow depressions should be avoided. Finally, a good site should also have a relatively low frequency of windy weather conditions.

Comparatively little has been written about the large-scale geographic factors that can influence local seeing conditions. It is known that seeing is affected by large-scale meteorological factors and can be influenced by proximity to the oceans but accounts are sketchy and are often contradictory. Nevertheless, some broad generalizations are possible and will be outlined here.

It is almost universally agreed that the passage of frontal systems produces poor seeing conditions. Cold fronts and, to a lesser extent, warm fronts are responsible for large-scale mixing of air masses of differing temperatures which, in turn, produces zones of atmospheric turbulence and poor seeing. The areas of the globe that are most commonly affected by frontal systems are the high and middle latitudes of both the northern and southern hemispheres. The equatorial and tropical latitudes are rarely influenced by frontal systems as these areas are largely dominated by the large-scale convective circulation set up by the subtropical highs and equatorial low.

There is less agreement as to the effects of proximity to the ocean on seeing. J.B. Sidgwick describes '. . . the ideal site as being a dry plateau of moderate elevation above sea level, covered with vegetation, located in a dry, clear climate, and far from valleys and the sea', and elsewhere states that '. . . (seeing) conditions in France are better away from the Atlantic coastal zone'. M.F. Walker, on the other hand, states '. . . the good seeing observed at Mt Hamilton, Mt Wilson, and Mt Palomar could result from the effect of the cold ocean current along the California coast'. In spite of these conflicting views it seems, on the whole, that the global pattern favors the marine influence as an ingredient for good seeing conditions.

While the influx of marine air onto land can create poor seeing, most notably by setting up local inversions, its greater effect is to moderate many of the temperature effects observed on land that set up atmospheric turbulence. Sidgwick's observations probably resulted from poor seeing experienced as a result of shallow inversions created by the influx of marine air over relatively flat terrain. If one can get above this inversion (e.g. by climbing a mountain) the situation can be improved considerably. Here the bad effects of the inversion are avoided with the added advantage of a stable laminar flow of air above since the cooler surface of the ocean will greatly reduce turbulence-causing convective currents. Indeed, the worldwide distribution of the larger, more recently established, observatories show most of them to be located on mountain sites that are moderately close to the ocean or other large bodies of water.

The jet stream is another factor in global patterns influencing seeing conditions. The mean path of the northern winter jet stream girdles the globe cutting a narrow path across Mexico, the southeastern USA, North Africa, northern India and southern China. During the summer its path moves northward to the US–Canada border and then meanders across northern Europe and central Asia. The primary effect of the jet stream is that it seems to establish the storm track for mid-latitude frontal movements. In addition, it also seems to be the major source of upper air thermal imbalances in lower mid-latitudes. The extremely high winds associated with the jet stream (up to 250 mph), even though it exists in the very thin air of the upper troposphere, can have a substantial effect on seeing at locations directly under its path.*

Given the global patterns outlined above, some general conclusions can be derived about where seeing conditions will be the best. Because of the movements of frontal systems and the associated turbulence in high and mid-latitudes, a position in the lower latitudes is preferred where frontal passages are less frequent. Although low latitudes have the advantage of small diurnal temperature changes, calm nights, and are largely outside the zones of cyclonic movement, the overall higher temperatures and strong convective air movements there tend to produce too much cloud cover to take advantage of any good seeing that may be encountered. An ideal site should also be located near an ocean or other large body of water on account of the reduced convective turbulence in these regions. Finally, in order to escape the bad effects of local inversions that are common to locations near water, the best sites will be found atop moderately high mountains.

Although it is recognized that most of the readers of this book do not have the option of locating their observing sites in accordance with the parameters given here, it is hoped that this information will provide a more complete picture of the factors that affect seeing. At this point there is no real understanding of the global patterns of atmosphere, topography, and ocean that make some regions better from the seeing standpoint than others. While local seeing conditions are easily tested and fairly well understood and a look at a world map produces some obvious patterns in observatory distribution, the reasons behind these broader patterns are as yet not understood. Part of the reason for this is the lack of any coordinated analysis effort over the years, but another is clearly caused by the enormity of the task. Such a concerted effort would be of value to professionals and amateurs alike for, although some astronomy is now moving into space, many of us will be here, beneath the atmosphere, for many years to come.

Seeing scales

The importance of excellent seeing for high-resolution photography cannot be overemphasized: excellent optics, steady mountings and perfect tracking are wasted if the seeing conditions are poor. An excellent method for determining the quality of atmospheric steadiness is the 10-point scale described by J.B. Sidgwick in the *Amateur Astronomers Handbook* which we reproduce here (Table 12.1). Designed for use with a small telescope (5-inch aperture), it is based on the appearance of the spurious disk and diffraction

* While a southern jet stream also exists, its path and characteristics are not well established.

rings of a relatively bright star as seen at high power (60 times or more per inch of aperture!).

When photographing with short focal lengths (20 to 36 inches) a seeing rating of 4 (often noted as 4/10 to indicate the scale height) is generally needed to obtain satisfactory results but photographs taken when seeing is rated 6/10 or more will exhibit markedly sharper star images. However, if the focal length is increased to 80 inches, a rating of 6/10 should be considered marginal and, even in better seeing, images will be degraded if motion of the image, caused by atmospheric effects, is greater than approximately 2 arc sec.* With the longer effective focal lengths that are required for lunar and planetary work, requirements become quite stringent as nearly perfect seeing becomes a must and some of the best planetary photographers consider ratings of 8/10 to be marginal!

For visual work, this 10-point seeing scale works quite well however, for photographic applications (deep sky and lunar/planetary work) image degradation will occur, even with a perfectly defined diffraction pattern, if the image is showing erratic motion. The degree of motion (and therefore the loss in resolution) can be easily estimated by comparing the amount of movement with the size of the Airy disk. The Airy disk, when well defined, is approximately equal to the visual resolution of the instrument in arc seconds. A final reminder: when assessing motions due to seeing be aware that the motion generally occurs in all directions. Therefore, if the motion is estimated to be 3–4 arc sec an exposure of any substantial duration will register a blur of some 6–8 arc sec in diameter.

Table 12.1. *Astronomical seeing scale*[a]

1.	Disk and rings undifferentiated; image usually about twice the size of the true diffraction pattern
2.	Disk and rings undifferentiated; image occasionally twice the size of the true diffraction pattern
3.	Disk and rings undifferentiated; image still enlarged but brighter at the center
4.	Disk often visible; also occasional short arcs of the rings
5.	Disk always visible; short arcs of the rings visible for about half the time
6.	Disk always visible, though not sharply defined; short arcs of the rings visible all of the time
7.	Disk sometimes sharply defined and distinct from the rings
8.	Disk always sharply defined; inner ring in constant motion
9.	Disk always sharply defined; inner ring stationary
10.	Disk always sharply defined; all rings virtually stationary

[a] This scale was designed for use with a 5 inch aperture refractor. The star should be of magnitude 1 or 2 and a magnification should be used which is at least 60 times the diameter of the objective.

* Image motion is only a concern with instruments less than about 6 inches of aperture. With larger apertures, image motion is not seen. This effect is discussed, in detail, in the previous discussion of seeing.

Lunar and planetary photography

As the heading would indicate, we are dealing with the photography of the Earth's moon and other planets of the solar system. While the brightness range and size range in this broad category are rather large, the principles leading to better-than-average work are essentially the same. While very nice photographs of the moon's disk can be obtained at prime focus with short focal lengths, high-resolution photographs of a small group of craters generally require that the image be enlarged considerably. We see these differences as being small variations on a theme within this category. We must also acknowledge that these areas are not ones to which we have applied ourselves over the years. While we do feel qualified to speak on basic principles we have obtained the aid of an acknowledged expert in this field to discuss some of the particulars that he has picked up over the years (see Chapter 13 for the contribution by James Rouse and see Appendix C for applicable formulas).

Since we are going to be dealing with objects that range in size from 10 arc sec to 40 + arc sec for planets and up to 1 or 2 arc min for crater groups on the moon, it should be obvious that focal lengths of 30–60 inches will not be adequate to obtain an image of reasonable size. By this, we mean an image that is large enough to produce a printed image of the object of interest (planet or crater) that is roughly 10–30 mm across and not overly grainy. In order to do this, requires an imaging system with a focal length on the order of 400–1200 inches! This, for an 8 inch aperture system, means working at focal ratios of about f/50–f/150! The obvious result of these necessary conditions is highly magnified but extremely faint images. Typical exposures on slow, fine-grain films may run from 10–20 + sec, while exposures on faster grainy films may be as short as 0.5–3 + s. We will discuss how to obtain such focal lengths after discussing the full implications of these requirements.

When dealing with the image of a planet, one would like to have the best resolution possible and would like to have an image that is not terribly disrupted by grain. For 6–12 inch aperture instruments the absolute limit of photographic resolution (given enough focal length) is about 2–4 arc sec. Since the largest angular diameter of any of the planets is about 45 arc sec, it should be clear that one must push hard to obtain the best possible resolution or all efforts will be wasted. This means that excellent seeing is an absolute must! If the seeing is not at least 8/10 there is no sense in taking out a camera (the Airy disk and diffraction rings around star images should be well defined almost continuously and the image should remain motionless and not jump around from second to second). If high-quality planetary photography is a prime goal, it is necessary to find sites that have exquisite seeing or else wait for those 4–5 nights per year when the seeing in the backyard is superb.

There is an interesting factor which comes into play when pursueing excellent planetary images. Since the planets are considerably brighter than nebulae and galaxies, it is not necessary to travel to dark sites which, in turn, allows a greater variety of sites to choose from. It also allows 'shooting' in weather conditions that would be less than ideal for deep-sky photography. A couple of commonly encountered examples may serve to illustrate the point.

1 One often finds that the air is quite stable when a thick, wet air mass settles in that knocks visibility down to only 1st or 2nd magnitude stars. Our friend, Jim Rouse, used to live in Naples, Florida and often found that his best work was obtained on nights when it was barely possible to keep the main mirror of his 8 inch f/7 Newtonian from dewing up. He rated the seeing as about 15 on a scale from 1 to 10 on such occasions. These, of course, were also the nights that the Alligators liked to wander about his backyard in search of snacks (rolls of film, cameras, feet, anything!). After moving from Naples, he has never seen similar conditions.
2 In Los Angeles, the area known as the San Fernando Valley is often enveloped in an inversion layer that traps the ground air (and smog, etc.) in a compressed layer several thousand feet thick (the thicker the better, as it turns out). This trapped layer is very stable thermally and, while one can barely see a 2nd magnitude star, the seeing is often excellent (a solar observatory was built in this area because of these atmospheric conditions. The seeing conditions at this solar observatory site were further enhanced by placing the observatory on a peninsula in the middle of an artificial reservoir).

Sites that consistently have excellent seeing are rare and one must conduct 'site surveys' to search them out. However, they are, in our opinion, well worth the effort.

With exposures running from a few seconds to perhaps 30 s and resolution goals on the order of 1–2 arc sec, another requirement for the photographic system should come to mind: a very stable, very accurate clock drive that can track to within an arc second for up to a minute and which is free of sporadic error motions. Without a precision drive the problem of getting a single good shot out of every 100 shots becomes one of sheer luck. With 'small' telescopes where the image must be magnified 10–20 times before falling on the film plane, it is impossible to 'guide out' 1 arc sec errors. The investment in such a drive becomes virtually mandatory if this is one's primary interest in astrophotography.

Again, since 1–2 arc sec resolution is the goal, it is essential that one have a very sturdy mounting. While there are several ways to get around the vibration created by the opening of a camera shutter, it is hard to be prepared for, or to stop, those small, brief breezes that you would never notice with a low-power eyepiece or a camera at prime focus. If your mounting is a little on the flimsy side, you will undoubtedly lose a number of images from time to time because of miscellaneous vibrations.

Getting down to the smaller details now, since the objects being photographed are small and since one is in need of a long focal length, it becomes more feasible to use Cassegrain-type optical systems for planetary work than it was for deep-sky work (although for lunar work, the system must be well baffled). In addition, since one is dealing with low-contrast objects, in most cases, it is advisable to use a small secondary to avoid scattering light out of the central diffraction disk. Finally, it should be noted that the elimination of a spider support makes a noticeable contribution towards increasing the contrast of the image.*

In order to obtain focal lengths of close to 1000 inches from a telescope of 50 inches focal length, some means of magnifying the image must be employed. The old standby for accomplishing this task has been applied for close to 100 years and is known as eyepiece projection. The oldest and most common form simply takes the light that is passing through the eyepiece and focuses it on a piece of film some distance away. There is another variation on this method which takes the beam of light projected by the eyepiece and passes it through the lens of a camera onto the film. The two systems are shown in Appendix C along with the equations used to derive the effective focal ratios of those systems. The contribution by Jim Rouse in Chapter 13 provides considerable information for deriving exposure values.

Of these two systems, some prefer the eyepiece + camera lens scheme because of the fact that they can mount the camera on a sturdy tripod, away from the telescope and thus avoid the effects of shutter vibration. This is true, but the same scheme would also work for the first system as well! If the camera is mounted on a separate mount, one must be sure that the film plane is square to the projected image and arrange to have the telescope's tracking motion carry the image into the film plane, and not in a track along the film or else the image will most certainly trail (even this arrangement is not fully satisfactory since the motion caused by the tracking of the telescope will result in a change of focus that will be noticeable over a 10 s exposure). Our preference is for the standard eyepiece projection method with the camera mounted on the telescope.

When using the telescope and such a long focal length, the vibration of the shutter or SLR mirror is magnified many times and becomes a serious problem for 'normal' applications. There are numerous methods of dealing with this shutter shake and we will describe several that have been particularly successful for various practitioners over the years.

1 The old 'hat trick' is the simplest, least expensive, and probably most frequently used method of eliminating shutter vibration. In this technique a 'shutter' that will cover the telescope aperture is held up in front of the telescope and the camera shutter is opened. Then, after 5–10 s, when the seeing looks particularly nice, the hand-held shutter is flipped out of the way and the exposure is counted down. At the end

* The standard four-vaned spider should be replaced with an optical window that eliminates scattered light and not a curved vane spider that simply scatters light about more uniformly!

of the exposure, the hand-held shutter is moved to cover the telescope aperture and the camera shutter is closed. The hand-held shutter can be a cardboard cutout, a Frisby, trash can lid, or anything else that does the job.

2 Similar in concept, but far more complex in design, is to build an additional shutter into the system and then isolate the vibration of that shutter from the telescope. Typically, a 'leaf shutter' is added into the eyepiece-projection system and then isolated from the telescope using elastic supports (rubber bands work well).

3 The long lever arm which the eyepiece-projection tube plus camera body provides is eliminated to remove much of the vibration. In this system, the image projected from the eyepiece is bent 90 deg using a mirror diagonal and the camera is mounted on the tube of the telescope. This is particularly useful when one needs to project the image over a distance greater than just a few inches. This technique results in a reversed image which must be projected with the emulsion side facing away from the lens for correct viewing or reproduction.

Finally, the projection system itself should be given some careful thought. The eyepiece used should be well corrected and should be of high quality, Orthoscopics and Plossls being commonly employed for this purpose. The focal length of the eyepiece normally falls in the range from 10 mm to 40 mm, depending on the projection distance and the effective focal ratio that is desired. We would like to point out, as an interesting bit of trivia, that eyepieces were not designed as projection lenses and that optical systems designed for one type of work do not usually function at their best when applied to a totally different application. An enlarger lens is designed for projection, and focal lengths of 25 mm and 50 mm are available. Since even 25 mm is rather a long focal length, it will be necessary to project the image over a longer distance which makes the third system listed above the ideal for such a projection system. If an enlarging lens is used, which is our position for optimum results, the highest-quality lens available should be employed.

When it comes to selecting a film, we tend to agree with Jim Rouse on the use of fine-grained film and longer exposures as being preferable to shorter exposures with coarser-grained films. We would recommend films like Kodachrome 25 and Kodachrome 64 for color work and 2415 Technical Pan for black-and-white work. With color films it is quite easy to judge when one has obtained the best exposure, the slide just looks good to the unaided eye and/or looks good when projected. However, with black-and-white films, the situation is a little different when it comes to evaluating a negative. If one were to take a series of increasingly longer exposures and make the best prints possible from negatives that were very thin up to negatives that were very dark, it would be possible to see that print quality does vary with image density. If the negative is too thin, one gets greys without details and often tries to compensate by printing on high-contrast papers. If the negative is too dark, the greys are all 'muddy' looking and show nothing but the grain. The negative needs to be very carefully exposed in order to obtain the best possible print. Whenever given a set of exposure values for any given subject, a series should be taken that brackets the recommended exposures and the negatives should be evaluated carefully to determine the correct exposure for your particular system.

Solar photography

We now come to the one body in the solar system that gives off enough light to allow for nice short exposures, where there is no need to worry about the adverse effects of reciprocity failure (and for those who detest the night life, it can be done in the daytime!). However, the problem here is that there is far too much light to work with safely. Warning: one should **not** look directly at the sun with (or without) a telescope without the aid of special protective filters to prevent **permanent** damage to the eye (one of the authors learned this the hard way; the other having worked at a solar observatory saw titanium shutters turned to ash by the focused light of the sun). Today, the problem of excessive light from the sun is cheaply dealt with by the use of lightweight, inexpensive mylar filters. These consist of a double layer of ultrathin mylar that is stretched to fit some type of ring structure that will fit over the telescope aperture. Another, more expensive, alternative is to purchase an aluminum-on-glass filter that serves the same purpose.

Since the density of these filters varies considerably (plus or minus a stop) around an optical density of roughly 5.0, it is difficult to predict exposures for any given filter without using a densitometer to read the filter density. Also, since the mylar filters are not neutral in color, the use of color film is generally precluded. Finally, since the exposures with these filters are on the order of 1/100 s, or shorter, the 'hat trick' method of making exposures will not work and one needs to rely on a stable mount to eliminate shutter-shake effects (locking up the mirror before making exposures is quite helpful).

Solar photography at long effective focal lengths also requires excellent seeing but with solar work the problem of defining excellent seeing is less precise than with night time work when stellar images can easily be examined. In the daytime, the matter of determining the 'seeing' becomes more qualitative (we should also note that sites having excellent night time seeing do not always have excellent daytime seeing, in fact there is little correlation at all, and each type of seeing needs to be assessed). In any case, there are several means of getting an idea of how good or how bad the seeing is but we are unaware of any published work that has attempted to present a rigorous 'solar seeing scale' such as those that have been developed for stellar estimates. However, there are solar seeing scales in use at various observatories. We present here the Mt Wilson solar seeing scale

(Table 12.2) as an example of the type of scales that are commonly used. We would like to note to our readers that high-resolution photographs of sunspot groupings can only be regularly obtained when the daytime seeing reaches level 4 or 5 on this seeing scale.

Perhaps the most interesting (and most costly) aspect of solar photography is photographing in the light of hydrogen alpha (656.28 nm). The light of calcium is also quite interesting. In hydrogen alpha (H-alpha) light, one can readily see the prominences that frequently line the edge of the sun's disk in addition to solar flares and the dark filaments that are caused by prominences that are seen floating above the sun's surface. The filters needed to do this work are costly but the price has come down considerably over the past decade. The higher-quality filters are generally housed in a small oven that maintains the filter at a constant temperature which keeps the central bandpass of the filter centered at the H-alpha wavelength. These filters can be manufactured with a variety of 'bandpass widths' (the bandpass width is the width, usually specified in Ångstroms or nanometers, of the spectrum over which the transmission of the filter is greater than 50 % of its peak transmission value; the 'bandpass width' is also sometimes called the 'half-power width' of the filter). The act of narrowing the bandpass of an H-alpha filter has a significant impact on the view obtained of the sun's disk.

1 With a 1 Å bandwidth, the prominences lining the edge of the disk can be easily seen and solar flares can be viewed but the finer details over the sun's disk are not well displayed.
2 With a 0.8 Å bandwidth, the prominences are seen very well, with better contrast than is provided by a 1 Å filter, and the details on the disk of the sun are readily seen in detail.
3 With a 0.6 Å bandwidth, the details on the disk are seen with excellent contrast and clarity however, the prominences are now difficult to see.

These filters are interference filters and thus are designed for use in a beam of parallel light. In a solar telescope, the focused beam from the instrument would be passed through a collimator and then into the H-alpha filter. In most amateur telescopes the problem of making a collimated beam of concentrated sunlight is avoided by stopping down the main instrument until the f/ratio reaches at least f/30. At such an f/ratio, the filter is able to

Table 12.2. *Mt Wilson solar seeing scale*

1. Limb of sun appears serrated, image in violent motion
2. Limb relatively steady, disk continuously blurry
3. Image varies from blurred to sharp over several seconds; limb motion about 3–4 arc sec
4. Image is sharp most of the time, but limb is still in motion; limb motion about 2 arc sec; fibrils in sunspots are visible
5. Image is absolutely steady; limb motion at 1 arc sec level; granularity visible on solar disk

perform close to its designed specifications. However, one should note that there will be a substantial resolution sacrifice in stopping an f/6 system down to f/30. Finally, in addition to the price of the filter one should also plan on purchasing an additional 'energy-rejection filter' that removes the light from the infrared and ultraviolet regions of the spectrum before it passes through the H-alpha filter. Without this filter, the useful lifetime of the H-alpha filter will be substantially shortened. The H-alpha solar filters made by DayStar have had an excellent reputation for many years (see Fig. 12.2).

At this point, many will be envisioning a beautiful photograph of the sun with the disk mottled by dark H-alpha clouds and streamers, sunspot groupings, flares, and a group prominences hanging on the edge of the sun's disk. We have all seen photographs like this, so that's the way it is! Wrong! The prominences are fainter, by a factor of 10 to 20, than any H-alpha features that can be seen on the sun's disk. Photographs that portray both have either been dodged or masked or else are a photocomposite made from two different exposures. Several articles have appeared on these techniques over the years, and our bibliography will be useful to those interested. As a final note, we should mention that color photography with an H-alpha filter is not exactly productive since everything is in one tone of red. If you really need slides and do not want to go through an extra copying stage, you can resort to color film or else use a special developing kit that will allow several different black-and-white films to be processed as a positive and apply a layer of red croseine dye.

Total solar eclipse photography

Now we come to the pinnacle of astronomical phenomena. In fact, it is doubtful if any other natural phenomenon can begin to compete with this

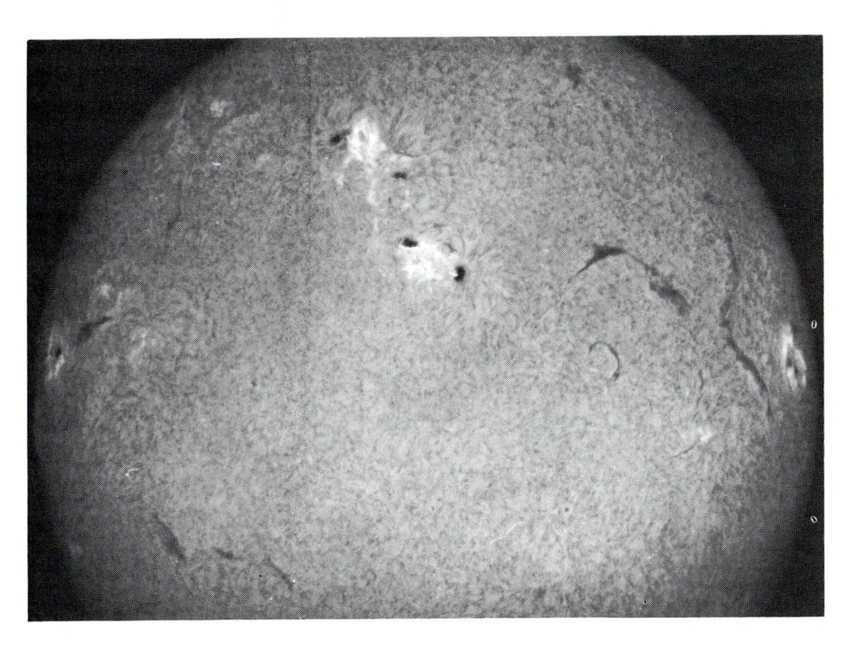

Fig. 12.2. An H-alpha photo of the Sun showing sunspots, bright activity centers and long dark solar filaments. Photo by Edwin Hirsch using a DayStar filter at f/30.

spectacular event. When seen for the first time, this event fills one with such awe that it is difficult to tear one's eyes away in order to make the exposures that have been planned. This feeling is even greater while alone in a relatively wild area. The reactions of the animal life to this event are surprising and must be considered an integral part of experiencing a total solar eclipse. If you have to travel with a group to the eclipse-viewing area, we strongly recommend that you wander away from the group in order to see and hear all of the effects of a solar eclipse.

Warning: one should **not** look directly at the sun **until** the total phase of the eclipse begins. After totality has begun (e.g. the sun's disk is entirely covered and prominences and corona are visible), it is safe to view the sun's outer atmosphere with the naked eye and with other viewing instruments. During the partial phases, special light-rejection filters must be used. Kodak Wratten filters do **not** reject the short and long rays that can cause eye damage.

In order to photograph the partial phases of an eclipse, filters and aperture stops are recommended to reduce the intensity of the sun's light. If one tries to photograph the sun using a SLR without such filters, it is quite possible and very likely that the camera will be severely damaged. In most SLRs, the shutter is made from a cloth material which can be incinerated if the mirror is left up for too long. Even the titanium shutters employed in a few SLRs can be damaged by prolonged exposure to the sun's light at prime focus. The full-aperture mylar filters that are rapidly becoming popular offer the best solution to this problem. However, it is difficult to predict the exact optical density of these filters and an exposure test needs to be conducted to know the exact exposures to use in photographing the sun's disk.

A number of information sources have been published which provide accurate information on photographing the sun's disk using filters and/or aperture stops. However, the information published regarding the photography of the total phases should generally be examined carefully. The intensity of the light coming from the sun's corona is variable and one may be very disappointed in the results obtained if some of the published guidelines are strictly followed. When studying the exposure data for the various regions of the sun's outer atmosphere, and then comparing these data for a number of total eclipses, it becomes quite evident that standard exposures for these phenomena just do not exist (Table 12.3). It is usually possible to eliminate some of the extremely brief exposures, but one should plan on trying a wide range of long exposures if planning to capture the sun's outer corona. If one is interested in photographing the outermost reaches of the corona, it is important to remember that this feature can extend out to 3 or 5 solar radii. It is therefore necessary to choose a lens that will contain a field of this size. This would mean that 400 mm f.l. lens is the longest that could be used in conjunction with 35 mm film.

The best strategy to adopt in photographing the phenomena visible during totality is to plan on running an entire series of exposures from the minimum range for the film–f/stop combination in use out to 10 or 15 s if you are interested in photographing the outermost reaches of the corona. It is very important to prepare and practice the exposure strategy that you plan to use during totality. The natural tendency at the start of totality is to stand gasping with mouth open in total awe of this phenomenon. If well prepared, it is possible to run off two rolls of 36-exposure film (including the time required to unload and reload the camera!) and then stand around enjoying the sight after you are finished, even if the duration of totality is only 3 minutes.

Photographing a total solar eclipse poses a special problem: the range of brightness from the chromosphere outwards to the outer corona is quite extreme. The intensity variation encountered is on the order of 500 to 1000 and if the outermost coronal streamers are included this range can easily be expanded by a factor of four to eight. This presents a range that would span 8 or more f/stops, which is far beyond the range of all color emulsions that are generally available. The only color film that even comes close is a special color emulsion known as Kodak XR Color film (XR = extended range). This film is not readily available to amateurs unless one plans on buying a large quantity and splitting it with a number of people. Black-and-white films can handle this range of light if exposed properly and processed properly. It is also possible to reconstruct an extended range color photograph by obtaining three black-and-white photographs through blue, green and red filters and then recombining the three images to form the final color image. One other approach to this problem is available which entails the use of what is known as a 'radial filter'. These are filters that are made specifically for the optical system that is to be used at the eclipse. The filter is very dense in the center and, at a radius slightly larger than the image of the sun's disk, decreases in density as one moves away from the disk of the sun. These filters must be placed in the focused beam very near the image plane and must therefore be of very good optical quality. In order to produce a perfectly neutral filter of high optical quality, these filters are constructed by depositing a thin inconel layer on optical glass. As one might imagine, these filters are very expensive and thus are not generally available to amateurs.

Table 12.3. *Solar eclipse exposure guide*[a]

Phenomenon	Neutral density filter	Exposure
Partial phases	5.0	1/60–1/125
Bailey's beads	4.0	1/60–1/500
Diamond ring	4.0	1/60–1/500
Prominences	None	1/100–1/500
Inner corona	None	1/25–1/500
Outer corona	None	1/25–3

[a]This shows the range of exposure that may be used to capture various eclipse phenomena. All exposures shown are for f/8 using ASA 64 film.

Lunar eclipse photography

Lunar eclipses occur much more frequently than solar eclipses. In fact, if you just stay in one general geographic location, you can expect to see a lunar eclipse every year on the average. Even though these events do occur frequently, they continue to be spectacular year after year. If you have only viewed these events from the confines of a city, you have missed much of the beauty that accompanies this phenomenon. Viewing a lunar eclipse under clear dark skies is a fantastic experience. As the moon glides into the Earth's shadow and begins to glow in beautiful shades of red, the sky changes from very bright with very few stars to a dark black sky studded with thousands of stars. As mid-totality approaches, the sky becomes darker and darker and, in the midst of this star-filled sky, looms a glowing red moon. If the lunar eclipse occurs while the moon is situated in the midst of the Milky Way, the view is even more beautiful. The only other natural phenomenon which can rightly claim to outdo a lunar eclipse is a solar eclipse or a great comet.

As you can well imagine, we would definitely recommend the use of color films for photographing a lunar eclipse. We generally use very little film on the partial phases of the eclipse and save our efforts for the total phase. Since the total phase is characterized by various shades and tones of red, we choose our film to best portray these tones. At present, Fujichrome 50, Kodachrome 25 and Kodachrome 64 are particularly well suited for this work. The exposures for the partial phases of a lunar eclipse can be calculated quite accurately. However, global atmospheric conditions present at the time of each eclipse can affect the brightness dramatically and the appearance of the moon during totality and a range of exposures should be used to ensure capturing this event.

Table 12.4 should be used as a general guide and the exposure provided should be bracketed due to the variability of eclipse circumstances. In particular, the exposures for the total phases of the eclipse should be widely bracketed to allow for the extreme range of variability that is encountered. Some eclipses may darken the moon to such an extent that it

Table 12.4. *Lunar eclipse exposure table*

	Exposure using ASA 64 (seconds)	
Phase	f/8	f/5.6
Moon 1/10 to 1/2 in umbra	1/8	1/16
Moon 1/2 to 3/4 in umbra	5	2
Portion in penumbra	1/8	1/16
Moon 3/4 in umbra to totality	5–10	4
Totality starts	5–15	2–10
Mid-totality	10–120 +	8–60 +

is barely visible to the unaided eye, while others may produce a bright orangish-red moon. If the eclipse is one in which the moon passes almost directly through the center of the Earth's shadow, one can usually expect a dark eclipse, while eclipses in which the moon comes close to the edge of the Earth's shadow will most likely produce a brighter eclipse. If there has been a sizeable amount of volcanic activity during the year or two prior to a particular eclipse, it is quite likely that a very dark eclipse will occur. One eclipse is seldom 'just like' any other eclipse that one has seen; each is unique and each is beautiful in its own way.

Just as the characteristics of each eclipse can vary, there are numerous approaches that can be taken in the photography of an eclipse. One can set out to just photograph the progression of the partial phases, or just photograph the total phases, one can photograph the star field in the vicinity of the moon during totality along with many other aspects of the eclipse. The lunar eclipse photographs we have included in the color section show just a few of the many projects that can be conducted. A list of possible projects would include:

1 Using a wide-angle lens, a series of images on a single frame can capture all of the various partial and total phases of an eclipse.
2 Using a good-quality telephoto lens, one can capture three images of the moon on a single frame that show the moon entering the Earth's shadow, mid-totality, and the moon leaving the Earth's shadow. This will show quite plainly that the Earth's shadow is responsible and produce a partial outline of its perimeter. It is necessary to keep the field fixed relative to a fixed star when each of the three exposures is taken (i.e. you don't have to guide during the whole hour-and-a-half, but the guide star has to be recentered before each of the exposures is made).
3 During totality, long exposures (i.e. 5–20 min) can be made that will show the star field surrounding the moon. These can be made with various focal length lenses to obtain a variety of photographs. If the eclipse occurs while the moon is among the Milky Way, this can yield some really spectacular results.
4 By taking a series of photographs of the moon during totality, a record is made that may show variations in the Earth's shadow during the course of totality. The eclipse of June 1982 showed a number of variations during the course of totality.
5 A very interesting effect can be captured by leaving the lens open (stopped down to f/8 or f/11) during the entire eclipse. This can be done using a stationary camera and allowing stars and moon to trail during the exposure, or one can guide the camera on the stellar background allowing only the moon's motion to show. The latter form of this process may yield some interesting information on the structure of the Earth's shadow.
6 Use your imagination!!!

A selection of fine astronomical photography

Leo C. Henzl

1
Horse Head Nebula and NGC2024, 25 min on Kodak 103a-F with 8 inch f/6 Newtonian.

2
The Double Cluster, 15 min on Kodak 103a-F with 8 inch f/6 Newtonian.

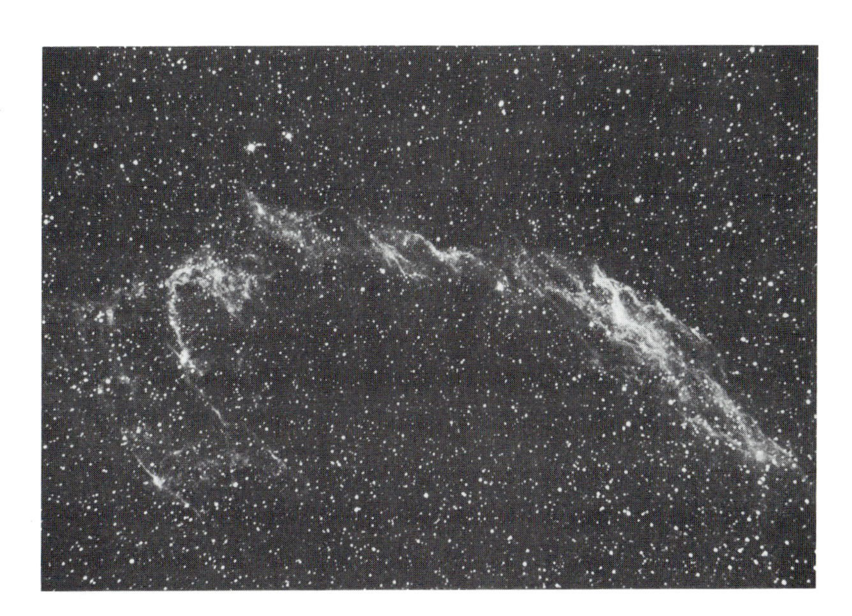

3
NGC6992–5, 25 min on hydrogen-hypered Kodak 103a-F with 10 inch f/6 Newtonian.

4
M22, 20 min on Kodak 103a-F with 10 inch f/6 Newtonian.

Lee C. Coombs

1
M104, 15 min on Kodak 103a-O with 10 inch f/5 Newtonian Processed MWP-2, 9 min.

2
M97, 15 min on Kodak 103a-O with 10 inch f/5 Newtonian. Processed MWP-2, 9 min.

3
M51, 15 min on Kodak 103a-O with 10 inch f/5 Newtonian. Processed MWP-2, 9 min.

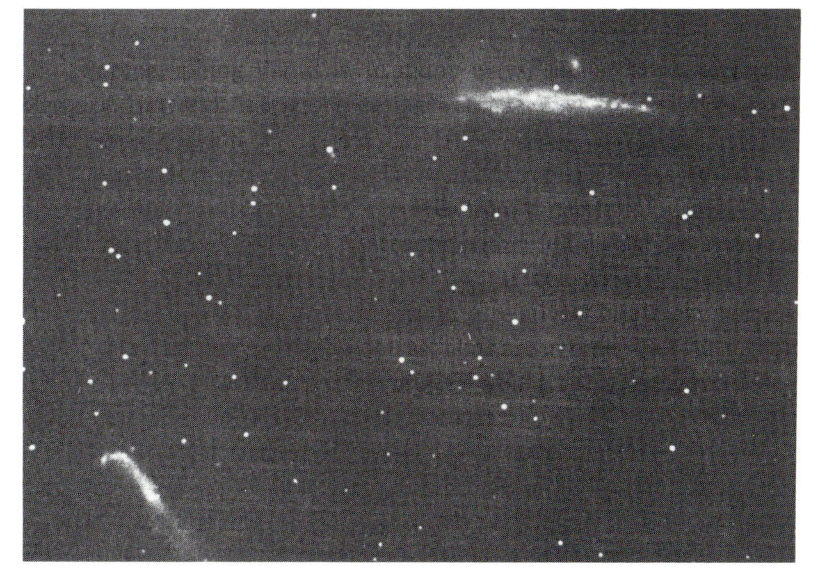

4
NGC4631 and 4656, 15 min on Kodak 103a-O with 10 inch f/5 Newtonian. Processed MWP-2, 9 min.

James K. Rouse

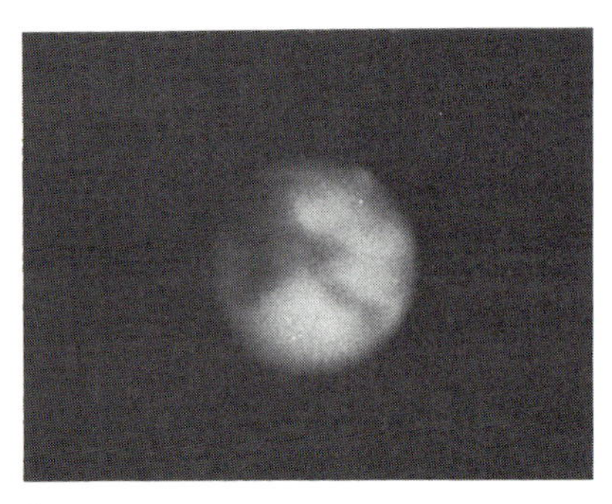

1
Mars, Major and Sabaeus Sinus, 5 s on Kodak High Contrast Copy film with 8 inch Newtonian at f/105. Processed in H&W developer.

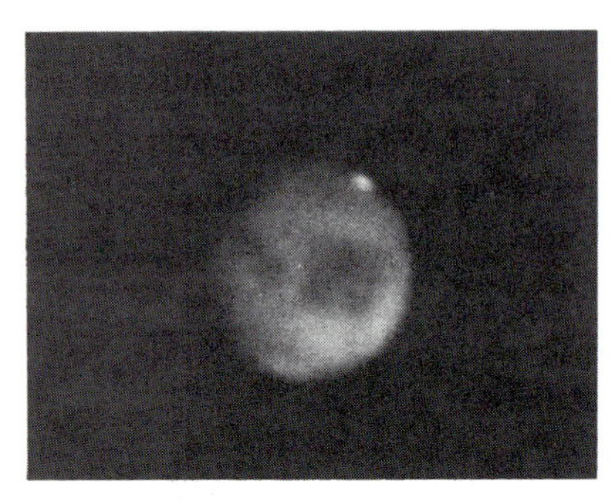

2
Mars, Meridiani Sinus and Mare Erythraeum, 5 s on Kodak High Contrast Copy film with 8 inch Newtonian at f/105. Processed in H&W developer.

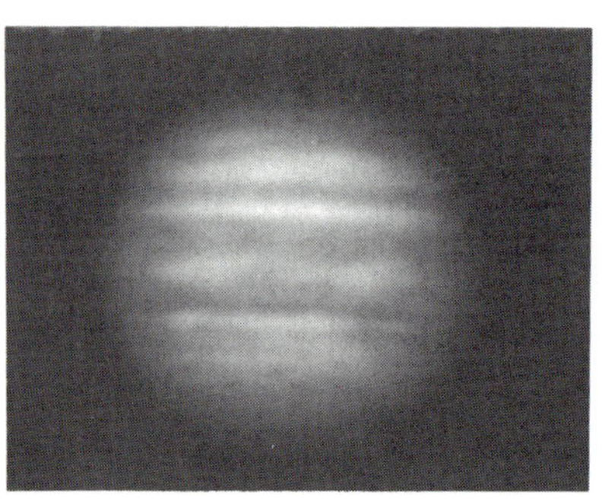

3
Jupiter, 5 s on Kodak High Contrast Copy film with 8 inch Newtonian at f/80. Processed H&W developer.

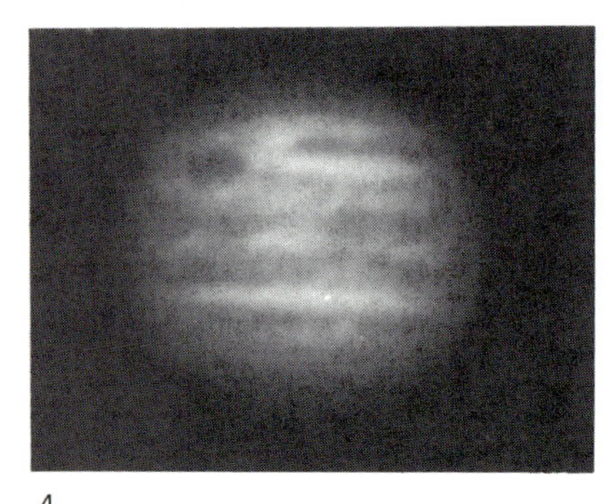

4
Jupiter, 5 s on Kodak High Contrast copy film with 8 inch Newtonian at f/80. Processed H&W developer.

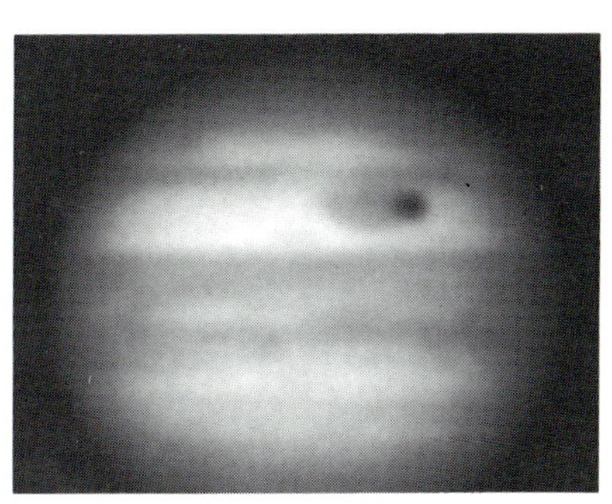

5
Jupiter and Ganymede with shadow transit, 5 s on Kodak S.O. 410 film with 8 inch Newtonian at f/105. Processed H&W developer.

6
Jupiter, 5 s on Kodak High Contrast Copy film with 8 inch Newtonian at f/80. Processed H&W developer.

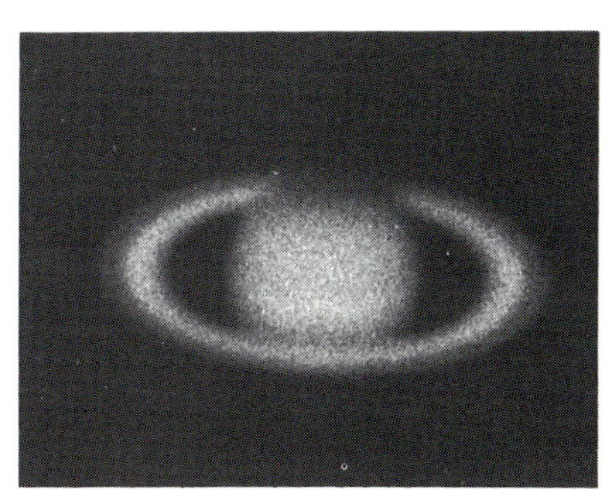

7
Saturn, 4 s on Kodak Plus-X Pan film with 6 inch Newtonian at f/135. Processed Edwal's FG7 (1 : 15) developer.

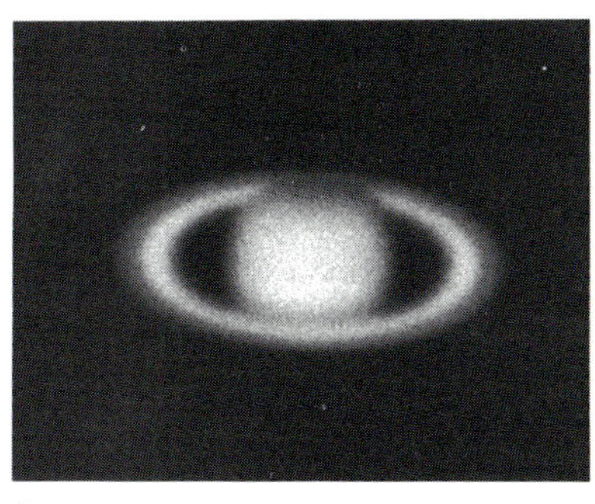

8
Saturn, print made from two sandwiched 3 s exposures on Kodak Plus-X Pan film with 6 inch Newtonian at f/135. Processed Edwal's FG7 (1 : 15) developer. Note improved grain over photograph at left.

Martin C. Germano

1
M83, 50 min on Kodak 103a-F with 8 inch f/10 Schmidt–Cassegrain. Processed Acufine 1 : 1, 10.5 min.

2
NGC253, 60 min on Kodak 103a-F with 8 inch f/10 Schmidt–Cassegrain. Processed Acufine 1 : 1, 15 min.

Jean Dragesco

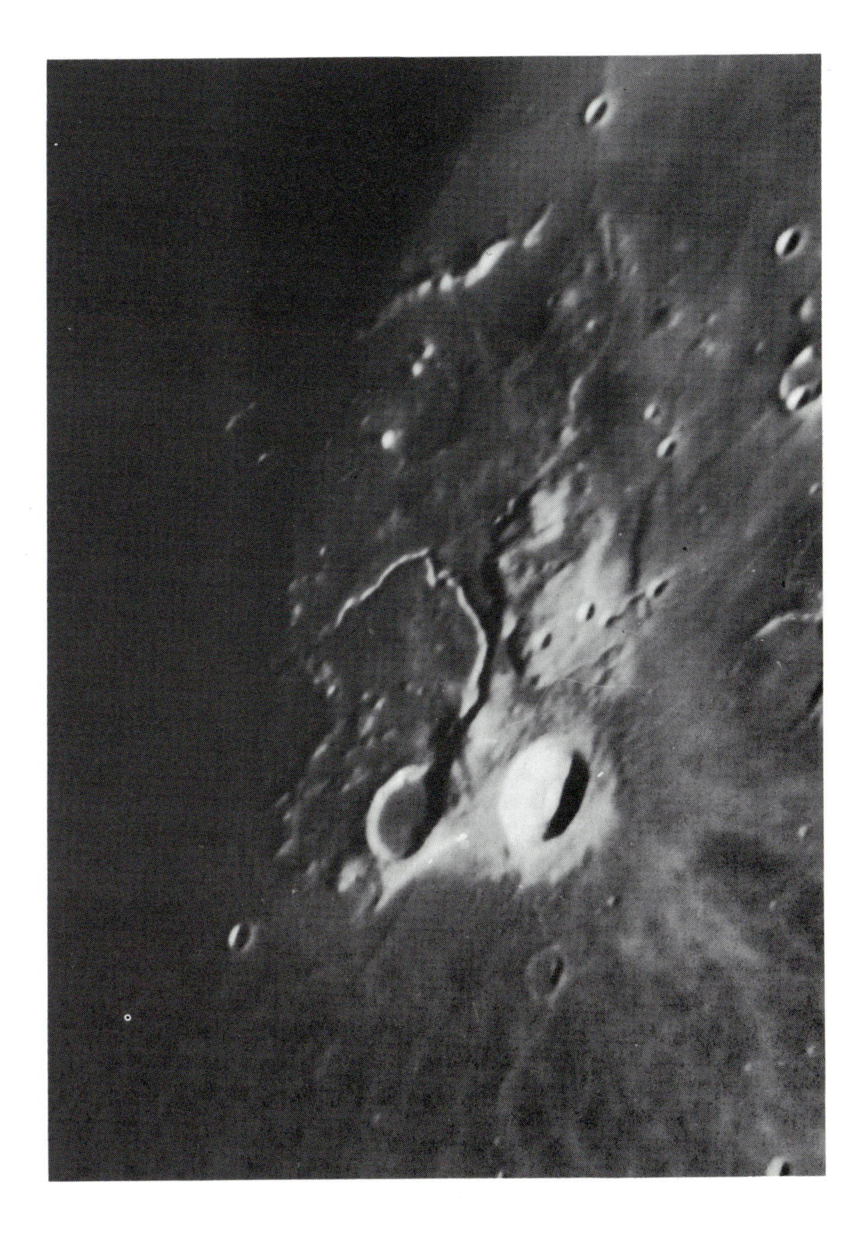

1
Schröter Valley, Kodak 2415 Technical Pan film with 14 inch Schmidt–Cassegrain at f/30. Processed in Ilford Microphen developer.

2
Stadius and Copernicus, 2 s exposure on Kodak 2415 Technical Pan film with 14 inch Schmidt–Cassegrain at f/30. Processed in Ilford Microphen developer.

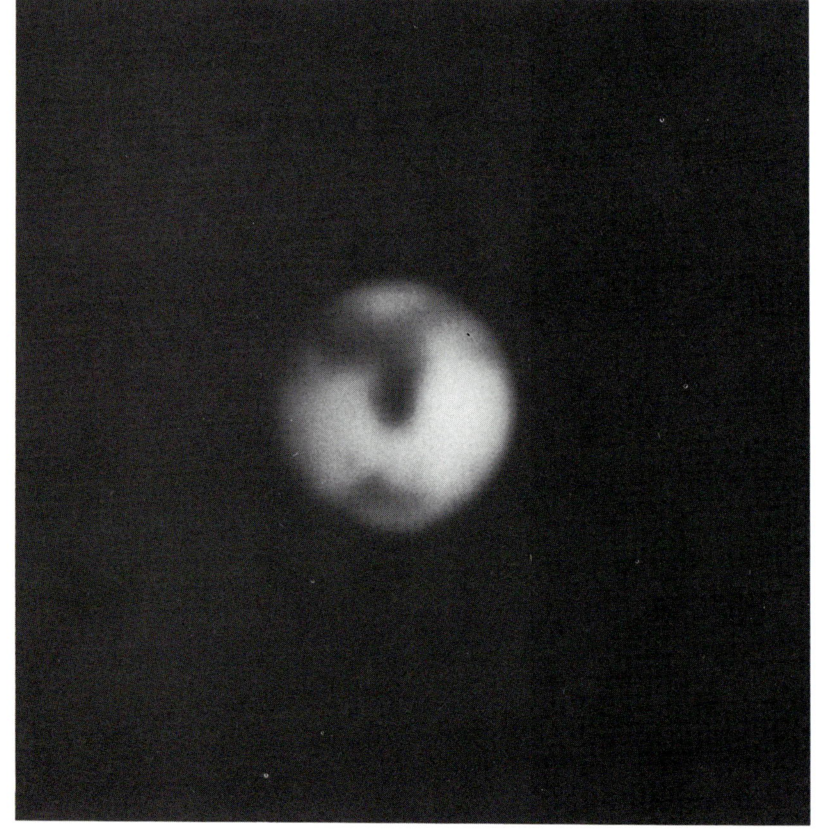

3
Mars, Syrtis Major, Kodak 2415 Technical Pan film with 14 inch Schmidt–Cassegrain at f/40.

4
Fracastorius, 2 s exposure on Kodak 2415 Technical Pan film with 14 inch Schmidt–Cassegrain at f/35. Processed in special Ilford developer.

Brad D. Wallis and Robert W. Provin

1
Comet West (1975 n), 40 min on Kodak 103a-O with 6 inch f/6 Newtonian. Processed MWP-2, 9 min. 24 May 1976.

2
M45, 30 min on Kodak 103a-O with 6 inch f/5 Newtonian. Processed MWP-2, 9 min.

3
M35 and NGC2158, 20 min on Kodak 103a-F with 6 inch f/4 Newtonian stopped down to f/5.6. Processed MWP-2, 9 min.

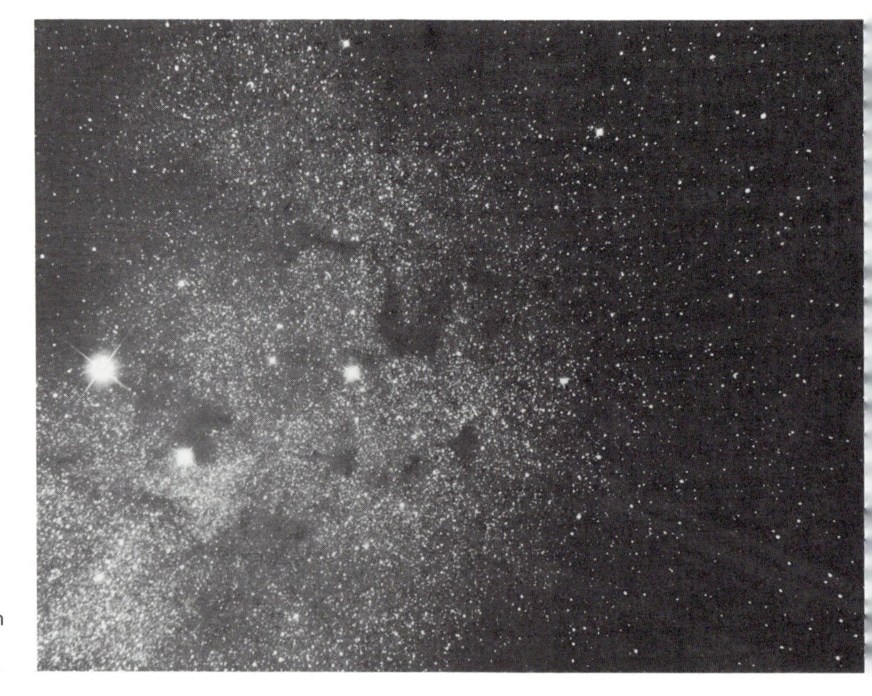

4
Barnard 72, 2 h on hydrogen-hypered Kodak 2415 Technical Pan film with 4.5 inch f/6.7 Newtonian. Processed MWP-2, 11 min. Bright star at left side of field is 44 Ophiuchi.

5
M31, M32 and NGC205, 45 min on Kodak 103a-F with 6 inch f/4 Newtonian stopped down to f/5. Processed MWP-2, 9 min.

6
NGC6992–5, 30 min on Kodak 103a-F with 6 inch f/4 Newtonian stopped down to f/5. Processed MWP-2, 9 min.

7
M65, M66 and NGC3628, 45 min on Kodak 103a-O with 8 inch f/6 Newtonian. Processed MWP-2, 11 min.

8
M101, 45 min on Kodak 103a-O with 8 inch f/6 Newtonian. Processed MWP-2, 11 min.

9
Comet West (1975 n), 15 min on Kodak 103a-O with 6 inch f/6 Newtonian. Processed MWP-2, 9 min. 7 March 1976.

10
Comet Bennett (1969 i), 15 min on Kodak Tri-X Pan with 135 mm lens at f/2.8. 8 April 1970.

11
Comet Kobayashi–Berger–Milon (1975 h), 10 min on Kodak 103a-F with 4.25 inch f/5.6 Newtonian. Processed MWP-2, 9 min. 12 August 1975. Note M106 in field.

12
M42, three 20 min exposures on Kodak 103a-O and one 5 min exposure on Kodak 103a-F were combined using integration and masking techniques. 6 inch f/4 Newtonian stopped down to f/5. Processed MWP-2, 9 min.

13
Pelican nebula (IC 5067–70), 2 h on hydrogen-hypered Kodak 2415 Technical Pan film with 6 inch F/5 Newtonian stopped to F/6.7. Processed D-19, 13 mins. Faintest star images near the center of the 5 × enlargement measured 18 μm on the negative.

K. Alexander Brownlee

1
M51, 95 min on gas-hypered Kodak 2415 Technical Pan film with a 16 inch f/5 Newtonian. Processed D-19, 5 min.

2
M63, 90 min on gas-hypered Kodak 2415 Technical Pan film with 16 inch f/5 Newtonian. Processed D-19, 5 min. (K.A. Brownlee)

Ron Potter

1
North American nebula (NGC7000), 15 min on gas-hypered Kodak 2415 Technical Pan film. 8 inch f/1.5 Schmidt camera with No. 92 filter.

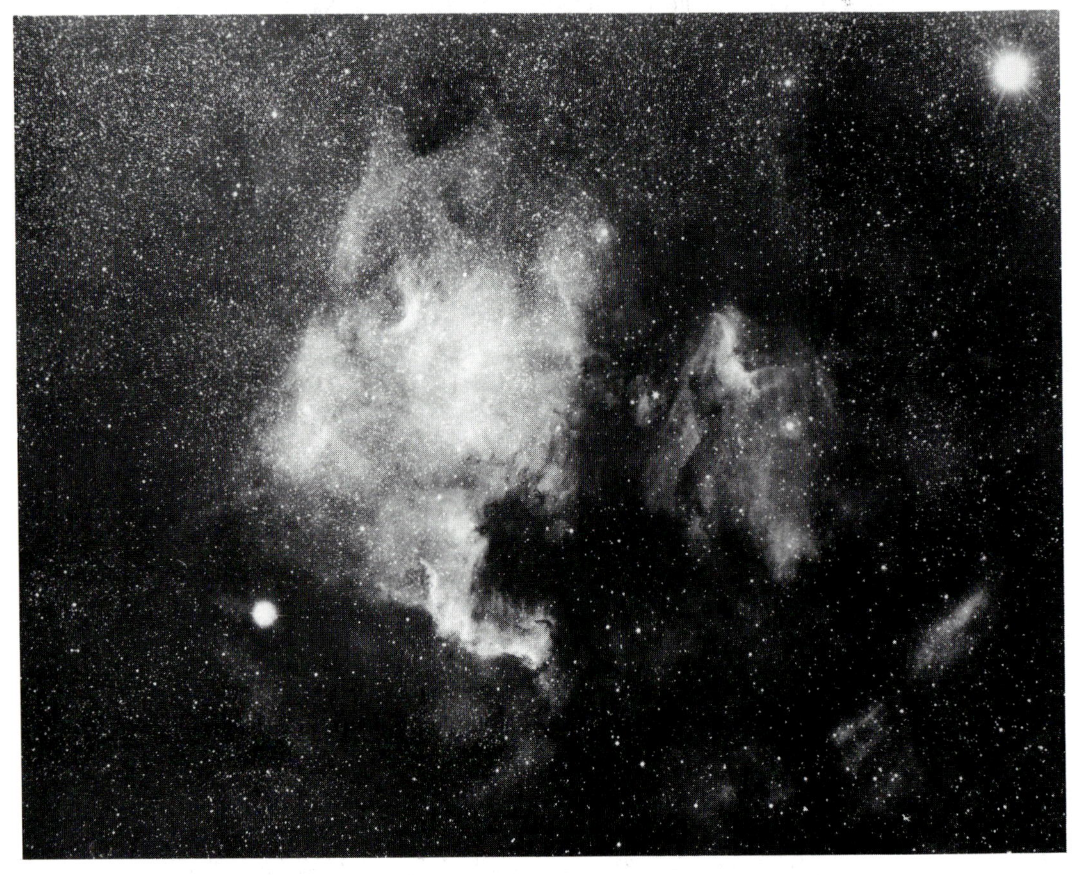

Alan McClure

1
Region around Rho Ophiuchi, 45 min on Kodak 103a-E with 7 inch f/7 Fecker triplet.

2
Comet Seki-Lines (1962c), 10 min on a panchromatic plate with a 5.5 inch f/5 Zeiss triplet. 9 April 1962.

13

Selected topics

Since it is impossible for any individual to master all of the various topics that constitute astronomical photography, we have gathered a collection of papers from a number of recognized experts in some of those areas where we do not have extensive experience. The people who have contributed to this work were all more than willing to cooperate in volunteering their knowledge and time. Before presenting the papers we will give a brief introduction to each of the contributors.

Planetary photography: Jim Rouse

For a great many amateurs lunar or planetary photography is their first encounter with astronomical photography. The first images of the moon or Jupiter are a real thrill and it seems quite easy at the start. For most, this introduction is as far as it goes since further effort has little meaning: the photographs from Mariner, Viking, and Voyager have surpassed anything that could ever be obtained by earthbound instruments. However, for a few, the challenge of getting better photographs is accepted and years of experimentation are devoted to the problem. Among those who have taken this challenge seriously are: Ed Hirsch, Howard Zeh, Robert Price, James Rouse, John Dragesco. To these practitioners there is more to high-resolution planetary photography than just using the proper exposure. The seeing must be perfect, the drive must be capable of near-perfect tracking, and the films used for each application are chosen carefully.

Jim Rouse has been quite active in the field of lunar and planetary photography for over a decade and has published numerous photographs and articles in *Sky and Telescope*, *Astronomy*, *Star and Sky*, and *The Astrograph*. He has learned to appreciate excellent seeing and takes photographs only when seeing conditions are nearly perfect. The techniques that Mr Rouse employs have produced some of the finest high-resolution planetary images seen in the amateur literature. Indeed, many practitioners would be hard pressed to duplicate his efforts using much larger aperture instruments (Mr Rouse uses an 8 inch f/7 Newtonian for most of his work). Like most devoted planetary photographers, Mr Rouse has spent hundreds of hours experimenting and tinkering to refine his techniques and many of the interesting results of this work are presented in this paper.

Filters for astrophotography: Jack Marling

As light pollution continues to encroach on the darkness of the night sky, amateur astronomers are turning more and more to the use of filters in an effort to stay ahead of this grim reaper. Not only are amateurs using filters that professionals have used for many years, they are also adopting a relatively new filter that has been designed to pass the light from particular astronomical objects while excluding the light from annoying pollution sources. Jack Marling founded a company called Lumicon with the aim of developing and marketing products designed to meet the special needs of the amateur astrophotographer. Mr Marling, through Lumicon, has developed light-pollution-rejection filters far beyond the point where earlier manufacturers thought they would be of any value and has been an active advocate of the use of such filters. He has also been active in promoting the use of gas hypering since shortly after the introduction of this technique to the amateur community. Lumicon has long been the only source of well-designed gas hypering equipment and it is also a source for information on the hypering characteristics of the most recently available films. In addition to his commercial work, Mr Marling has published numerous articles in *Sky and Telescope*, *Astronomy*, *Deep Sky Monthly*, and the *AAS Photo-Bulletin* and has presented papers at most of the important amateur astronomical meetings. Despite his busy schedule, we managed to talk him into contributing a paper and are pleased to present it here.

Automatic guiders for astrophotography: Ron Potter

Over the past two decades, electronics have evolved rapidly and have gradually infiltrated the world of the amateur astronomer. The drive correctors of today bear little resemblance to those of the 1960s and bear no resemblance whatsoever to the mechanical drive regulators used in earlier days. Today, the computer is moving into the world of the amateur. It is used to make calculations to plan the construction of telescopes and their accessories. It can be used at the telescope to guide and plan an evening's work, to reduce data in real-time as it is being gathered or even to run the telescope itself. The dream of an automatic guider has certainly tantalized most astrophotographers (especially on cold winter nights) yet most of these devices have generally fallen short of expectations. Automatic guiders have been made by amateurs and manufactured commercially but the photographs resulting from their use generally show signs that betray the failings of the guider: stars are often trailed and will often show small spikes indicating that the star had moved some distance over a brief time during the exposure.

The photographs of Ron Potter first came to our attention some years ago. His work often appeared in *Astronomy* and *Sky and Telescope* and were typically of several-hours duration, employing instruments of very long focal lengths, yet the guiding was always perfect. At first we were impressed with what seemed to be superhuman stamina in guiding such long exposures but when it was discovered that he used an automatic guider our interest was whetted further. Since we were interested in doing our own long-exposure work, we hoped that Mr Potter had discovered a commercial unit that really worked. When we found that he had designed and built the unit himself we invited him to write a paper describing his autoguider so that others (ourselves included) could benefit from his obvious skills. Mr Potter felt that schematics would not be of much use published in a book format. Hopefully we will see more of his work and the plans for this device in the literature of the near future.

Cooled-emulsion photography: Jack Newton

Cooled-emulsion photography was one of the first methods of fighting reciprocity failure, and of increasing apparent film speeds during long exposures, to become popular among amateur astronomers. The first articles on this technique appeared in the early 1960s and the excellent work of Evered Kreimer helped to lure a number of amateurs into using this process. Since Kreimer's work, other top astrophotographers have kept the cold camera in the spotlight for the past two decades: Orien Ernest, Leo Henzl and Jack Newton. Once the cold camera was made available commercially by Celestron many amateurs were able to try this interesting approach. Until hydrogen hypering was shown to be effective on color emulsions in 1977–8, the cold camera was the only process available for amateurs that wanted to use color emulsions for long-exposure work. Today, cold camera photography is approaching a new era when solid-state thermoelectric cooling 'chips' will replace the dry ice that is currently used to cool film.

Jack Newton has been working with the cold camera for many years and has built and used each of the three basic cold camera types: vacuum, dry gas, and optical plug models. Beginning with a 12 inch reflector, he has built larger and larger telescopes culminating in the 20 inch f/5 Newtonian which he now operates at his Victoria, Canada observatory. Over the past decade he has published three books (the *Cambridge Deep Sky Album* being his most recent), and has published numerous fine photos and articles in *Astronomy*, *Sky and Telescope*, and the *Journal of the Royal Astronomical Society of Canada*. In addition to all of this, he has served as the photographic expert on five solar eclipse expeditions. The paper presented here outlines some of the techniques and processes that Mr Newton employs in producing his superb photographs.

Planetary photography*

JAMES K. ROUSE

Introduction

Even with the advent of recent spacecraft photographs displaying remarkable lunar and planetary detail, many who have a love for the night sky find great satisfaction in photographing some of the details they have seen at the eyepiece.

Mother Nature, the limitations of optical and emulsion science, and observer imperfections band together to make lunar, solar, and planetary photography a challenging, but highly rewarding experience.

We will attempt here to eliminate many of the variables encountered when trying to produce high-resolution photographs of the moon, sun, and planets.

Mother Nature

The astrophotographer has the misfortune of living below a seething ocean of air whose ever-changing flow can at times ruin the chance to see or photograph fine lunar or planetary detail. Depending on where you live (in the mountains, on the plains, or on the shore) and depending on the season and the prevailing weather patterns, periods of calm air may range from less than a second to several hours. Generally, humid, somewhat hazy, wind-free nights will produce a steadier image than will a very clear, dry night sky.

To illustrate, you have probably noticed how dark and clear the sky is right after the passage of a cold front. This is great if you want to observe galaxies or nebulae, but it is usually a disaster for lunar or planetary observations because the air is much too turbulent. It is better to wait two or three days and photograph just before the next cold front comes.

High-resolution astrophotography requires the best of 'seeing' conditions. On a scale of 1 to 10, I have found a seeing of at least 7 is necessary for any chance of a really good photograph. A 'seeing' level of 7 might be

* Some of the films and developers recommended in this chapter are special-order items and will not be found in the average photography supply store or discount store. Any competent photography supplier can special-order Kodak 2415 Technical Pan film in 36-exposure cassettes, or Edwal FG-7 Developer, or Cibachrome print materials. H & W Control Developer should be ordered directly from the H & W Company, Box 332, St Johnsbury, VT 05819, USA.

defined as: 'moderate-to-high power images steady for periods up to 5 s with intermittent image breakup due to turbulence'. This level of clarity allows one to take 20 or 30 exposures of up to 5 s with the probability that at least one exposure will be sharp.

Geography plays a vital role in the quality of the image you will see, so it is best to do plenty of experimenting in your area to determine when the best 'seeing' occurs – before or after midnight, after a thunderstorm, just after dusk, or just before dawn, etc.

Optics

For observing or photographing fine lunar, planetary or solar detail, one must first have a telescope of good optical quality. This usually is not a problem with most commercially made telescopes, mirrors and lenses; however, it is a good idea to check around and even 'test drive' a few different instruments before making a large purchase.

Optical alignment (collimation) is probably as important as optical quality, although no amount of alignment will produce a good image if your primary mirror has gross irregularities (turned-down edge, astigmatism, inaccurate parabolization, etc.) or if you have introduced strains into the optics by improperly clamping or glueing them into their cells. Study a book on telescope making or follow your manufacturer's guidelines for collimation of optics.

If your telescope is a portable one, be sure to check for proper collimation each time you get ready for serious observation or photography. Sometimes it does not take much of a bump to misalign optics just enough to affect image quality.

For high-resolution observing or photography, it is generally best to use a telescope of f/6 or higher for reflectors, f/12 or higher for refractors. With shorter f/ratios, fine collimation becomes much more critical and optical quality also will need to be the absolute best. In addition, with anything other than prime-focus photography you will need to magnify the image by projection through an eyepiece. It is much easier to magnify an image from a long-focus telescope than it is from a short-focus one.

If you already own a short-focus telescope (f/4 or f/5), don't despair. Fine photos can still be taken; however, you will need to spend extra time making sure your optics are strain-free and properly aligned.

Assuming one has high-quality optics, fine collimation, and good seeing conditions, the diameter of your objective lens or mirror will determine how much fine detail you can see or record on film. Generally, a 6 inch or larger telescope is needed to do any worthwhile observing of fine detail; however, smaller instruments of good quality can also be used very effectively. Even 2.4 inches of aperture is enough to record several cloud bands on Jupiter (see Fig. 13.1). Keep in mind, though, that the smaller your aperture, the less light you will have to focus with. So, with small-diameter telescopes, once you have decided how much magnification to use, it will probably be best to focus the camera on a bright star, then to line up the telescope on the planet you want to photograph.

Mounts and drives

High-magnification photography puts an extremely stringent requirement on the mechanical aspects of your instrument. If you are serious about astrophotography, buy a heavy-duty mount. Some telescope manufacturers will sell just a tube assembly, allowing one to build or buy a heavy, precision mount.

There is also the concern regarding mount alignment on the celestial pole. Unless you have critical polar alignment, you will not be able to take 5 s or longer exposures at high magnification without some image blur. A good way to tell if your alignment is acceptable for photography is to use your highest power eyepiece, center a star or planet in the field, and (with the drive running) see whether the image stays centered for at least 10 min. Do not forget – Polaris is almost a degree (twice the diameter of the moon) away from true north, so alignment on Polaris is not quite good enough.

For high-magnification lunar photography, or for long-exposure, prime-focus photos of the lunar crescent or lunar eclipses, you will need a drive corrector to slow down your clock drive to the lunar rate. This is because the moon moves across the sky from west to east at the rate of about 0.5 deg/h (one lunar diameter per hour). That translates into a movement of 5 arc sec for a 10 s exposure – too much movement to record really fine detail.

The camera

Another important aspect of good astrophotography is selection of a camera body which has the features desirable for high-resolution photography. Generally, any single-lens-reflex (SLR) camera can be made to work; however, if you have the funds available, invest in a SLR with interchangeable focusing screens, an eyepiece magnifier and a low vibration mirror/shutter. Any of the better camera brands should meet this last criterion; however, the amount of vibration that will be transferred to the telescope when the mirror slaps up and the shutter clicks open will be inversely proportional to the mass of your telescope and mount.

Fig. 13.1. Photographic resolution test using apertures of 2.4, 3.0, 4.25 and 6.0 inches. This series shows the resolution increase one might expect when photographing through telescopes of various standard apertures. All exposures were 15 s at f/100 on Kodak high-contrast copy film, using an 8 inch f/7 Newtonian with various aperture stops.

If your telescope is not very massive then it will probably be best to begin each exposure by first placing a piece of black cardboard in front of the telescope, open the shutter, wait a few seconds for vibrations to die down, and remove the black cardboard to begin your exposure. An air-bulb shutter release works much better than a cable release (less movement and vibration is transferred from the photographer's hand to the camera).

Interchangeable focusing screens are desirable because the standard screen in most SLRs consists of a microprism or split image surrounded by ground glass. It is very difficult to obtain accurate focus at high magnification on such a screen. A clear, aerial image screen is great for very-high-magnification photography, while a screen with a clear center spot surrounded by a ground glass field is a good all-purpose one, allowing high-magnification focusing on the clear spot and prime-focus focusing on the ground-glass area. This is mentioned because it is often difficult to achieve critical focus at prime focus with a perfectly clear focusing screen.

If you already own a SLR with a non-interchangeable focusing screen, then focusing at high magnification should be done on a star rather than on the moon or planets.

For all close-up photography of the moon, sun or planets, some form of lens must be inserted in the optical path to enlarge the image on the film. This is most commonly done by using a high-quality eyepiece inserted into an adapter between the camera and the telescope focusing mount.

For most high-resolution work, you will need to increase the effective focal length (EFL) of your instrument to at least 300 inches EFL. Many amateur lunar and planetary photos are taken at 800–1000 inches EFL or more. Keep in mind that the more you magnify the image, the longer your exposure will have to be. A practical guideline is to use just enough magnification so that an exposure of 5–8 s gives an image of proper density on the film. Normally, if you try to expose longer than this, clock-drive irregularities, atmospheric turbulence, or slight misalignment on the pole will cause degrading of the image. The actual magnification you achieve is dependent on the focal length of the eyepiece chosen and on the distance of the eyepiece from the film.

To calculate magnification, use the formula:

$$\text{Magnification} = \frac{\text{Distance from center of eyepiece optics to film}}{\text{Focal length of eyepiece}} - 1$$

Once the magnification has been calculated, the EFL can be determined by multiplying magnification by the focal length of the telescope. Similary, the effective focal ratio (EFR or F/number) can be calculated by multiplying magnification by F/number of the telescope. For example, let us assume you are using an 8-inch f/10 telescope of 80 inches focal length. You have determined the eyepiece to film distance to be 110 mm and you are using a 10 mm eyepiece for projection. Magnification, EFL and EFR are then calculated as follows:

$$\text{Mag} = \frac{110}{10} - 1 = 10\times$$

$$\text{EFL} = 10 \times 80 \text{ inches} = 800 \text{ inches}$$

$$\text{EFR} = 10 \times \text{F}/10 = \text{F}/100$$

You will now be using a system equivalent to a telescope with a focal length of 800 inches, or nearly 67 feet!

Keep in mind that the length of exposure required is in direct proportion to the squared values of the F/ratios. For example, a 4 s exposure at f/80 is comparable to a 1 s exposure at f/40: $40^2/80^2 = 1600/6400 \times 4\,\text{s} = 1\,\text{s}$; or about a 6 s exposure at f/100: $100^2/80^2 = 10\,000/6400 \times 4\,\text{s} = 6.25\,\text{s}$.

Since focus, seeing, and drive accuracy are less critical at lower EFRs, you may find it easier to work with an EFR of less than f/80 at the outset.

Films

Astrophotography is a unique branch of photography and, as such, it requires specialized films and processing for optimal results. Nearly any film can be made to produce a pleasing image of whatever object you are photographing; however, for capturing the greatest detail possible on film, careful film selection (and, in the case of black and white photography, careful developer selection) is of the utmost importance. Two schools of thought exist concerning the best way to achieve the greatest amount of image detail:

One school suggests that fast films, coupled with high EFRs (f/150 or so) and short exposures, will produce the best results. The thinking here is that short exposures (2–3 s) will preserve image details that might be lost owing to atmospheric turbulence if the exposure were too long. Fast film is required for short exposures and, since resolution goes down as film speed goes up, greater magnification is required. This technique works well and some very good photographs have been produced in this way. School Two teaches that slow, fine-grained, high-resolution film, coupled with longer exposures at a more moderate EFR, will produce the best image. In order to use this method, the seeing must be very good for periods of 5 s or more.

My experience has been that School One works best on color films and School Two on black-and-white. An inaccurate or unstable clock drive may force one to adopt the first school.

With color photography, we are concerned not only with high resolution, but also with color balance. Some fine-grained color films like the Kodachromes and Ektachrome 64 produce very detailed images, but color balance is frequently off by quite a bit. This is because the film is designed for correct color balance with short exposures. Long exposures shift the color characteristics significantly, as well as slowing the effective speed of the film. For this reason, a 10 s exposure probably will not be twice as dense as a 5 s one. High-speed films like High Speed Ektachrome

(ISO 160), which has been replaced by Ektachrome 200, seem to produce a much better color balance, plus, its higher speed allows one to use a higher EFR, with a larger image. Color plates 4 & 5 show a comparison of planetary images taken on various color films.

In black-and-white photography, resolution, image contract, and density are the chief concerns. Very-high-resolution films are best, such as Kodak's 2415 Technical Pan film. Kodak's High Contrast Copy film (now discontinued) has probably produced the most spectacular results and is a vivid testimonial to the validity of 'School Two' principles. For lunar, solar, and planetary photography, a developer must be chosen which will modify the high contrast of films like 2415. H & W Control developer, Edwal FG-7, or Kodak HC 110 or D-76 give good results, both in image resolution and in contrast control. With H & W Control, try diluting 15 ml of developer with 8 ounces (237 ml) of water and process for 5 min at 75 °F. With Edwal FG-7, the same processing parameters produce good results. 'Fine-grain' developers like Microdol-X should be avoided, as they tend to soften the image, thereby reducing resolution and sharpness.

Lunar photography

Since someone first hooked a camera to a telescope, the moon has been a favorite photographic subject. Even with today's incredible photographic technology, it is still a real challenge to produce a picture with all the fine detail visible through the eyepiece.

Prime-focus photography of the moon has initiated many astrophotographers. Rather nice photographs are possible with the smallest of telescopes, or even with common telephoto lenses. With prime-focus photography there is normally plenty of light available from the moon, so a film having the finest grain possible should be chosen. Two exceptions are lunar crescents with earthshine (color plate 6) and lunar eclipses, both of which require fast color films. Tables 13.1 and 13.2 list some recommended films and exposures for shooting various lunar phases at prime focus and at an EFR of f/80. Since the moon's brightness varies considerably with

Table 13.1. *Data for prime-focus lunar photography*

Lunar phase	Film	Exposure at f/8 (s)
Crescent	Tri-X Ektachrome 200/400 Kodacolor 400	1–30 depending on thickness of crescent
First or third quarter	2415 Technical Pan Kodachrome 25/64 Kodacolor 100	1/15 to 1/4 1/125 to 1/60 1/250 to 1/125
Full moon	2415 Technical Pan Kodachrome 25/64 Kodacolor 100	1/250 to 1/60 1/125 to 1/60 1/500 to 1/250

changing phases, any recommended exposures will require some adjustment if the moon's phase is more than a day or so on either side of first or third quarter. In black-and-white photography of the moon, whether at prime focus or with eyepiece projection, use very fine-grained high-resolution films like Kodak 2415 Technical Pan. Also, a developer which will moderate the contrast, like H & W Control or Edwal FG-7, is necessary so that your final print will have good tonal quality and not the super contrasty blacks and whites we see on so many amateur photographs (see Figs. 13.2 and 13.3). Plates 6 & 7 show the results that can be achieved with both small and large telescopes in color photography of the moon, both at prime focus and at high magnification.

Solar photography

Nearly every telescope manual contains precautionary warnings of the dangers involved in viewing the sun, and the very same cautions apply in solar photography. There is only one method generally recognized as safe for viewing the sun through eyepiece or camera – the use of a vacuum metalized sun filter directly over the aperture of the telescope. Such filters serve to protect not only the viewer's eyes, but also the telescope, eyepiece, and the sensitive metering systems of many cameras. The use of any other type of filter – including neutral-density filters, cross polarization, exposed film and even the popular green-glass filters supplied with some telescopes – is strictly for those willing to risk serious eye damage.

The same basic film recommendations for lunar photography also apply when shooting the sun, but exposures will vary considerably with the brand and size filter you decide to use. Table 13.3 lists some starting exposure values, but there's no substitute for generating you own exposure data by trial and error.

Planetary photography

Utilizing the techniques already discussed, let us now consider how each of the planets represents a unique challenge to the photographer.

Mercury

Photographing Mercury is one of the toughest assignments for the astrophotographer. Not only is the planet almost as small as our moon, but it is confined to a tight orbit around the sun that prevents it from being

Table 13.2. *Data for high-magnification photography of the moon near first or third quarter*

Film	Exposure at f/80 EFR (s)
2415 Technical Pan	2–4
Kodacolor 100	2–4
Kodachrome/Ektachrome 64	3–6

Fig. 13.2. Black-and-white photographs of the full moon – one very high contrast, one normal contrast: (*a*) was processed using a high-contrast film developer (Kodak D-19), while (*b*) used H & W Control Developer. Both negatives were printed on a #2 grade paper with identical enlarger settings and exposure times.

Table 13.3. *Data for sunspot photography*

Film	EFR	Exposure (s)
2415 Technical Pan	f/8	1/500–1/250
	f/50	1/60–1/30
Kodacolor 100	f/8	1/500–1/250
	f/50	1/60–1/30
Kodachrome/Ektachrome 64	f/8	1/500–1/250
	f/50	1/30–1/15

Fig. 13.3. High-magnification lunar photography. Tri X and 2415 compared. When the 'seeing' is good, high-resolution films produce sharper, less grainy photographs. Compare these two high-magnification lunar photographs: (*a*) on Tri-X film; (*b*) on 2415 Technical Pan film.

(*a*)

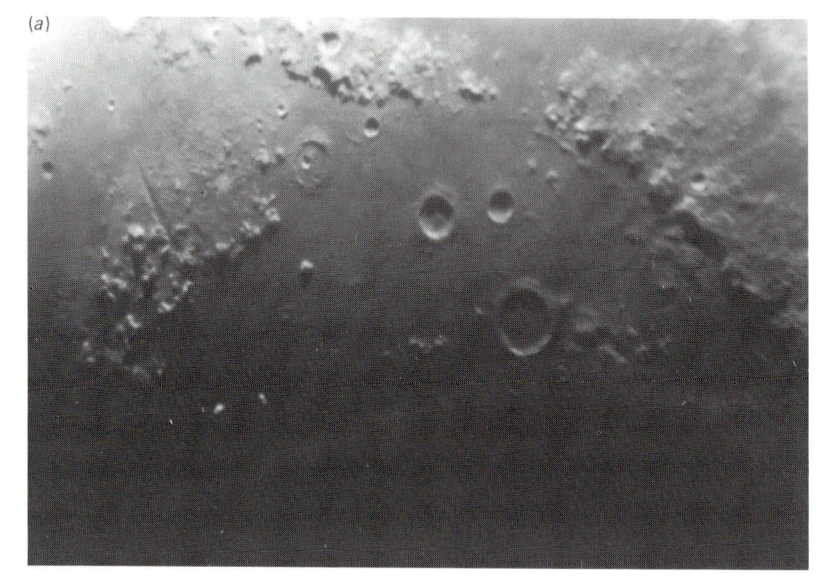

(*b*)

visible well above our horizon in a relatively dark sky. Because of this, we are forced to view Mercury through many miles of our own turbulent atmosphere and, consequently, we rarely have an undistorted view of it.

To an observer, Mercury runs through phase changes similar to those of the moon. It is best seen at eastern or western elongation (maximum separation east or west of the sun) when its phase resembles the first or third quarter moon; half of Mercury's surface will appear illuminated. This is the ideal time for photographing Mercury because it is as far as it can ever get from the blinding glare of the sun.

If you are fortunate enough to live in an area of high elevation which typically has good seeing conditions, you may want to try photographing Mercury in a twilight sky during one of its favorable eastern or western elongations. Start searching for Mercury right after sunset or about half an hour before sunrise and, as soon as you locate the planet, begin photographing (even if the sky is quite bright).

Since Mercury is small and you will often be shooting through a lot of atmosphere in a relatively bright sky, it will probably be best to stick with a high-resolution, high-contrast emulsion such as Kodak 2415 Technical Pan film. Processing should be in a high-contrast developer like Kodak D-19. If you want to use color, try Ektachrome 64 and have your laboratory 'push' it about two extra stops for increased contrast.

When shooting Mercury, use just enough magnification to see the phase and shape of the planet on the camera viewscreen. Plan to use a whole roll of film at whatever magnification you have chosen and 'bracket' your exposures widely; e.g. from 1/30 s to 10 s. When shooting in a bright sky, a red or yellow filter may also help to increase the contrast between Mercury and the sky.

Venus

Photographing Venus is, in many ways, similar to photographing Mercury. However, it is much easier because Venus' elongations carry it farther from the sun.

Nearly all of Venus' phases can be photographed. Over a period of just a few months Venus gradually changes from a nearly full disk to a very large, thin crescent and, as it approaches either full-disk phase or crescent phase, it will appear lower and lower in our sky; we must therefore be prepared to photograph these phases in a relatively bright sky just after sunset or before sunrise. High-contrast films are in order then, just as they are for Mercury. An interesting project would be to see how close to conjunction with the sun you can photograph Venus. A clear view of the eastern or western horizon will be essential in such a project.

As Venus approaches Earth and grows larger in angular diameter, you will need less and less magnification to produce an acceptably large image on film. An EFR of about f/50 with 2415 or Ektachrome 64 film and exposures of around 1/8 to 1 s should produce good results, but do not hesitate to bracket your exposures even more than this, especially if you shoot in bright sky conditions.

Mars

In attempting to photograph the fine, low-contrast details on Mars, generally the slower, fine-grained films perform better; however, when Mars is near its largest apparent size, the higher-speed films such as Ektachrome 200 also work well.

With black-and-white photography of Mars, the standard films such as Tri-X, Plus-X, and even Panatomic-X prove too grainy and do not exhibit enough image contrast or resolution. Of all the black-and-white films tried, High Contrast Copy film produced the best results. Since this film is no longer available, Kodak 2415 Technical Pan is the next-best choice and should produce nearly equivalent results. Since Mars displays very low-contrast details, it may be tempting to beef-up the contrast by using a high-contrast developer such as Kodak D-19. However, with this procedure, the details on the Martian terminator tend to be lost, making Mars look more gibbous in the final print than it really is. For this reason, H & W Control Developer or Edwal FG-7 should be used, diluted 15 ml of developer in 8 ounces (273 ml) of water, and processed for 5 min at 75 °F. Optimum exposure on 2415 film should be about 2–4 s at f/100.

Jupiter

Jupiter is probably the most-photographed planet. It is large enough and exhibits enough detail that nearly any size telescope can be used to yield pleasing photographs.

Whatever size telescope is used, a minimum of about 250 inches EFL will be needed to produce an acceptably large image on the film. Using 2415 film, an EFR of about f/100, with an exposure of 4–8 s should produce optimum results. As with Mars, H & W Control Developer or Edwal FG-7 (same dilution and development) should be used for true-to-life images with moderate contrast.

Jupiter is an excellent subject for color photography with telescopes of 6 inch or larger aperture. Cloud bands of various hues, as well as the Great Red Spot, can be captured in fine detail and beautiful color. Some color films produce a color shift with long exposures that can be very undesirable. To avoid this, Ektachrome 200 is a good film choice and will produce accurate color images, as well as ample resolution, at an EFL of 800 inches or greater. Exposures of 2–4 s at f/100 or 6–8 s at f/150 should give excellent results on Ektachrome 200. To preserve as much fine detail as possible, it will probably be best to stay with films of index ISO 200 or less.

Saturn

Over a period of about 14.75 years Saturn's position, relative to ours, makes its ring system appear to open or close so that, at times, we have a beautiful view of the rings and at other times they all but disappear.

Saturn is smaller than Jupiter and twice as far from the sun, so it appears both dimmer and smaller than the Giant Planet and, consequently, it is more difficult to photograph successfully. As with Jupiter and Mars, the standard black-and-white films are just too grainy and do not produce enough image resolution to be the films of choice for Saturn (Fig. 13.4). Kodak 2415 Technical Pan again wins out as the best all-round film. At an EFR of f/80, 4–8 s using H & W Control Developer or Edwal FG-7 should produce optimum results. For color work on Saturn, Ektachrome 200 works best. Exposure should be 4–8 s at f/100 for a very nice image.

Uranus

Uranus presents a unique challenge for the serious planetary photographer. Because of its distance from us and from the sun, Uranus presents a disk only 4 arc sec in diameter and around 6th magnitude brightness. So, to capture an image of Uranus of any dimensions, a very long EFL is needed, which in turn, forces one to use very long exposures.

Photography of the pale, green disk of Uranus is probably best left to those with access to a 12 inch, or larger, telescope. If you want to meet the challenge, try coaxing about 1500 inches EFL out of your telescope. Use Ektachrome 400 or Tri-X developed in Edwal FG-7 and expose for 30–60 s at f/200.

Neptune and Pluto

Because of their dimness and small angular diameters, both Neptune and Pluto are best photographed at prime focus amid the starry background. Of course, no visible planetary disk will be seen; however, photographing on successive nights will show the movement of these wanderers amid the stars.

Darkroom techniques

With proper attention paid to the preceding techniques, you should now have a first-class image on film of your favorite solar system object. At this point, what you do with that image in the darkroom in making a black-

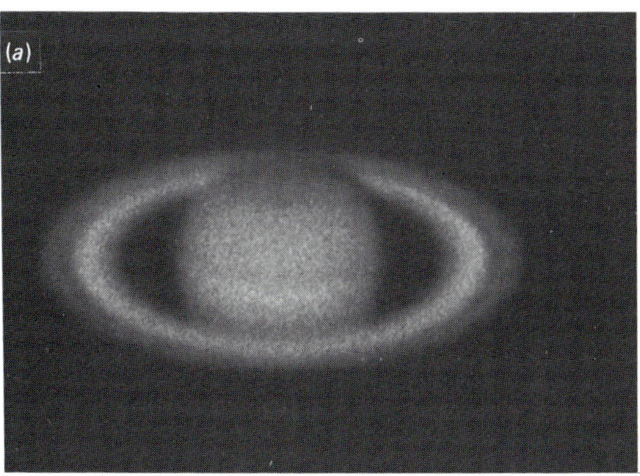

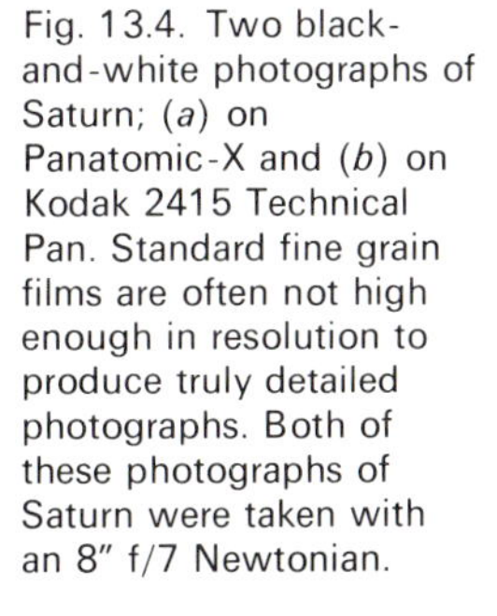
Fig. 13.4. Two black-and-white photographs of Saturn; (*a*) on Panatomic-X and (*b*) on Kodak 2415 Technical Pan. Standard fine grain films are often not high enough in resolution to produce truly detailed photographs. Both of these photographs of Saturn were taken with an 8″ f/7 Newtonian.

and-white or color print can mean the difference between a great portrait or a mediocre print.

With black-and-white printing, be sure to plan ample time for experimenting with various degrees of enlargement and various contrast grades of paper and exposures in order to get a print of proper contrast and detail. Often, a change of one contrast grade can have quite an impact on the quality of the image. With experience, you will be able to produce film exposures with just enough density to print well on a #2 grade of paper. Do not hesitate to invest in a set of polycontrast filters (or use the magenta filters in your enlarger if you have one so-equipped) and experiment until the perfect print is achieved.

Sometimes 'sandwiching' two or more negatives of the same object together will produce a better-looking image on paper. This is true because of the grain-cancellation effect the two negatives have on one another. Also, if you find you have not given the film enough exposure and you wind up with very light images on the negative, sandwiching can produce an image dense enough to print well on paper of a normal contrast grade.

In color printing, I have found that it is best to take color slide photographs and to make direct positive prints on reversal papers because, when you take color negative photographs, you will have no slide for projection.

Choose the process of your choice; however, if you want the best, the Cibachrome process is highly recommended. All color planetary photographs in this chapter were reproduced from Cibachrome prints. This process is easy to use, requires a minimum of filtration changes, no color analyzer, and produces prints with great color saturation and highly fade-resistant images.

Table 13.4. *Planetary exposure table*

Planet	Film	EFR	Exposure (s)	Developing
Mercury	2415 Technical Pan	Variable	Variable	D-19
	Ektachrome 64	Variable	Variable	Pushed 1 stop
Venus	2415 Technical Pan	f/50	1/8–1	D-19
	Ektachrome 64	f/50	1/8–1	Pushed 1 stop
Mars	2415 Technical Pan	F100	3–5	H & W Control or FG-7
	Kodachrome 64	f/100	4–8	Standard
Jupiter	2415 Technical Pan	f/100	4–8	H & W Control or FG-7
	Ektachrome 200	f/100	2–4	Standard
Saturn	2415 Technical Pan	f/80	4–8	H & W Control or FG-7
	Ektachrome 200	f/100	4–8	Standard
Uranus	Tri-X	f/200	30–60	Edwal FG-7
	Ektachrome 400	f/200	30–60	Pushed 1 stop

Filters for astrophotography

JACK B. MARLING

Let's face it, most of us live in or near urban locations with only fair skies that have substantial artificial light pollution. We need good filters to give us a strong advantage in our battle against the ever-encroaching pollution of the night sky. In visual astronomy it is not too hard to drive to a dark location and set up a large altazimuth Dobson-type telescope; however, for photography such trips require a lot of extra effort because of the extra necessary equipment. Then, once at a dark-sky site, one still needs to align the scope accurately on the celestial pole. This is most accurately done by 'drift alignment', which takes an hour or more.

Whenever possible, it is best to use a permanent site for astrophotography. Thus, I have a small observatory in my backyard in Livermore, California with two permanent piers that carry a 14.5-inch f/7.1 telescope and either a 10-inch f/5 or an 8-inch f/4.5. However, backyard photography requires filters in order to obtain beautiful photographs in or near big cities. This is well illustrated by the beautiful photograph (Fig. 13.5) of the Rosette Nebula which was taken from my backyard observatory using a deep-sky filter. This dim nebula would have been impossible to photograph without a filter. A skilled astrophotographer at an urban site can duplicate the quality of photographs taken at dark-sky sites without filters. The only price you pay is in exposure duration.

Red filters

In general, black-and-white photography with red filters should be reserved for emission nebulae which have a strong red emission near 656 nm as shown in Fig. 13.6. A No. 25 red filter has an orange cut-off wavelength of 590–600 nm. This means that it absorbs all light at shorter wavelengths and transmits all light at longer (red) wavelengths. This is adequate to suppress most blue, green, and yellow light pollution from mercury streetlights (Fig. 13.7). However, some of the light from the new high-pressure sodium streetlamps (Fig. 13.7) passes through a No. 25 filter and stronger filters may be needed like a No. 29, No. 92, or an H-alpha pass filter (discussed below).

The only photographic components of emission nebulae passed by a red filter are the ionized nitrogen emissions at 655 and 658 nm and the H-alpha emission at 656 nm. Since these lines carry much of the emission energy and are at essentially the same wavelength, the only requirement of a black-and-white photographic filter for nebulae is that it pass light near 656 nm with the highest possible transmission. The closer the cut-off wavelength is to 656 nm, the more light pollution will be rejected and the attainable contrast boost will be greater using such a filter to photograph emission nebulae.

The No. 29 dark red filter has a cut-off wavelength (the point where transmission has fallen to 50 %) of about 610 nm. The No. 92 filter has a cut-off wavelength of about 635 nm. The Lumicon H-alpha pass filter has a cut-off wavelength of 640 nm yet has 90 % transmission for the H-alpha and ionized nitrogen lines and is the strongest sharp cut-off glass filter available.

Fig. 13.5. This photograph of the Rosette nebula was taken using an 8 inch f/4.5 reflector and a Light pollution rejection (LPR) filter to reduce light pollution from nearby Livermore, California.

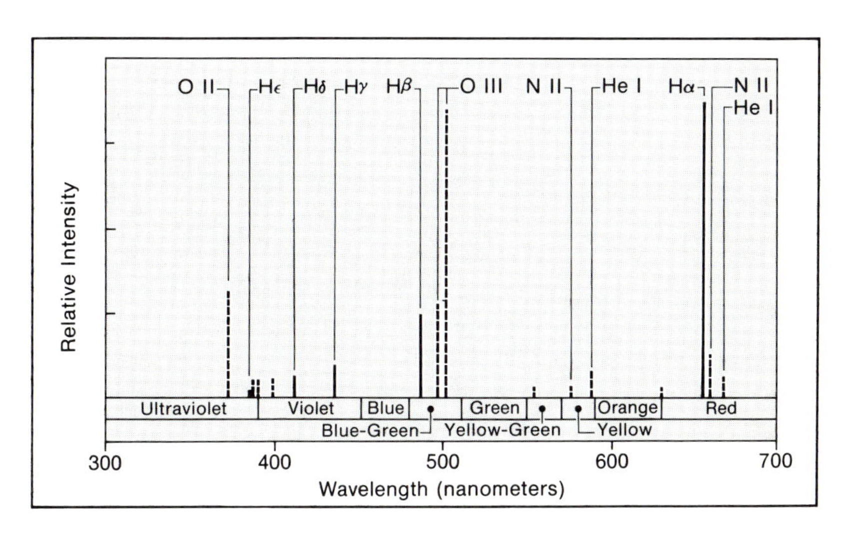

Fig. 13.6. The spectra of a typical emission nebula is shown with spectral lines identified. The regions of the spectra corresponding to commonly used color names are labeled in accordance with accepted color standards. The nebula used for this example is the Trapezium region of M42. Other nebulae spectra vary widely from this example; the lines that generally vary in intensity from nebula to nebula are shown with dashed lines.

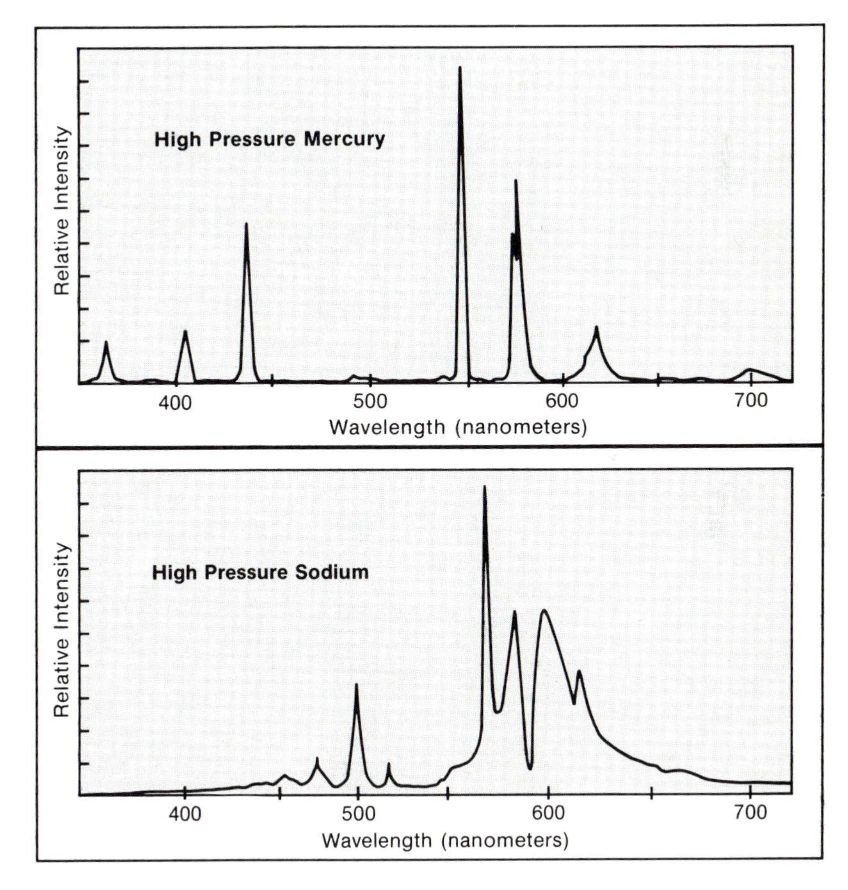

Fig. 13.7. The spectra of high-pressure mercury and sodium lights, recorded by a low-resolution scanning spectrometer and thus the singular spectral lines appear as broadened peaks. In the case of the high-pressure sodium light, the discrete spectral lines are superimposed upon a broad continuum in the 550–650 nm range (spectra courtesy of Jack Marling).

The Schott RG645 filter has a 650 nm cut-off wavelength but has only 60 % transmission at 656 nm yielding lower contrast than the Lumicon H-alpha pass filter (see Fig. 13.8).

The use of a very dark red filter for prime-focus photography is difficult, since it becomes very hard to find a star bright enough to achieve camera focus.* Thus a compromise is often needed in selecting the filter for such applications. The No. 29 dark red filter transmits enough light to see some stars for camera focusing, yet yields good rejection of unwanted light pollution. However, the strongest possible filter should be used if focusing does not require examination of a star image, such as when one is using telephoto lenses.

The H-alpha pass filter

Many of the largest deep-sky objects in the night sky are diffuse emission nebulae. These are photographed best with wide-angle camera lenses or telephoto lenses. Examples are Barnard's Loop in Orion, the California nebula in Perseus, the North American and Veil nebulae in Cygnus, and many others. Unfortunately, many of these nebulae are so faint that it is almost impossible to photograph them well without using strong filtration to remove artificial light pollution. Even from the darkest sites, natural airglow due to atomic oxygen at 558 nm and 630 nm interferes with the extremely long-duration photographs necessary for the faintest nebulae. In all cases, without exception, the best filter is the H-alpha pass filter (Lumicon trademark). It absorbs all visible light up to 640 nm yet transmits 90 % of the light at the 656 nm H-alpha line. It is very dark red in appearance and removes all artificial and natural light pollution. It is especially useful in areas with many high-pressure sodium streetlights which emit a lot of undesirable yellow, orange, and red light pollution into

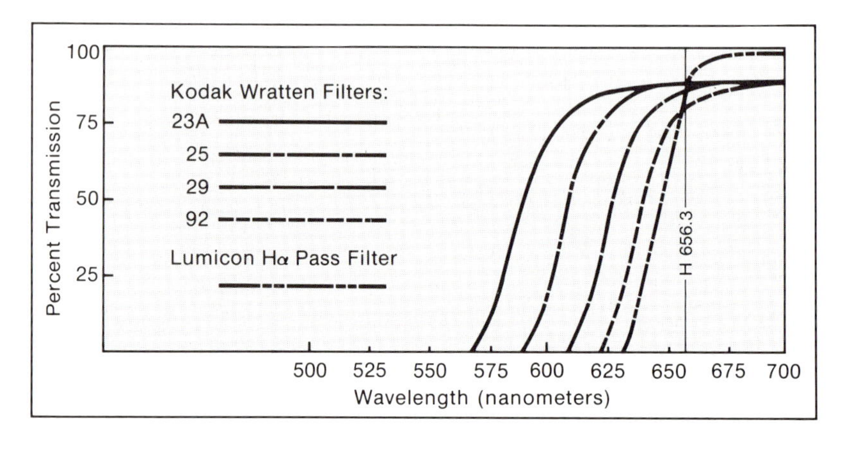

Fig. 13.8. The transmission curves of four commonly used 'red' filters are shown along with that of a Lumicon H-alpha pass filter to illustrate how the cut-off wavelength approaches the 656.3 nm hydrogen line as one moves from one filter to another.

* Authors' note: another alternative is to use a clear filter (glass or gelatin) in place of the dark red filter for focusing purposes. This filter must be the same optical thickness as the filter that will be used to take the photograph.

the sky that cannot be removed entirely with conventional red filters like the No. 25 and No. 29 filters.

The H-alpha bandpass filter has 90% transmission for both the H-alpha line at 656 nm and the ionized nitrogen line at 658 nm that are important in emission nebulae. Light pollution is reduced 10–20 times, which means that contrast can increase about 10-fold on photographs exposed to the sky fog limit. This is illustrated in Fig. 13.9, showing two photographs of the North American nebula taken from my observatory site in Livermore, California. The exposure for Fig. 13.9 (left side) was 16 min in duration without a filter. Light pollution was not extremely bad and some nebular detail is shown. Fig. 13.9 (right side) was taken the same night but with the H-alpha pass filter for a 32 min exposure. Light pollution was completely removed, contrast is better and the star images are greatly improved.

Light-pollution-reduction filters

Some of the new light-pollution-reduction filters (LPR filters) used to improve visual astronomy can also be used to improve prime-focus astrophotography of planetary and diffuse nebulae. The primary and

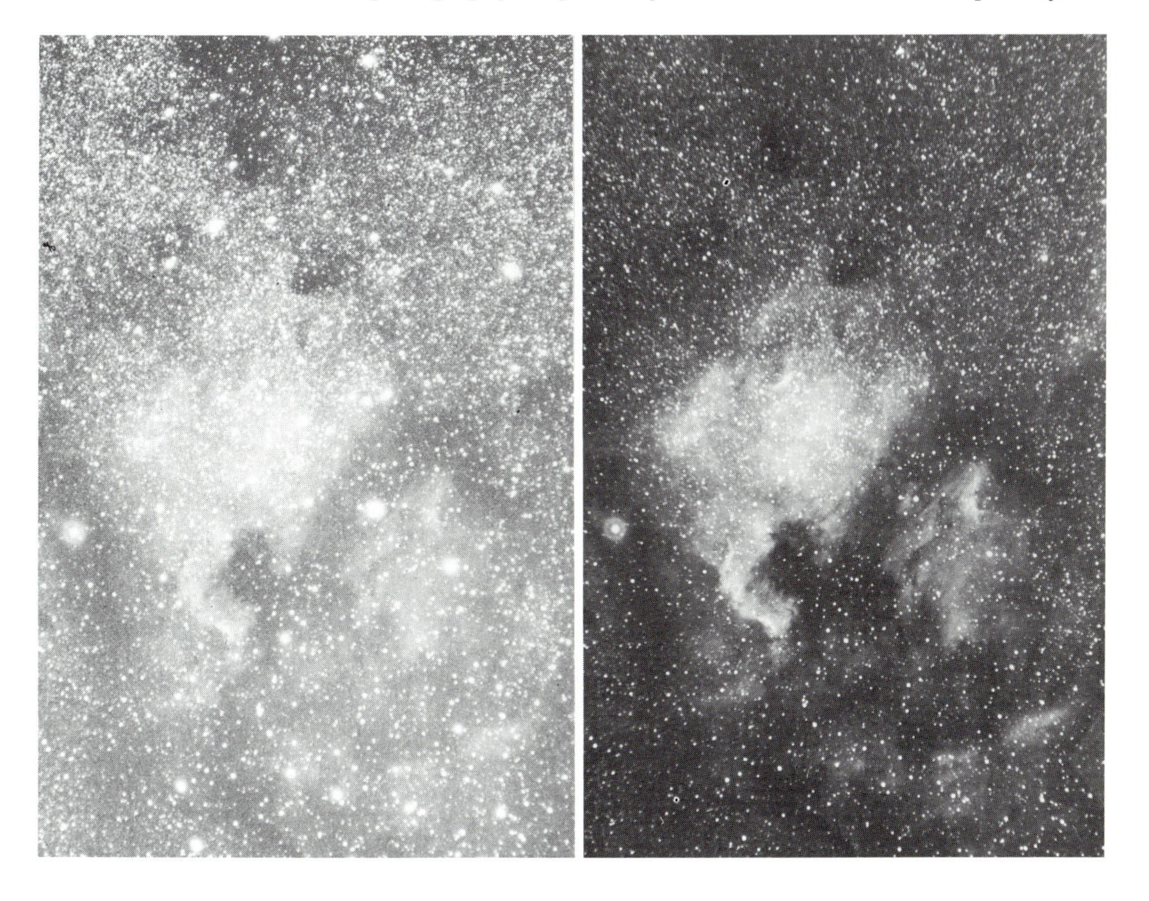

Fig. 13.9. These photographs illustrate the dramatic improvement that can be obtained by using a red cut-off filter on hydrogen emission nebulae. The exposure on the left was 16 min without any filtration while the exposure on the right was 32 min using a Lumicon H-alpha pass filter.

essential requirement is the highest possible transmission near 656 nm in the red to record the H-alpha and ionized nitrogen emissions. Since a well-made photovisual LPR filter has 80–95 % transmission on nebular emission lines, the 'filter factor' is quite small and previously established exposures for given objects need only be multiplied by about 1.2 to obtain the equivalent exposure using an LPR filter. Figure 13.10 shows the transmission curve for the Lumicon deep-sky filter.

A most dramatic illustration of the potentials of the LPR filter can be seen in the photograph of the beautiful Rosette Nebula, NGC 2237, in Monoceros shown in Fig. 13.5. This was taken in Livermore, California which is a typical urban location plagued by light pollution. From this location, I reach the sky fog limit in 20 min using hypered 2415 TP in an f/4.5 telescope. For a very dim nebula like the Rosette, 20 min is too short an exposure to record it well. With a deep-sky LPR filter, I was able to expose hypered 2415 TP for 60 min before reaching the sky fog limit. This resulted in a dramatic increase in both detail and contrast, as illustrated in Fig. 13.11 which shows a comparison of these two photographs. Both photographs were taken on the same night, using the same film and were developed and printed identically. Thus, even from backyard locations plagued by urban light pollution, superb astrophotographs are possible.

Filters *can* be used in color photography. The only requirement is the need to preserve an adequate color balance. Thus a dark red filter cannot be used in color photography for obvious reasons. A well-made LPR filter for color photography must pass light in the blue, green, and red regions to yield satisfactory color balance. Thus, a wide bandpass filter like the deep-sky filter is ideal since Fig. 13.10 shows that these essential colors are passed to the film. Again, high transmission of the H-alpha line is essential. In color photography, the sky fog limit is reached when light pollution causes a *slide* to lose contrast or to have a light background. Exposures 2–4 times longer are possible using an LPR filter. Beyond that, the color balance of the

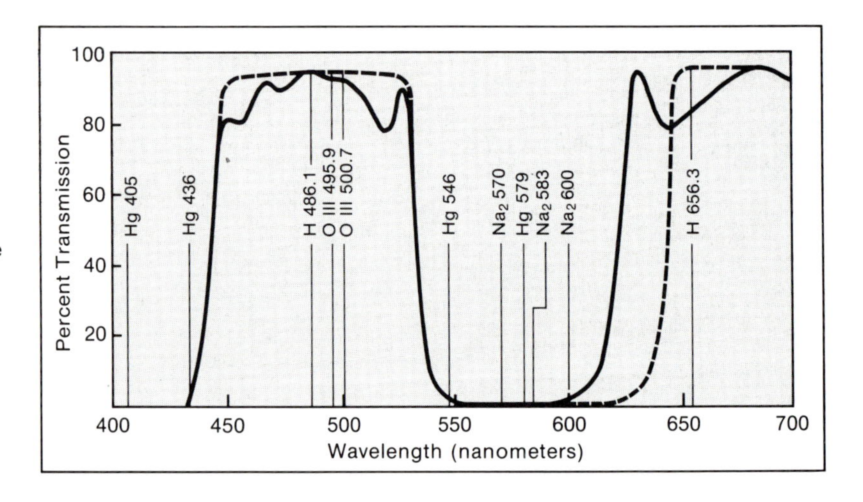

Fig. 13.10. The transmission curve of an 'ideal' nebula filter is shown (dashed line) along with that of an actual nebula filter (solid line). In addition, prominent lines of hydrogen nebulae and light pollution sources are shown.

background is shifted and becomes purplish in very long-filtered color photographs.

Schmidt-camera photography

These specialized cameras are characterized by both high resolution and high speed. Currently available commercial Schmidt cameras have photographic speeds ranging from f/1.5 to f/2.4 and custom units are sometimes encountered with speeds as fast as f/0.5. The sky fog limit with such instruments will typically be from 2 to 8 min using hypersensitized films. Filters are almost essential for these fast optical systems. However, the commercial units are only set up for use with thin gelatin filters such as the Kodak Wratten filters.

Only two filters are optimum for use in these commercial Schmidt cameras: they are the red No. 92 Wratten filter and the blue–green No. 44 Wratten filter. Both the No. 29 and No. 92 filters have the same 80–85 % transmission at the H-alpha line, but the No. 92 rejects about twice as much light pollution and thus yields far higher contrast.

For galaxies and other blue objects, a red filter is unsuitable. For these objects, the blue–green No. 44 filter is far superior. It rejects artificial and natural light pollution and can yield 2–3 times higher contrast than an unfiltered exposure. Exposure durations with systems in the f/1.7–f/2.4 range can be extended to 15–30 min using this filter with hypered 2415 Technical Pan.

Fig. 13.11. These photographs illustrate the dramatic improvement that can be obtained by using a light–pollution–rejection filter on emission nebulae. The exposure on the left was 20 min without any filtration while the exposure on the right was 60 min using a Lumicon deep-sky filter.

Minus violet filters for pinpoint star images

In most telephoto lenses, the usual astronomical telescope mirror is replaced with a number of lenses which perform the image-forming task. This introduces an important disadvantage, namely residual chromatic aberration. Commercial lens systems are corrected to provide relatively color-free or achromatic images in the blue and yellow (or red) regions of the spectrum. Unfortunately, most stars give off a lot of violet and ultraviolet radiation which will be recorded by most films. Thus a star's image comes to a pinpoint focus for visible light but is out of focus in the violet and ultravoilet. For pinpoint star images, when telephoto lenses are being used, it is generally necessary to use a filter which absorbs both violet and ultraviolet light.

Fig. 13.12 shows three 5 min exposures of the Pleiades taken on hypered 2415 TP using a 300 mm f.l. f/4, telephoto lens. All were taken on the same night using the same film and all were developed and printed identically, in order to facilitate comparisons. Figure 13.12 (left side) was taken without any filter and has the characteristic bloated star images created by residual chromatic aberration. Figure 13.12 (center) was taken with a standard commercial 'UV filter', which removes ultraviolet light below its 380 nm cut-off limit. This filter provides only a slight improvement and the stars are still rather large. Fig. 13.12 (right side) was taken using a minus violet filter which removes both ultraviolet and violet light below its 440 nm cut-off wavelength. Now the star images are much sharper and higher resolution is obtained. Note the small triplet of stars that is resolved near the bright star in the center. Note also that there was virtually no loss in the limiting stellar magnitude. The light that is removed by a minus violet filter does not come to a focus so it contributes almost nothing toward the

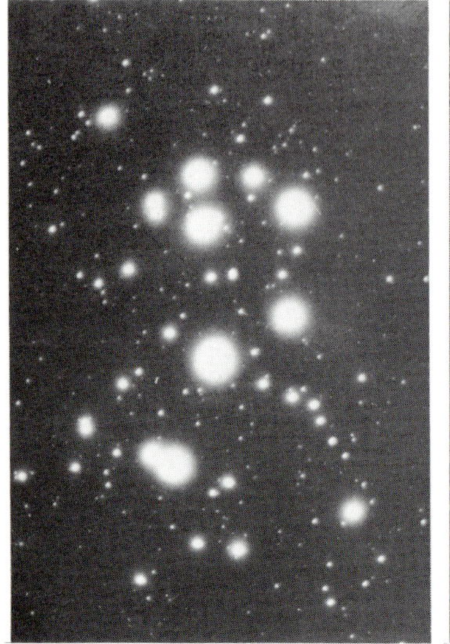
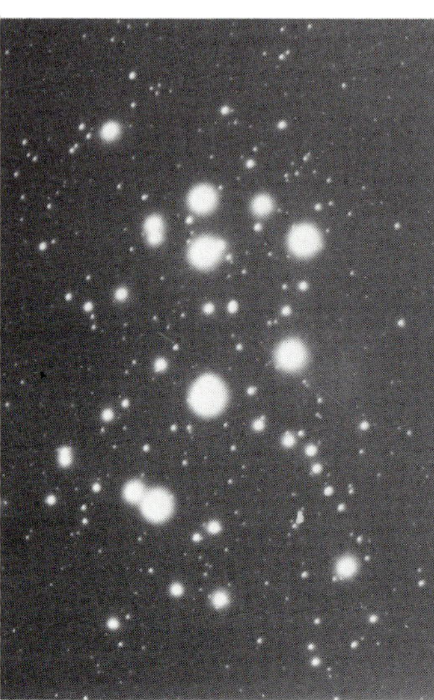

Fig. 13.12. Marked image improvement with telephoto lenses is often obtained by removing light from the ultraviolet as this series of photographs shows. The image at left was taken using no filtration. The center image was made using a standard photographic 'UV filter'. The image on the right was taken using a Lumicon minus violet filter.

formation of a faint pinpoint image. Therefore, the image improvement is obtained without any cost to the photographer.

In color photography the minus violet filter will reduce or eliminate the blue halos that form around bright star images. In the photography of emission nebulae, it also enriches the red color by removing the light from the ultraviolet oxygen emission line at 373 nm. That line records as blue on color film and causes nebulae to look 'pink' (or magenta). The minus violet filter with a 440 nm cut-off wavelength is a pale green color and does cause a color shift. If light pollution is present, the green and yellow light from sodium and mercury streetlights causes exposures to appear greenish. This color shift can be eliminated by using a minus violet filter with a 420 nm cut-off wavelength. This filter is clear in appearance and yields better color renditions. For black-and-white photography, the 440 nm minus violet filter is to be preferred as it removes the violet light pollution from mercury vapor lines at 405 nm and 435 nm.

Conclusion

Since we have to live with light pollution, filters give us the ability to remove selectively unwanted skyglow while passing the desirable light from a deep-sky object. Visually, a good high-contrast LPR filter makes faint nebulae much easier to see and make it possible to see some nebulae that are totally invisible when viewing is attempted without the LPR filter. Since film is more affected by light pollution than the human eye, filters yield a greater improvement in astrophotography. The use of red cut-off filters with films like 103a-F and 2415 Technical Pan can yield dramatic increases in the amount of detail that can be recorded in some faint nebulae. The special combination of the H-alpha pass filter and hypered 2415 Technical Pan yields remarkable contrast gains in nebula photography. The red cut-off filters and UV cut-off filters also help to reduce or eliminate the effects of chromatic aberration when using telephoto lens systems. LPR filters with high H-alpha and high visible light transmission allow both color and black-and-white astrophotography. With filters, remarkable photographs of faint deep-sky objects are possible even from areas suffering from severe urban light pollution. The increased use of filters in several areas has contributed significantly to the revolution in amateur astronomy.

Automatic guiders for astrophotography

RON POTTER

Introduction

Those who have spent a few nights taking long-time-exposure photographs through a telescope by visual guiding will sympathize with anyone who proposes to automate this procedure. The requirement becomes even more compelling when a long focal length (and hence slow) telescope is used to increase the image size for improved resolution. Four to six hour exposures are often needed to obtain adequate film image density, especially if filters are used.

Unfortunately, most commercially available trackers are currently not sensitive enough to be very useful. Owing to differential atmospheric refraction between the guide star and the object of interest, and owing to slight misalignment of the telescope polar axis, it is necessary to choose a guide star that is close to the object being photographed, as well as one that is oriented properly in position angle to minimize errors due to these effects. For example, a guide star should be chosen at nearly the same elevation above the horizon as the object of interest to minimize tracking rate differences due to differential atmospheric refraction. As a consequence of these restrictions, the tracker must be able to guide on a wide variety of stars, especially dim ones, that might appear at this optimum location. Experience has shown that it is necessary to be able to track on an 11th or 12th magnitude star (as measured through an 8 inch diameter reflector) reliably for several hours to satisfy this sensitivity requirement.

Some of the advantages and a few of the problems associated with automatic guiding are described here, including some ideas on the design of an automatic guider to reduce or eliminate these problems. The author has designed and built a guider that will track reliably on a 12th magnitude star (through an 8 inch telescope), and has logged over 700 h of guiding to date, in photographing over 180 different deep-sky objects. Most exposures last for 3 or 4 h, although several 6 h exposures have been made, with good results.

The photograph shown as Fig. 13.13 is an example of a 6 h exposure taken through a No. 92 red filter on hypersensitized Kodak 2415 film. The telescope was a 14 inch Schmidt–Cassegrain type, having a focal length of 3.91 m (f/11 system). The object is the Bubble nebula (NGC 7635) located at 23 h 19.6 min in right ascension, and 61.05 deg in declination (1975

Fig. 13.13. A 6 h exposure of the Bubble nebula (NGC7635) through a No. 92 filter on hypersensitized Kodak 2415 film, using a 14 inch f/11 telescope. Guide star is approximately 0.4 deg below the object.

coordinates). The guide star is about 0.4 deg below the object. The elongated star images above the object are due to differential atmospheric refraction. For this object, the use of a guide star to one side causes considerably more distortion of the star images than shown in this photo.

Advantages of guiding automatically

Even with some of the newer fine-grain films, the ultimate resolution limit on most astrophotographs is set by the film grain. Thus, the amount of detail in a photograph is generally dictated by the image size. Every time the image size is doubled, the area is quadrupled, and the amount of visible detail increases by a factor of four. Unfortunately, larger image sizes mean reduced light intensity for each unit of area on the film, and hence the exposure time must be increased to maintain the same contrast.

Most amateurs can tolerate visual guiding for 30 min or so, but 1 h exposures begin to resemble drudgery, and 4 h shots are really not feasible. An automatic guider allows the photographer to engage in more constructive activities, such as reading, or even sleeping! The net results are many more photographs of higher quality each season, and a much more comfortable photographer.

Some potential problems

As any astrophotographer well knows, there are a multitude of things that can and do go wrong during most exposures. This can be a very frustrating hobby, since none of the exposures ever turn out to be perfect. The use of an automatic guider does very little to eliminate these problems, and a few new problems are created. However, the one consolation is that very little time and effort has been wasted when a photograph comes out poorly, since the automatic guider can simply be utilized to try one more time.

In order to obtain the required sensitivity, an aperture must be used to eliminate as much as possible the background skylight around the guide star. This implies that any perturbations of the guiding telescope can move the guide star out of this aperture rather easily. Hence, gusts of wind can be very annoying. Also, some telescopes tend to jerk occasionally, as the unit is rotated by the clock drive over the duration of an exposure. Problems with the guider due to small earthquake tremors have even been experienced on several occasions. If the optical elements or camera are loose, then these may move during an exposure. In many of these cases, the cause is difficult to determine, and there is a tendency to blame the automatic guider for the poor performance.

The effect of atmospheric scintillation (seeing) is normally not a problem, unless it is so bad that the photograph itself is degraded. If the scintillation is fast, then the tracker simply averages the star position. If it is very slow, then the tracker can correct (at least partially) for these apparent changes in star position.

The problems caused by differential atmospheric refraction are generally ignored by most astrophotographers, but become very important for long-time exposures, or for wide-angle shots using a Schmidt camera. The Earth's atmosphere refracts all of the incoming light (except for light entering at the zenith), in a manner such that all images seem to have a greater elevation angle above the horizon. This effect is of the order of 30 arc min at the horizon, and decreases rapidly for higher elevation angles. However, the apparent tracking rate required to follow an object varies with the elevation of the object above the horizon. Hence, if a guide star is chosen with an elevation different from the object being photographed, then the object will move slightly with respect to the guide star over the duration of the exposure, and a blurred image will result. Computer simulations show that the trajectories of stars around the guide star can be rather complicated, and there really is no way to compensate for these effects in general.

This problem is particularly apparent when photographing objects that are low in the southern sky (for photographers in the northern hemisphere), and therefore it is difficult to obtain good long-time exposures of such objects. The problem is also very severe in Schmidt camera photos, even for short-time exposures (15 min), owing to the wide field of view. Again, the parts of the sky having low elevation angles cause the most trouble. The best approach is to pick a guide star as near as possible to the object of interest, and having nearly the same average elevation angle. It is also best to straddle the meridian during the exposure, so that the beginning and ending elevation angle are the same.

Fig. 13.14 shows a plot of contours of constant error due to differential atmospheric refraction for an optical system with a focal length of 305 mm, such as might be typical for a Schmidt camera, or an ordinary telephoto

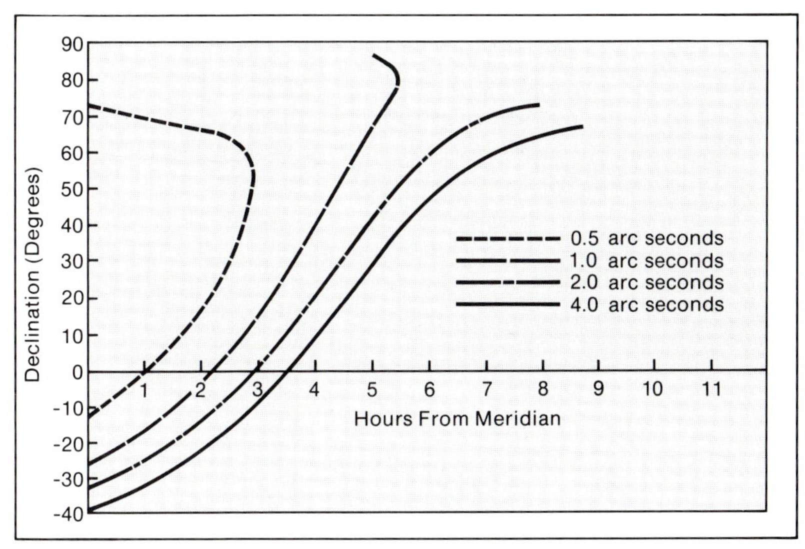

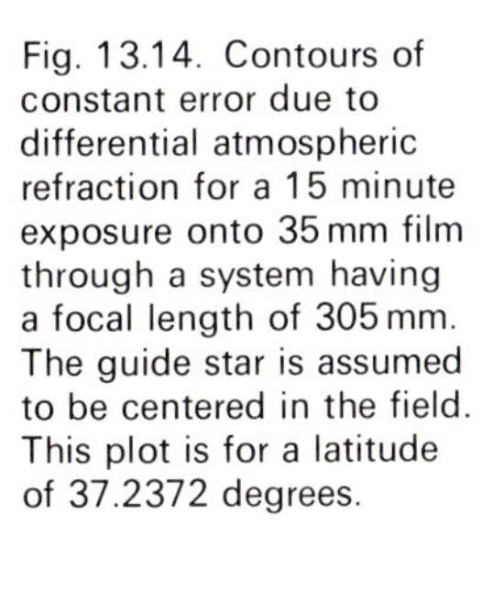
Fig. 13.14. Contours of constant error due to differential atmospheric refraction for a 15 minute exposure onto 35 mm film through a system having a focal length of 305 mm. The guide star is assumed to be centered in the field. This plot is for a latitude of 37.2372 degrees.

camera lens. The guide star is assumed to be in the center of the field, and the contours show the error (in arc seconds) at the corners of the image on a 35 mm frame of film for a 15 min exposure. The error is minimal at the zenith, and near the meridian.

Although an automatic guider will generally reduce the effects of clock drive error, it is best to remove this burden from the guider if possible. Any effort needed by the guider to reduce clock drive error also tends to reduce the sensitivity of the guider, since this error acts like 'noise' to the unit. Also, to maximize sensitivity, the servo system built into the guider will typically exhibit a position error that is proportional to the velocity error between the telescope and the guide star. Thus, velocity errors in the clock drive will translate into position errors on the photographic film, even with a good guider design. This problem is compounded if the guide star is dim, since it is necessary to reduce the servo loop gain to reduce the system bandwidth (and hence the noise), and this reduction in gain causes an increase in position error due to these velocity errors.

As a consequence of longer time exposures, the usual problems with airplane lights, satellite traces and meteor trails are compounded with automatic guiding. In some parts of the sky, and at certain times, it is nearly impossible to obtain a 4 h exposure without at least one of the above effects. In addition, problems with clouds, fog, dew, and wind are more noticeable over these longer times.

Another potential problem is poor polar alignment, which can introduce an apparent field rotation during the exposure. This alignment is particularly difficult for portable equipment, although it is not completely trivial even for permanently mounted instruments. The effects of polar misalignment are most noticeable near the pole itself, but may still cause some problems elsewhere. For fixed telescopes, it is not too difficult to adjust the polar alignment to within a 1 arc min, or better, although some telescopes are sufficiently flexible so that the polar axis position depends upon the rotation angle from the meridian, and hence the polar axis direction changes during an exposure.

The ingredients needed for a good guider

Probably the single most important requirement for a practical guiding system is sensitivity. In order to track on a 12th magnitude star in an 8 inch telescope reliably, it is really necessary to design a system that will track on a 14th magnitude star under ideal conditions. A 14th magnitude star is fairly difficult to see in an 8 inch telescope, and is particularly difficult to locate in a tiny aperture, so simply testing the guider sensitivity is somewhat challenging. Furthermore, it is important essentially to eliminate all clock drive errors, including periodic errors due to the driving gears or worm. It is possible effectively to cancel these periodic errors by generating electronically a phase-modulated clock signal incorporating the correct amplitude and phase of the modulation, although the adjustment of this cancellation system is very time consuming.

There are two main requirements for obtaining the best sensitivity. First, it is necessary to use a comparison or chopping scheme, as opposed to a direct coupled measurement of the star light intensity. All photodetectors have too much DC drift to be used in the direct coupled mode. Instead, an average of the photocurrent with the guide star both covered and uncovered is measured, and this value is used as a reference with which to compare the photocurrent when the star is 'half' covered in each direction. The errors from this comparison in each direction are used to drive the telescope motors on each axis to center the guide star. In the present guider, this cycle of four measurements of the guide star is repeated 25 times per second. See Fig. 13.15 for a schematic diagram of the motion of a pair of orthogonal knife edges to accomplish this chopping function.

The second main requirement for good sensitivity is the use of a vacuum photomultiplier tube for the detector. Solid-state detectors, such as avalanche photodiodes, do exist, and have very high quantum efficiency. However, the photomultiplier has a much better signal-to-noise ratio (by a factor of at least 100), even though its quantum efficiency is only around 10%. This occurs because the photomultiplier has many orders of magnitude more built-in gain than does the avalanche photodiode, so that the pre-amplifier noise can be completely neglected when the tube is used. The solid-state detectors are more sensitive to red light than to blue light, while the photomultiplier has the opposite characteristic. Thus, the photomultiplier will generally 'see' hot blue stars better than the eye, but may not see the cooler red stars quite as well. A conventional 1P21 photomultiplier is currently being used. It is important to use a well-filtered and regulated power supply for the photomultiplier. These tubes tend to be very microphonic, and must be well shock mounted. In addition, both electrostatic and magnetic shields are recommended.

As mentioned earlier, it is also important to mask as much of the background skylight as possible, by imaging the guide star through a small aperture, thereby reducing the background noise level. The aperture in the current guider is about 0.1 mm^2 in area. This aperture area must not be modulated by the technique used to cover and uncover the star, or else a false star signal may result.

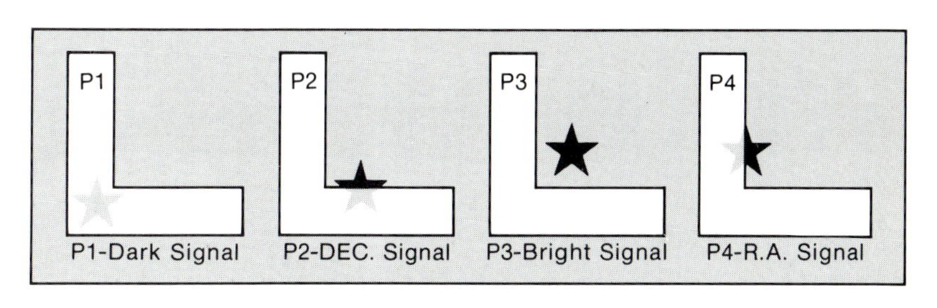

Fig. 13.15. The guide star light is chopped by means of four positions of two orthogonal knife edges, to eliminate the effects of drift and gain variations in the photodetector. The two half-covered signals are compared with the average between the bright and dark signals, giving error signals for the two telescope axes.

Fig. 13.16 shows a diagram of the optical arrangement of the tracker in an off-axis guiding configuration. Note the aperture stop just after the pair of knife edges, in a plane where the star field is essentially still in focus.

The amplitude probability density (or histogram) plot of the noise current from a photomultiplier is decidedly asymmetrical, since all of the shot noise pulses from the arrival of discrete photons are always of the same polarity (each photon triggers a burst of electrons, which results in a negative pulse of current). The impulse response characteristic of the pre-amplifier determines the shape of the probability histogram. For asymmetrical noise of this type, it is possible to improve the signal-to-noise ratio by adding a non-linear circuit element of the appropriate shape. In the present guider, this ideal non-linear element is approximated by back-to-back Schottky clipping diodes, which are metal-semiconductor junctions that allow current to flow in only one direction. This technique improves the signal-to-noise ratio by about a factor of two.

The total loop gain (or amplification) of the servo system should be adjustable for each axis, and should be set to the lowest possible value consistent with adequate compensation for clock drive errors. When the loop gain is reduced, the noise bandwidth is also reduced, and hence the noise power is reduced. Some attention should be paid to the design of the servo system, and to the compensation used to stabilize the loop. The servo should have a good transient response with good damping for a large range of loop gain values. The servo bandwidth (or frequency response) must be large enough to handle the periodic clock drive errors, including any harmonics of this basic periodicity, and must be wide enough to handle gross motion of the guide star caused by atmospheric turbulence. The present instrument has a minimum bandwidth of around 15 mHz (the periodic clock error fundamental is 2.5 mHz).

Finally, it is important to use good engineering practice in the selection of components, and in their layout and construction, to minimize

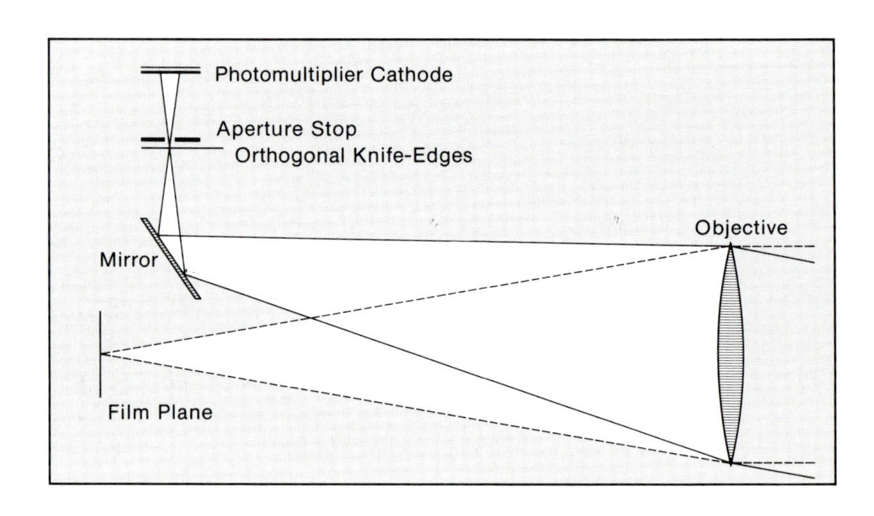

Fig. 13.16. This shows the general optical configuration for off-axis guiding, and shows the relative locations of the knife edges, the aperture stop, and the photomultiplier tube.

the noise. Care is needed in the routing of low-level ground currents in the amplifiers and filters, and any digital logic circuitry must have a separate grounding system, and a well-filtered power supply that does not affect the power supplied to the low-level circuits. All low-level analog circuits must be well shielded. Beware of tribo-electric effects (transient voltages induced by the physical change in a capacitance in the presence of a DC voltage) in the cable connecting the photomultiplier output to the pre-amplifier, since any motion of this cable will induce large amounts of noise into the system.

The difference between the guide star light plus background light and background light alone can be readily measured with a chopping technique of this kind, so a very sensitive photometer is built into the guider. This is very useful, both in the acquisition of a guide star and in monitoring the guider performance during an exposure. If the instrument is tracking properly, then the guide star magnitude will be very stable. For initial star acquisition, it is also helpful to monitor the error signals in each direction, to allow manual operation of the telescope motors until the star is centered in the aperture.

Since the error signals are readily available, they can be monitored on a strip chart recorder, giving a record of the performance of the system during an exposure. This is also useful in the measurement of periodic clock drive error, since any periodic variations will be compensated by the guider, and will show as a periodic error component.

Fig. 13.17 shows a block diagram of the guider currently used by the author. The modulated star light from the two knife edges impinges on the photomultiplier tube, driven by a -900 V regulated power supply. The resulting current variations from the tube are AC coupled to a transresistance pre-amplifier (and attenuator) that converts signal currents to

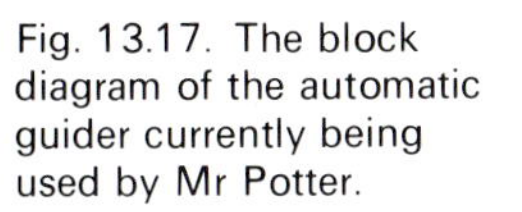

Fig. 13.17. The block diagram of the automatic guider currently being used by Mr Potter.

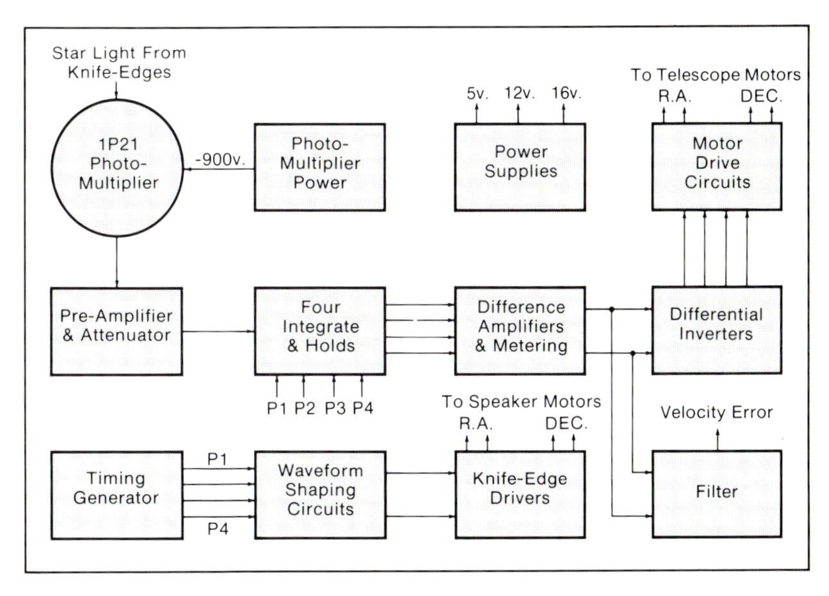

voltages. These voltages are sampled at the four times denoted by P1 through P4, corresponding to the four positions of the knife edges (shown in Fig. 13.20), and the samples are stored on four capacitors (denoted as integrate and hold circuits). The appropriate differences between these signals are used to drive the telescope motors, and the monitoring meters. A timing generator, in conjunction with some shaping circuits form the proper voltage waveforms to drive the knife edges. This pair of orthogonal knife edges are driven by a pair of small (1.5 inch diameter) speakers with most of the cone removed to increase their compliance. Knife edge motion is 0.25 mm peak-to-peak in each direction.

A 12th magnitude star (through an 8 inch telescope) induces a change of about 100 picoamperes in the 1P21 photomultiplier tube, and this sensitivity varies about 0.5 magnitude over the visible color range of stars (spectral types O through N).

Figs. 13.18 through 13.20 show some additional examples of deep-sky photographing using the tracker. These exposures range from 2 to 4 h in duration, and include different parts of the sky. Notice that the film grain is still rather obvious, even though hypersensitized Kodak 2415 film was used. Also, note the effects of differential atmospheric refraction on NGC 253. This object is very low in the southern sky, and 4 h is about the maximum practical exposure time, even under optimum conditions.

Conclusions

The use of an automatic guider for astrophotography allows routine exposures of several hours which, in turn, allows the image size to be enlarged to improve film resolution. In addition, instead of wasting valuable time guiding visually, the telescope and camera can be left unattended, while more productive activities can be pursued. It is possible to build a guider that will work reliably on a 12th magnitude star through an 8 inch telescope. This sort of sensitivity is needed for a practical guiding system.

Fig. 13.18. M64, the Blackeye galaxy, 2 h exposure on hypersensitized Kodak 2415 film, with no filter. A 14 inch f/11 telescope was used.

Fig. 13.19. M1, the Crab nebula, 3 h exposure on hypersensitized Kodak 2415 film, through a Wratten No. 25 (red) filter. The red filter helps bring out the filament structure of this supernova remnant. A 14 inch f/11 telescope was used.

Fig. 13.20. NGC253 4 h exposure on gas hypered Kodak 2415 Technical Pan through a Wratten No. 4 filter (pale yellow). This object is very low in the southern sky (−25° 26′ dec.) and shows some elongated star images due to differential atmospheric refraction. A 14 inch f/11 telescope was used.

Cooled-emulsion photography

JACK NEWTON

Cold camera photography actually developed quite by accident. Astronomers noticed that photographic plates taken on cold winter nights recorded more information than identically exposed and developed plates taken on warmer nights. The colder conditions had, in fact, retarded the reciprocity failure of the film.*

Most emulsions suffer from reciprocity failure when exposed for more than a few seconds. The nature of this effect is that the accumulated image density of the film ceases to be related to time in a linear manner. As light strikes the emulsion, a chemical change in its photosensitive crystals occurs, rendering the particles developable. In long exposures at low light levels, the crystals are unable to retain some of the light energy that they capture and thus the crystals cease to grow at the same pace throughout the exposure.

It is possible to enhance the ability of the emulsion crystals to store the incident light by chilling the film using dry ice at −78 °C (−108 °F). At this temperature the crystals are better able to hold the energy accumulated from low-level incident light and thus the resulting image reveals fainter details than an equal but uncooled exposure on the same film.† The arresting of reciprocity failure becomes even more significant when dealing with color film. Color film is made up of three emulsion layers which may each show differing reciprocity effects. For example, if during a long exposure the red layer is less affected by reciprocity failure than either the blue or green layers, the resulting developed negative would be badly shifted to the red. The cold camera keeps the three layers in a much closer color balance during a long exposure and the end result is a color photograph that not only shows more nebulosity or stars but also retains a color balance that is close to that observed in 'normal' (short exposure) applications.

The primary function of a cold camera is to facilitate cooling the film while preventing it from frosting over during the exposure. With the aid of George Ball, of the Royal Astronomical Society of Canada, two very

* E.S. King discovered, investigated, and published accounts of this phenomena in 1912.

† While the effective speed of a film used for long exposures is greater when the film is cooled because of the fact that reciprocity failure has been arrested, the actual instantaneous film speed for short exposures is decreased by cooling.

successful cameras have been designed and built. The first camera is of the optical plug variety and was designed to handle 35 mm film in the standard cassette form. The optical plug is the medium used to prevent moisture from condensing on the film. The plug is cut from a 50 mm thick plastic or plexiglas sheet and is used to press the film against the copper endplate of the dry-ice chamber. This system will remain frost-free for over 30 min of exposure depending on the ambient temperature and humidity conditions. Eventually, as the optical plug is cooled, the face opposite the film will cool to the point where it will frost over and cause the light striking the film to be diffused. The camera body was fabricated from black 5 mm plastic sheet in such a way that the body containing the film and the cold chamber fits over a shutter assembly so that the camera can be removed without altering the focus (see Figs 13.21 and 13.22).

The dry-ice chamber was constructed using PCB plastic pipe with a threaded cap. A spring-loaded plunger is fitted to the cap to keep the sublimating dry ice firmly pressed against the copper plate at the bottom of the chamber. Copper is used because it is an efficient thermal conductor and thus transfers the cold quickly and evenly to the film. A shutter release was fabricated using brake cable from a 10 speed bicycle. The whole camera can be constructed for less than US$20.

The second cold camera is a dry-gas variety. A shutter assembly was built with a heated optical window at one end and an opening to the film at the other. A length of plastic tubing is used to carry the dry gas (nitrogen) to

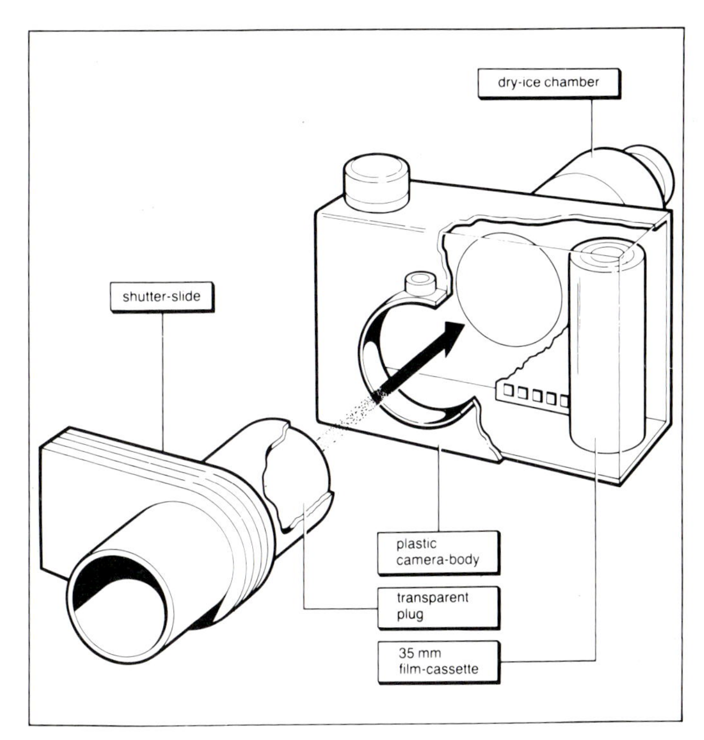

Fig. 13.21. The simpler of two cold camera designs by Jack Newton and George Ball is shown here. Both designs use dry ice to chill 35 mm roll film in order to lessen the effects of reciprocity failure during long exposures. In this model a simple clear plastic plug fills the space between shutter and film. (From *The Cambridge Deep Sky Album*, © 1983, Cambridge University Press.)

the chamber in order to purge the chamber of moist air. The film is situated in the camera as before but now it is the 35 mm opening in the shutter assembly that pins the film against the copper cold plate (see Figs 13.21 and 13.22). The dry-gas method offers a number of advantages over the plastic plug design. Optical plugs tend to acquire a static charge that attracts dust which inevitably shows up on the photograph. The plexiglass is also easily scratched. These problems are avoided with the dry-gas design since the film does not come into contact with any optical surface.

Both of these cold cameras were designed for use with standard 35 mm or 120 roll film which does not have to be cut into short strips. This makes processing quite easy since any color lab can now handle the film. Since the cold camera stops the film from losing much of its speed, exposures are short and the film does not need to be push processed. In fact, pushing increases grain and may even destroy the color balance. Processing Fujichrome 400 as a negative only increases the grain and is not recommended when using the cold camera.

The optical plug camera is by far the easiest system to use in the field. The edge of the plug should be threaded and then painted flat black to prevent stars at the edge of the field from reflecting back onto the film surface. Cans of compressed air can be used to direct blasts of air onto the plug to remove dust from the surface that will be pressed against the film. Small scratches on the polished ends of the plug do not seem to have a bad effect on the photographs since one end is pressed firmly against the film and

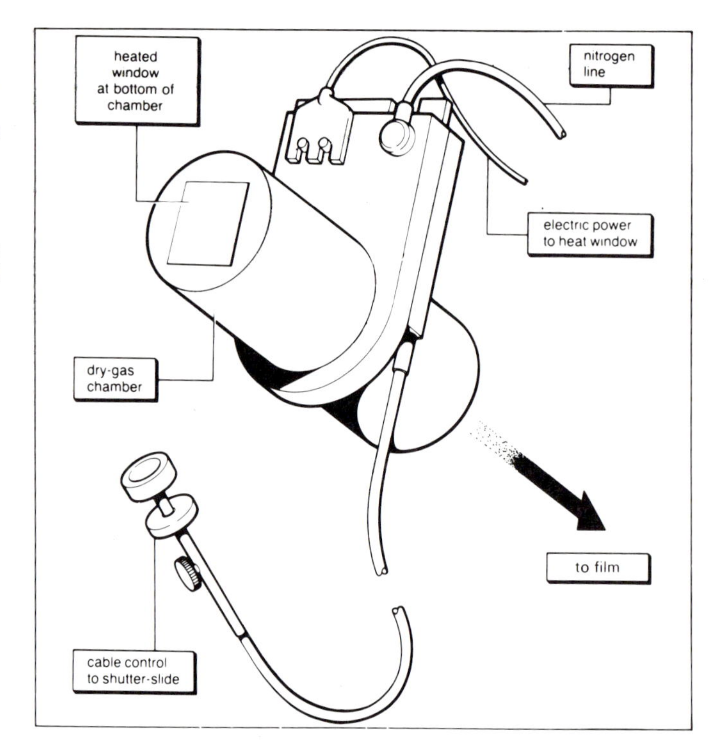

Fig. 13.22. The dry gas chamber of the more sophisticated cold camera design is shown here. This chamber is filled with dry nitrogen and the optical window is heated to prevent frosting. The bottom window presses against the film and keeps it in contact with the cold platten of the dry ice chamber shown in Fig. 13.21. (From *The Cambridge Deep Sky Album*, © 1983, Cambridge University Press.)

the other end is so far from the focus that the scratches have no effect. If glass plugs are used, they will require a heater at the telescope end in order to keep frost from forming and care must be taken to see that too much heat is not applied. If excessive heat is applied to the plug, the star images may actually be degraded.

The one advantage of the dry-gas camera is that it eliminates the dust problem created by the plastic cameras. The dry-gas camera requires the use of a nitrogen source and thus is difficult to use in the field. It also requires the use of a heated window. The least desirable camera (from the portability viewpoint) is the vacuum model. Film needs to be cut and placed in the vacuum chamber and the optical window must be heated to prevent frost from forming (but less heat need be applied than with the dry-gas model).

A typical photographic run begins with the shutter portion of the camera attached to the telescope with an eyepiece placed in the optical plug compartment. The object to be photographed is centered in the field of view and a suitable guide star is located with the guiding head on the telescope. Next, the camera is focused using an optical plug fitted with a knife edge at the film plane. A nearby star of 3rd or 4th magnitude is best for focusing. The object is then recentered and the guide star positioned on the crosshairs. The shutter is closed, the focusing plug is removed, and a clean plexiglas plug is inserted. The camera body is now placed over the plug and secured. Crushed dry ice is removed from the thermos where it has been stored and is packed into the camera. Finally, the shutter is opened and the exposure is made.

On completion of the exposure, the shutter is carefully closed and the camera removed and capped. Excess dry ice is returned to the thermos bottle and a hairdryer is used to warm the camera back. After about 4 min of warming with the hairdryer the film is warm enough to be advanced for the next exposure. After a plug has been used it should also be warmed using the hairdryer. Several plugs should be kept on hand and rotated during the night in order to minimize the time between exposures (and the chances of a used plug getting cold enough to frost over). The whole cycle averages about 1 hour.

Exposure times for a particular location are found by actual experimentation to find the sky fog limit. The sky background changes from night to night and can affect exposures by as much as 20 %. On nights when I can easily see a 5.7 magnitude star located between Eta and Zeta Ursa Minor I can photograph for 20 min using Fujichrome 400 in a 50 cm f/5 Newtonian. If the star is not visible at all I will cut my standard exposure down to 16 min. Each film has its own characteristic sky fog sensitivity. Tri-X is the best black-and-white film for the cold camera while 2415 TP is not affected at all by cooling.

Ektachrome 400, Fujichrome 400, and VR 1000 color negative film seem to work best for cooled color photography. Ektachrome 400 is about 4 times faster when cooled and shows little color shift. However, the sky background shows a slight green tint. Fujichrome is slightly faster and shows deeper (more saturated) reds and blues than Ektachrome 400. The grain of both these films is about the same. A skylight filter (UV cut-off filter) will improve Fujichrome 400 by increasing the contrast and the background becomes a natural blue-black. Ektachrome 400 is also improved by this process but the sky retains its green tint. Kodak VR 1000 is about 3–4 times faster when chilled and seems well suited for the cold camera. This film is very blue-sensitive at ambient temperatures and the blue background is saturated by fog before faint details can be recorded. However, when VR 1000 is used cold, in conjunction with a good light-pollution filter, the results are stunning. The filter/film combination is a near perfect match. The weaker blue transmission of the filter compensates for the film's blue sensitivity and the end result is a well-balanced color rendition. The filter permits exposures to be extended by a factor of almost four. Fujichrome 400 is also quite effective when used with light-pollution filters. It is slightly slower than VR 1000 but the grain is much finer.

Appendix A

Catalogue of interesting photographic objects epoch 1950.0

RA HH	RA MM	Dec. DD	Dec. MM	NGC(IC) catalog number	Messier catalog number	Type[a]	Name [b]
00	01	68	20	7822		DN + Cl	
00	11	−23	27	45		G	
00	12	−39	30	55		G	
00	22	−72	21	104		Gl	* 47 Tucanae
00	39	85	03	188		G	* Dwarf Galaxy
00	40	40	36	221	M32	G	
00	40	41	00	224	M31	G	* Andromeda Galaxy
00	45	−12	09	246		Pl	*
00	45	−21	01	247		G	
00	45	−25	34	253		G	*
00	50	−26	52	288		Gl	
00	50	−73		SMC		SMC	* Sm. Magellanic Cloud
00	53	−37	58	300		G	
00	54	60	48	IC 59–63		N	* Gamma Cass. Region
01	16	58	04	457		Cl	
01	19	05	00	488		G	
01	26	63	02	559		Cl	
01	30	60	27	581	M103	Cl	
01	31	30	24	598	M33	G	
01	34	15	32	628	M74	G	*
01	39	51	19	650–1	M76	Pl	
01	43	61	01	663		Cl	
01	55	37	25	752		Cl	
02	17	56	54	869/884		Cl	* Double Cluster
02	29	61	13	IC 1805		DN	*
02	19	42	07	891		G	*
02	24	33	22	925		G	*
02	39	42	34	1039	M34	Cl	
02	40	−00	14	1068	M77	G	
02	41	01	10	1073		G	
02	44	−30	29	1097		G(SB)	*
03	44	23	58	Mel 22	M45	RN + Cl	* Pleiades
03	00	−23	04	1187		G	
03	08	−20	46	1232		G	*
03	18	−19	35	1300		G(SB)	*
03	21	−37	25	1316/1317		G	Fornax A
03	32	−36	18	1365		G	
03	33	−25	06	1371		G(SB)	
03	37	−26	30	1398		G	*

Catalogue of interesting photographic objects epoch 1950.0 (cont.)

RA HH	RA MM	Dec. DD	Dec. MM	NGC(IC) catalog number	Messier catalog number	Type[a]	Name[b]
03	40	−47	24	1433		G	
03	42	−44	48	1448		G	
04	00	36	17	1499		N	* California nebula
04	02	60	47	1501		Pl	
04	06	30	38	1514		Pl	
04	11	51	07	1528		Cl	
04	19	−55	04	1566		G	
05	01	23	44	1746		Cl	
05	19	33	28	IC 410		Cl + N	
05	13	34	16	IC 405		R + DN	* AE Aur
05	22	−24	34	1904	M79	Gl	
05	25	35	48	1912	M38	Cl	
05	31	21	59	1952	M1	Pl	* Crab nebula
05	32	34	07	1960	M36	Cl	
05	33	−05	25	1976	M42	DN	* Orion nebula
05	33	−05	18	1982	M43	DN	
05	26	−69		LMC	LMC	LMC	* Lg. Magellanic Cloud
05	44	00	02	2068	M78	RN	
05	40	−69	04	2070		DN	* Tarantula neb.
05	40	−02	20	IC 434		DN	* Horsehead
05	49	32	33	2099	M37	Cl	
06	00	10	26	2141		Cl	
06	05	24	20	2168	M35	Cl	
06	06	20	31	2174–5		DN + Cl	*
06	11	12	50	2194		Cl	
06	14	22	48	IC 443		N	
06	19	−27	14	2217		G	
06	30	04	40	2237		DN + Cl	* Rosette
06	36	08	46	2261		N	* Hubble
06	38	09	57	2264		N + Cl	* Madonna
06	45	−20	42	2287	M41	Cl	
07	01	−08	16	2323	M50	Cl	
07	03	−10	29	IC 2177		N	
07	16	−13	10	2359		R + DN	
07	24	69	08	2366		G	
07	26	21	01	2392		Pl	
07	32	65	43	2403		G	*
07	34	−14	22	2422		Cl	
07	39	−31	32	2439		Cl	
07	40	−14	42	2437–8	M46	Cl + Pl	*
07	42	−23	45	2447	M93	Cl	
07	44	−37	51	2451		Cl	
07	54	53	33	2474–5		Pl	
07	51	−38	25	2477		Cl	
08	00	−60	44	2516		Cl	
08	08	−12	41	2539		Cl	
08	09	−49	07	2547		Cl	
08	10	−37	29	2546		Cl	
08	11	−05	38	2548	(M48)	Cl	
08	30	−44		—		DN	Gum
08	37	19	52	2632	M44	Cl	Beehive
08	48	12	00	2682	M67	Cl	

Catalogue of interesting photographic objects epoch 1950.0 (cont.)

RA		Dec.		NGC(IC) catalog number	Messier catalog number	Type[a]	Name[b]
HH	MM	DD	MM				
08	50	33	38	2683		G	
09	14	−36	24	2818		Cl + Pl	
09	19	51	12	2841		G	*
09	26	−56	45	I.2488		Cl	
09	29	21	44	2903		G	*
09	44	−30	58	2997		G	
09	52	69	18	3031	M81	G	
09	52	69	56	3034	M82	G	
09	59	55	57	3079		G	
10	01	−25	55	3109		G	
10	01	−59	53	3114		Cl	
10	03	−07	28	3115		G	
10	17	45	49	3198		G	
10	41	11	58	3351	M95	G(SB)	
10	43	63	30	3359		G	
10	44	12	05	3368	M96	G	
10	43	−59	25	3372		DN	* Eta Carina
10	45	12	51	3379	M105	G	
11	03	−58	24	3532		Cl	
11	09	55	57	3556	M108	G	
11	12	55	18	3587	M97	Pl	* Owl
11	15	−62	26	I.2714		Cl	
11	16	13	23	3623	M65	G	
11	18	13	17	3627	M66	G	
11	18	13	53	3628		G	
11	34	−61	19	3766		Cl	
11	34	−62	44	IC 2944		N + C1	
11	55	53	39	3992	M109	G	
11	59	−18	35	4038–9		G	Antennae Galaxy
12	11	15	11	4192	M98	G	
12	13	13	25	4216		G	
12	14	69	45	4236		G	*
12	15	38	05	4244		G	
12	16	14	42	4254	M99	G	
12	16	47	35	4258	M106	G	*
12	17	29	53	4274		G	
12	19	18	40	4293		G	
12	19	14	53	4298/4302		G	
12	19	04	45	4303	M61	G	
12	20	30	10	4314		G	
12	20	16	06	4321	M100	G	
12	21	−61	37	4349		Cl	
12	23	−72	24	4372		Gl	
12	23	13	10	4374	M84	G	
12	23	18	28	4382	M85	G	
12	23	33	49	4395		G	
12	24	13	13	4406	M86	G	
12	25	13	17	4438		G	
12	27	08	16	4472	M49	G	
12	28	41	58	4485–90		G	
12	28	12	40	4486	M87	G	
12	30	14	42	4501	M88	G	

Catalogue of interesting photographic objects epoch 1950.0 (cont.)

RA HH	RA MM	Dec. DD	Dec. MM	NGC(IC) catalog number	Messier catalog number	Type[a]	Name[b]
12	32	08	28	4535		G	
12	33	14	46	4548		G	
12	33	12	50	4552	M89	G	
12	34	28	14	4559		G	
12	34	26	16	4565		G	* Needle
12	34	11	32	4567/68		G	
12	34	13	26	4569	M90	G	
12	35	12	05	4579	M58	G	
12	37	−26	29	4590	M68	Gl	
12	37	−11	21	4594	M104	G	* Sombrero
12	40	11	55	4621	M59	G	
12	40	32	29	4631		G	*
12	41	11	49	4649	M60	G	
12	42	32	26	4656		G	* Fishhook
12	48	25	46	4725		G	
12	48	−06	08	4731		G	
12	49	41	23	4736	M94	G	
12	51	−60	05	4755		Cl	
12	54	21	47	4826	M64	G	Black Eye
13	02	−49	01	4945		G	
13	10	18	26	5024	M53	Gl	
13	11	36	51	5033		G	
13	13	42	17	5055	M63	G	
13	16	−20	47	5068		G(SB)	
13	22	−42	45	5128		G	*
13	24	−47	03	5139		Gl	* Omega Centauri
13	28	47	27	5194–5	M51	G	* Whirlpool
13	34	−29	37	5236	M83	G	*
13	39	28	38	5272	M3	Gl	
13	54	05	15	5364		G	
14	01	54	35	5457	M101	G	*
14	42	02	10	5746		G	
15	01	−54	09	5822		Cl	
15	15	56	31	5907		G	*
15	16	02	16	5904	M5	Gl	
15	20	05	15	5921		G	
16	09	−54	05	6067		Cl	
16	14	−22	52	6093	M80	Gl	
16	20	−26	24	6121	M4	Gl	
16	22	−40	35	6124		Cl	
16	26	−26	20	IC 4606		DN	* Alpha Sco Region
16	29	−52	31	6152		Cl	
16	30	−12	57	6171	M107	Gl	
16	38	−48	40	6193		Cl	
16	39	36	33	6205	M13	Gl	* Hercules Cluster
16	45	−01	52	6218	M12	Gl	
16	49	−40	18	IC 4628		N + Cl	
16	51	−41	43	6231		Cl	
16	53	−40	38	H 12		Cl	
16	54	−04	02	6254	M10	Gl	
16	57	−44	36	6259		Cl	
16	58	−30	03	6266	M62	Gl	

Catalogue of interesting photographic objects epoch 1950.0 (cont.)

RA		Dec.		NGC(IC) catalog number	Messier catalog number	Type[a]	Name[b]
HH	MM	DD	MM				
17	00	−26	11	6273	M19	Gl	
17	10	−37	03	6302		Pl	
17	16	−18	28	6333	M9	Gl	
17	16	−23	30	B 72		DkN	* Elephant's Trunk
17	17	−36	01	6334		RN	
17	16	43	12	6341	M92	Gl	
17	21	−49	54	IC 4651		Cl	
17	30	07	06	6384		G	*
17	37	−53	39	6397		Gl	
17	35	−03	13	6402	M14	Gl	
17	37	−32	11	6405	M6	Cl	
17	51	−34	48	6475	M7	Cl	*
17	54	−19	01	6494	M23	Cl	
17	49	70	10	6503		G	
17	59	−23	02	6514	M20	N + Cl	* Triffid
18	02	−24	20	6523	M8	DN	* Lagoon
18	02	−22	30	6531	M21	Cl	
18	15	−18	27	6603	M24	Cl	
18	16	−13	48	6611	M16	DN + Cl	* Eagle
18	17	−17	09	6613	M18	Cl	
18	18	−16	12	6618	M17	DN	* Omega
18	25	06	32	6633		Cl	
18	28	−32	23	6637	M69	Gl	
18	29	−19	17	IC 4725	M25	Cl	
18	33	−23	58	6656	M22	Gl	*
18	37	05	26	IC 4756		Cl	
18	40	−32	21	6681	M70	Gl	
18	43	−09	27	6694	M26	Cl	
18	48	−06	20	6705	M11	Cl	*
18	52	−30	32	6715	M54	Gl	
18	52	32	58	6720	M57	Pl	* Ring
19	05	−63	56	6744		G	
19	06	−60	04	6752		Gl	
19	15	30	05	6779	M56	Gl	
19	37	−31	03	6809	M55	Gl	
19	16	06	26	6781		Pl	* Bubble
19	42	−14	53	6822		G	
19	52	18	39	6838	M71	Gl	
19	57	22	35	6853	M27	Pl	* Dumbell
20	03	−22	04	6864	M75	Gl	
20	04	35	38	6871		Cl	
20	11	38	16	6888		N	*
20	22	38	21	6913	M29	Cl	
20	33	28	08	6940		Cl	
20	34	59	58	6946		G	
20	18	40				DN	* Gamma Cyg Region
20	44	30	32	6960		N	* Veil (Eagle)
20	47	44	11	IC 5067		N	* Pelican
20	51	−12	44	6981	M72	Gl	
20	54	31	30	6992–5		N	* Veil (Cirrus)
20	56	−12	50	6994	M73	Cl	
20	57	44	08	7000		N	* North American

Catalogue of interesting photographic objects epoch 1950.0 (cont.)

RA		Dec.		NGC(IC) catalog number	Messier catalog number	Type[a]	Name[b]
HH	MM	DD	MM				
21	01	−11	34	7009		Pl	Saturn
21	28	11	57	7078	M15	Gl	
21	31	−01	03	7089	M2	Gl	
21	30	48	13	7092	M39	Cl	
21	38	−23	25	7099	M30	Gl	
21	38	57	14	IC 1396		N	
21	51	47	02	IC 5146		N	Cocoon
22	02	46	16	7209		Cl	
22	27	−21	06	7293		Pl	* Helix
22	34	33	41	—		G	* Stephan's Quintet
22	35	34	10	7331		G	*
22	55	−41	20	7424		G	
23	12	61	14	7538		N	
23	19	60	54	7635		Pl	*
23	20	40	35	7640		G	
23	22	61	20	7654	M52	Cl	
23	55	56	26	7789		Cl	

[a]Object type is indicated by the following codes: N or DN, Nebular; DkN, Dark Nebular; RN, Reflection Nebular; Cl, Cluster; Gl, Globular Cluster; G, Galaxy; Pl, Planetary; SB, Barred Spiral Galaxy.

[b]Asterisk indicates an object of particular interest.

Appendix B

Computers and astrophotography

As the years have passed since we first started dabbling in this black art form, known as astrophotography, newer and fancier toys have been introduced. Some are only toys and their luster passes after a brief gleam while others are truly useful and find a permanent use in the amateur community. One 'toy' that has been more than just a little welcome is the microcomputer. While its direct uses for the casual worker are limited (e.g. one cannot compute optimum exposures for a particular object) its peripheral uses are only limited by the imagination and talent of the user (and by the amount you want to pay).

The microcomputer field is really just beginning to expand as a computer that was purchased in 1982 has already become obsolete by 1986 owing to the rapidly changing technology. Once the growing pains of a new technology have passed, one would certainly hope that a home microcomputer would still be useful and compatible with machines built 5 or 6 years later. The key here, to a non-programmer, is the word 'compatible'; will one still be able to purchase software and accessories for an older machine in the not-too-distant future? The machine will more than likely be functioning perfectly in 10 years but you may not be able to expand its capabilities much if it does not have new software and hardware available for more than just a year or two after it is purchased.

The statement above has two immediate implications: try and insure that the company that built the machine will still be around in 5 years and learn to write your own programs in whatever languages you can purchase for the machine! The corporate size is no guarantee that your machine will still be supported in 5 years (as owners of the IBM PC Jr found out early in 1985) and a bargain buy on a machine may not really be such a bargain if the machine in question is being 'closed out'. In order to be certain of a long future with any machine that you buy, it would be well to be certain that the micro can do everything that you really would like to be able to do in the next 2–4 years (within a reasonable budget, of course). In short, do not buy a machine that you will outgrow in just a few months.

As far as languages go, Basic, Fortran, C, and Pascal all have good 'number crunching' capabilities. Of these four, Basic is probably the most easily transported from one machine to another ('C' is supposed to have this same advantage but it is still undergoing major growth and refinements). Fortran is the 'good old standby' for number crunching and is quite like Basic in a number of ways but does not (usually) have the string handling capabilities that are found in most versions of Basic. There is no driving factor to choose one language over another when starting out other than that it meets your own personal requirements. Can it be compiled? Can you access the computer's graphics? Is it fast enough? Does it have a reasonable set of diagnostic errors that can be interpreted easily? Is the manual understandable or can you get other sources to help you understand the manual?

It would also be a good idea to try and learn some of the basics of the 'assembly code' for the central processing unit (CPU) that resides in your machine (e.g. Z80, 8086, 8088, etc.). This may sound like a type of torture (and when you start to read the manual you will think worse thoughts than that!) but it will quite probably prove useful over the period that you have to deal with your machine. Subroutines can be built requiring smaller memory and they will also run faster when written in assembly code. In addition, you can access more of the micro's built-in capabilities than you could just using a high-level language like Fortran.

We cannot tell anyone what machine to buy or what machines not to buy. It is even hard to give guidelines as to what features are needed and which are not. We can only pass on some information that has filtered in to us over the past few years since the introduction of the home computer. However, we can pass on a number of possible projects that may prove useful and we can even toss in a couple of quick-and-dirty programs that you can try out on your own machine.

1 Calculate zonal settings for Foucault tests of mirrors.
2 Calculate the size of a fully illuminated field for a system with a secondary or with light baffles.
3 Calculate the illumination across the focal plane (see Program 'ILLUM').
4 Determine the RA and Dec. of objects on a photographic plate.
5 Raytrace various optical systems.
6 Determine the magnitudes of stars on a plate from their diameters.
7 Determine the position of a comet or asteroid on any night at any time (given the orbital elements).
8 Determine orbital elements of a body given three (or more) observations of its position.
9 Reduce photometric data.
10 Keep a database logbook of all observations or photographs for easier future reference (use Dbase II or Dbase III for this one).
11 Write a program that will allow your machine to become a real-time clock and observer's helper. By reading the micro's internal clock and doing some fast calculations your micro can display Local Time, UTC,

Siderial Time, and the Julian Date (with loads of time to spare!). In addition, you can build in a photographic catalogue of interesting objects and get the machine to list all that are within 2 hours of the meridian whenever a certain key is struck. This one is really fun to write and very useful too!

Program: 'Illumination'

The following sample Basic program will compute the illumination across the focal plane, from the center outward, and plot the data as well as print it. The effects of different baffle apertures in the system can also be overplotted to check for vignetting by either the focuser of the T-mount.

This program has proved most useful in planning a system. With this program, not only knows that the system is fully illuminated out to a certain radius, but one also knows how much vignetting is occurring over the entire film plane to within a few per cent. For multi-element systems like Cassegrains, the user simply pretends that one is looking through two circular holes suspended at the appropriate distances (use the effective focal length and not the distance from primary to focal plane when the program asks for the focal length of the system). As more elements are added, the approximations used will become less accurate but these are still considerably better than a 'guestimate'.

```
1000 REM  ...  I= ILLUMINATION  .....  B= DISTANCE OFF AXIS
1050 REM  ...  D= PRIMARY DIA   .....  F= FOCAL LENGTH
1100 REM  ...  A= SEC.MINOR AXIS  ...  L= DIST.FROM OPT.AXIS TO FOC.PLANE
1150 PI=3.1415926#:LL=2
1200 REM --  CLEAR SCREEN W/ "GRAPH" COMMAND
1250 GRAPH
1300 REM  ---  PRINT OUT HEADER INFORMATION  ---
1350 PRINT TAB(13)"PROGRAM TO COMPUTE ILLUMINATION ACROSS THE FOCAL PLANE"
1400 PRINT TAB(12)"OF A NEWTONIAN SYSTEM ... ALL INPUTS SHOULD BE IN INCHES"
1450 REM          -----    COPYRIGHT 1985, BRAD D. WALLIS  -----
1500 GRAPH L(12,1)        '  POSITION CURSOR ON LINE 12
1550 INPUT "INPUT DIAMETER OF PRIMARY MIRROR";D
1600 INPUT "INPUT FOCAL LENGTH OF SYSTEM";F
1650 INPUT "INPUT SECONDARY MINOR AXIS";A
1700 INPUT "INPUT DISTANCE FROM OPTICAL AXIS TO FOCAL PLANE";L
1750 DMIN=.2 +(L*((D-.2)/F)):DMIN=CINT(100*DMIN)/100:PRINT
1800 PRINT "THE SMALLEST DIAGONAL FOR VISUAL USE IN SUCH AN INSTRUMENT IS"
1850 PRINT TAB(20);DMIN;"INCHES (MINOR AXIS)":PRINT
1900 D2MIN=1!+(L*((D-1)/F)):D2MIN=CINT(100*D2MIN)/100
1950 PRINT"THE SMALLEST DIAGONAL FOR PHOTOGRAPHIC USE IS ";D2MIN;"INCHES"

2000 IF A>DMIN THEN GOTO 2300
2050 PRINT
2100 REM ---- IF DIAGONAL INPUT IS TOO SMALL, RESET TO MINIMUM  -----
2150 PRINT "THE DIAGONAL SIZE ENTERED IS TOO SMALL FOR VISUAL APPLICATIONS"
2200 PRINT "THE DIAGONAL SIZE WILL BE CHANGED TO EQUAL THE MINIMUM USEFUL"
2250 PRINT "SIZE FOR VISUAL USE: ";DMIN;" INCHES":A=DMIN
2300 FOR II=1 TO 500: KL=II^2: NEXT II
2350 REM   -----  CLEAR SCREEN THEN SET UP FOR PLOTTING  -----
2400 GRAPH:PLOT B(8,8)(308,218):GRAPH H(1,2)
```

The graphics commands above first clear the screen, then draw a rectangle with opposite corners at (8,8) and (308,218) and then set the active screen to only lines 1 and two to avoid overwriting any graphics with future input requests. This will undoubtedly need to changed for other micros.

```
2450 REM   -----   OUTPUT TO PRINTER (IF ON)  -----
2500 LPRINT "ILLUMINATION ACROSS THE FIELD OF A ";D;"-INCH  F/";F/D;" SYSTEM"
2550 LPRINT "WITH A ";A;" INCH MINOR AXIS SECONDARY WHICH IS ";L;" INCHES "
2600 LPRINT TAB(10)"FROM THE FOCAL PLANE":LPRINT:LPRINT
2650 LPRINT TAB(8);"DISTANCE";TAB(29);"PERCENT";TAB(45);"MAGNITUDE"
2700 LPRINT TAB(8);"OFF-AXIS";TAB(27);"ILLUMINATION";TAB(48);"LOSS":LPRINT
2750 FOR J=0 TO 30
2800 REM ---  COMPUTE  I  AT .05 in SPACING W/ TICKS AT .10 in  ---
2850 REM        IF BAFFLE DATA, PLOT LINES AT HALF SPACING
2900 B=J*.05
2950 IF J=0 THEN GOTO 3900
3000 R2=A*F/(L*D):X=2*B*(F-L)/(L*D)
3050 A2=(X^2+1-R2^2)/(2*X)
3100 B2=(X^2+R2^2-1)/(2*X*R2)
3150 IF A2>1 OR B2>1 THEN GOTO 3900
3200 RCOSA=-ATN(A2/SQR(-A2*A2+1))+1.5708
3250 RCOSB=-ATN(B2/SQR(-B2*B2+1))+1.5708
3300 I=(RCOSA-X*SQR(1-A2^2)+R2^2*RCOSB)/PI
3350 I=CINT(I*1000)/10:NUM=100/I:GOSUB 5050:DM=2.5*NSTRT
3400 DM=CINT(100*DM)/100:FLAG=1
3450 X=B*200+8:Y=I*2+8
3500 IF J=0 THEN PLOT (X,Y) ELSE PLOT TO (X,Y)

  The "PLOT" commands above serve to do different tasks.
  "PLOT (X,Y)"  will plot a point at (X,Y), while
  "PLOT TO (X,Y)" will plot a line from the current (last)
  graphics position to the point (X,Y).  These commands will
  probably need to be changed for various machines.

3550 IF J MOD LL <> 0 THEN GOTO 3750
3600 IF J MOD 2 = 0 THEN TMRK=3 ELSE TMRK=6
3650 REM ---  PLOT ILLUMINATION DATA  ---
3700 PLOT (X,TMRK):PLOT TO (X,Y)
3750 LPRINT TAB(10);B,TAB(30);I;TAB(50);DM
3800 NEXT J
3850 GOTO 4150
3900 IF FLAG=1 THEN I=0 ELSE I=100

3950 IF I=0 THEN DM=999
4000 IF I=100 THEN DM=0
4050 GOTO 3450
4100 REM ----  PLOT TICK MARKS ON Y-AXIS  -----
4150 FOR K=0 TO 20
4200 Y=K*10+8
4250 IF K MOD 2 <>0 THEN GOTO 4400
4300 PLOT (2,Y):PLOT TO (8,Y)
4350 GOTO 4450
4400 PLOT (5,Y):PLOT TO (8,Y)
4450 NEXT K
4500 REM   -----  ASK FOR 'BAFFLE' DATA  -----
4550 INPUT "DO YOU WANT TO SEE ANY BAFFLE PLOTS "; P$
4600 IF LEFT$(P$,1)="Y" THEN GOTO 4650 ELSE GOTO 5000
4650 INPUT "WHAT IS THE DIAMETER OF THE BAFFLE";A
4700 INPUT "HOW FAR IS THE BAFFLE FROM THE FOCAL PLANE";L
4750 LPRINT:LPRINT:LPRINT:LPRINT
4800 LPRINT "ILLUMINATION ACROSS THE FIELD OF A ";D;"-INCH  F/";F/D;" SYSTEM"
4850 LPRINT "WITH A ";A;" INCH DIAM. BAFFLE STOP WHICH IS ";L;" INCHES "
4900 LPRINT TAB (10)"FROM THE FOCAL PLANE":LPRINT:LPRINT
4950 FLAG=0:LL=1:GOTO 2650
5000 GRAPH H(1,24):END
5050 REM ------   SUBROUTINE TO CALCULATE BASE 10 LOGARITHMS   ------
5100 REM  ******      NUM = INPUT NUMBER, NSTRT = LOG10(NUM)
5150 NDEL=.25
5200 IF NUM<10 THEN NSTRT=0
5250 FOR IQ=1 TO 35
5300 TST=10!^NSTRT:RTST=ABS(TST-NUM)
5350 IF RTST<.0001 THEN GOTO 5700
5400 IF TST>NUM THEN GOTO 5550
5450 NSTRT=NSTRT+NDEL
5500 GOTO 5600
5550 NSTRT=NSTRT-NDEL
5600 NDEL=NDEL*.66667
5650 NEXT IQ
5700 RETURN
```

Output from 'Illumination'

Graphics. The graphics produced by 'Illumination' is a plot of illumination v. off-axis distance. The illumination axis has ticks at 10 % intervals from 0 to 100 and the off-axis distance for the main system draws a vertical line each 0.10 inches off-axis. For baffle problems, the spacing is 0.05 inches so as to clearly show the vignetting created by the baffle.

Fig. A.1. This plot shows the illumination across the field of an 8 inch f/6.0 system which has a 2.6 inch minor axis secondary that is 8 inches from the focal plane. The region that is 'double ruled' shows the vignetting caused by a 1.5 inch diameter T-mount located 2 inches from the focal plane.

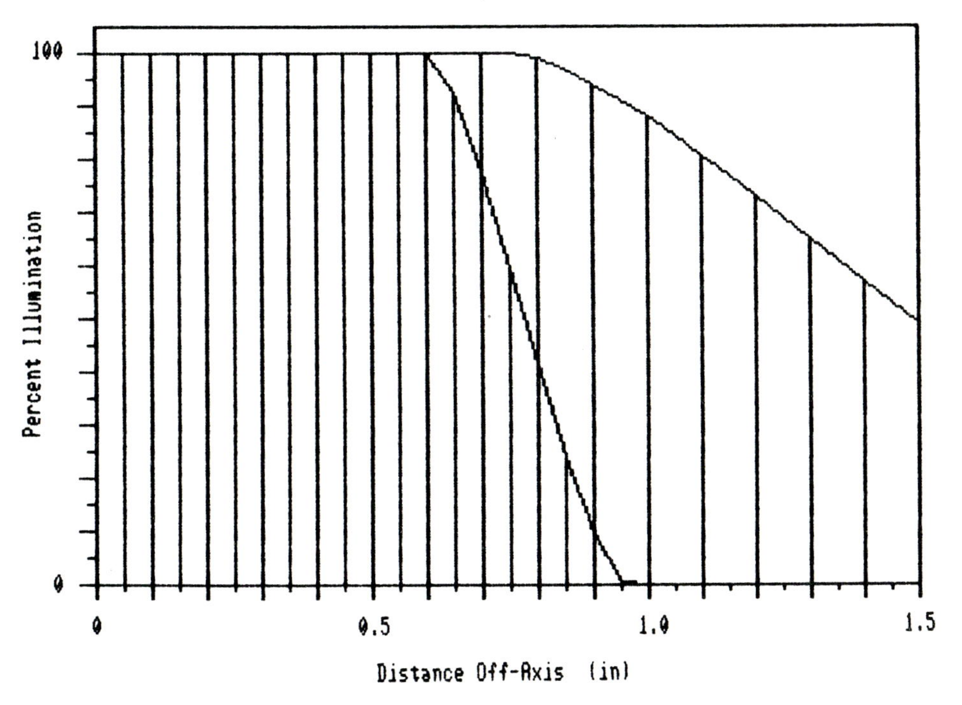

Appendix B continued overleaf.

Printer. The printed output of 'Illumination' is a tabulation of off-axis distance, percentage illumination and magnitude loss due to any vignetting. We only show the printout for the illumination due to the effects of the secondary.

Illumination across the field of an 8 inch f/6 system with a 2.6 inch minor axis secondary which is 8 inches from the focal plane

Distance off-axis	Percentage illumination	Magnitude loss
0	100	0
0.05	100	0
0.1	100	0
0.15	100	0
0.2	100	0
0.25	100	0
0.3	100	0
0.35	100	0
0.4	100	0
0.45	100	0
0.5	100	0
0.55	100	0
0.6	100	0
0.65	100	0
0.7	100	0
0.75	100	0
0.8	99.1	0.01
0.85	96.9	0.03
0.9	94.2	0.06
0.95	91.2	0.1
1.0	87.8	0.14
1.05	84.3	0.19
1.1	80.6	0.23
1.15	76.8	0.29
1.2	72.9	0.34
1.25	69	0.4
1.3	65	0.47
1.35	61	0.54
1.4	57.1	0.61
1.45	53.1	0.69
1.5	49.2	0.77

Appendix C

Useful formulas

Image size at prime focus

$$h = s\,F/k$$

where

s = Size of object in the sky
h = Size of object on film (in).
F = Focal length in inches
k = 57.3 if s is in degrees
k = 3438 if s is in minutes
k = 206265 if s is in seconds

Image scale at prime focus

$S = 57.3/F$ deg/inch
$S = 135/F$ min/mm
$S = 8100/F$ sec/mm
(F is in inches).

Visual limiting magnitude (approx.)

$$M_v = 9.1 + 5.0 \cdot \log(D)$$

where: D is the diameter in inches.

Photographic magnitude (approx.)

$$M_p = 5 \cdot \log(D) + 2.50 \log(E) + 9.5$$

(for 103a-O, epoch 1976)
where: E is the exposure in minutes.

Limiting photographic exposure (approx. assuming a dark sky)

$$\log(E) = 0.201 + 2.0 \log(f)$$

(for 103a-O, epoch 1976)
where: f is the f/ratio.

Photographic resolution

$$R_p = \mathrm{MAX}\left[\frac{k}{F}, \frac{24.8}{A}, (S)\right]$$

A = Aperture in inches

F = Focal length in inches

(S) = Seeing in arc sec

k = Film constant (103 for 103a-emulsions, 26 for 2415 Technical Pan)

MAX $[A,B,C]$ = Greatest value of A, B, or C.

Size of Newtonian diagonal required to illuminate a given field

$$d = I + S[(D - I)/F]$$

where

d = Minor axis of the secondary

S = Distance from the optical axis to the film plane

D = Diameter of the primary

F = Focal length of the primary

(Keep all units consistent).

Required offset from optic axis for a Newtonian Diagonal

$$x = d(D - I)/4F$$

where

x = offset from optical axis

(All other parameters previously defined. Keep all units consistent.)

Effective focal ratios of photographic systems (*see Fig. A.2*)

Prime focus or direct objective

$$\text{E.F.R.} = \frac{F}{D}$$

A focal

$$\text{E.F.R.} = \frac{F \times F_c}{D \times F_e}$$

Positive projection

$$E.F.R. = \frac{F \times L_c}{D \times F_e}$$

Negative projection

$$E.F.R. = \frac{F \times L}{D \times B}$$

where

D = Diameter of primary
F = Focal length of the primary
L = Distance from 'eyepiece' to film plane
F_e = Focal length of eyepiece
F_c = Focal length of the camera lens
B = Distance from 'Barlow lens' to primary image plane
(Keep all units consistent, do not mix inches and mm.)

Fig. A.2

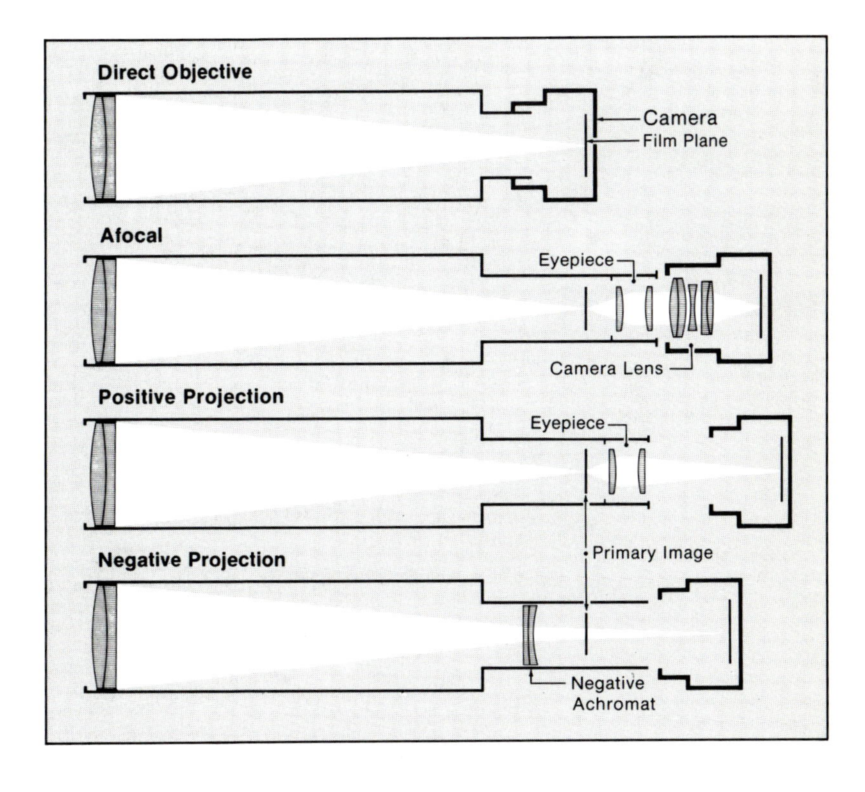

Bibliography

ATLASES AND CATALOGUES

Arp, H.C. *Atlas of Peculiar Galaxies*. Pasadena, CA, California Institute of Technology, (1966).

Barnard, E.E. *A Photographic Atlas of Selected Regions of the Milky Way*. Washington DC, Carnegie Institution of Washington,(1927).

Becvar, A. *Atlas of the Heavens–1950.0*. Sky Publishing Corp., Cambridge, Mass. (1964).

Becvar, A. *Atlas of the Heavens Catalogue–1950.0* Sky Publishing Corp., Cambridge, Mass. (1964).

Jones, K.G. (ed.) *Webb Society Deep Sky Observer's Handbook:* Vol. 2 *Planetary and Gaseous Nebulae* (1979); Vol. 3 *Open and Globular Clusters* (1980); Vol. 4 *Galaxies* (1981); Vol. 5 *Clusters of Galaxies* (1982). Hillside NJ., Enslow Publishers.

Mallas, J. & Kreimer, E. *The Messier Album*. Cambridge, Mass., Sky Publishing Corp, (1978).

Ross, F.E. & Calvert, M.R. *Photographic Atlas of the Milky Way*.Chicago, Ill., The University of Chicago Press, (1934).

Sandage, A. *The Hubble Atlas of Galaxies*. Washington DC, Carnegie Institution of Washington, (1961).

Tirion, W. *Sky Atlas 2000.0*. Cambridge, Mass., Sky Publishing Corp., and Cambridge, England, Cambridge University Press, (1981).

Vehrenberg, H. *Atlas of Deep Sky Splendors*. Cambridge, Mass., Sky Publishing Corp., (1967).

Vehrenberg, H. *Atlas of Deep Sky Splendors*. Cambridge, Mass., Sky Publishing Corp., (1978).

Publications of the Lick Observatory, Vol. 8. Sacramento, CA, University of California Publications, (1908).

Publications of the Lick Observatory, Vol. 11. Sacramento, CA, University of California Publications, CA (1913)

Publications of the Lick Observatory, Vol. 13, Berkley, CA, University of California Press, (1918).

Palomar Sky Survey. Pasadena, CA, California Institute of Technology, (1954).

Smithsonian Astrophysical Observatory Star Catalogue. Washington, DC. Smithsonian Institution, (1966).

GENERAL

Covington, M. *Astrophotography for the Amateur*, Cambridge University Press (1985).

Dittmer, H.R. A northwesterner and his astrophotography equipment. *Sky and Telescope*, **49** (6) p. 399 (1975).

Gordon, B. *Astrophotography*. Richmond, VA, Willman-Bell Inc. (1985).

Henzl, L.C. Astrophotographic routes. *Astronomy*, **3** (5) p. 59 (1975).

Keene, G.T. *Star Gazing with Telescope and Camera*. New York, NY Amphoto, (1962).

McClure, A. Photographing Comet Ikeya. *Sky and Telescope*, **25** (5) p. 263 (1963).

Mayall, R.J. & Mayall, M.W. *Skyshooting – Photography for Amateur Astronomers*. New York, NY Dover, (1968).

Newton, J. Some elements of astrophotography. *Astronomy*, **2** (7) p. 42 (1974).

Paul, H. *Outer Space Photography for the Amateur.* New York, NY Amphoto, (1960).
Provin, R.W. & Walliş, B.D. Astrophoto Gallery. *Star and Sky*, **2** (2) p. 40 (1980).
Rihm, K. Some deep-sky photographs. *Sky and Telescope*, **43** (1) p. 22 (1979).
Roth, G.D. (ed.) *Astronomy: a Handbook.* Cambridge, Mass., Sky Publishing Corporation, (1975).
Schur, C. Experiments with all-sky photography. *Sky and Telescope*, **63** (6) p. 621 (1982).
Sidgewick, J.B. *Amateur Astronomer's Handbook*, 2nd edn. London: Faber & Faber (1961).
Sussman, A. *The Amateur Photographer's Handbook*, 8th edn. New York, NY: Thomas Y. Crowell Company, Inc. (1973).
Texereau, J. *How to Make a Telescope*, 2nd edn. Richmond, VA: Willman-Bell, Inc. (1984).
Vehrenberg, H. Photographs of deep-sky objects. *Sky and Telescope*, **55** (4) p. 295 (1978).

Journal abbreviations
AJ: Astronomical Journal
ApJ: Astrophysical Journal
JOSA: Journal of the Optical Society of America
MNRAS: Monthly Notices of the Royal Astronomical Society
PASP: Publications of the Astronomical Society of the Pacific
Pub AAS: Publications of the American Astronomical Society
Proc. Am. Phil. Soc.: Proceedings of the American Philosophical Society
PSE: Photographic Science and Engineering
AAS Photo-Bulletin: American Astronomical Society Photo-Bulletin

Chapter 1

Ashbrook, J. Herschel's 'large 20-foot' telescope. *Sky and Telescope*, **54** (3) p. 174 (1977).
Barnard, E.E. The development of photography in astronomy. *Popular Astronomy*, **6** (8) p. 425 (1898).
– Photographs of the Milky Way. *ApJ*, **1** (1) p. 10 (1895).
– Photograph of the nebula NGC 1499, near the star ksi Persei. *ApJ*, **2** (5) p. 350 (1895).
– Photograph of the Milky Way near the star theta Ophiuchii. *ApJ*, **9** (3) p. 157 (1899).
– Diffused nebulosities of the heavens. *ApJ*, **17** (1) p. 77 (1903).
– The Bruce photographic telescope of the Yerkes Observatory. *ApJ*, **21** (1) p. 35 (1905).
– A great photographic nebula near pi and delta Scorpii. *ApJ*, **23** (2) p. 144 (1906).
– On a nebulous groundwork in the constellation of Taurus. *ApJ*, **25** (3) p. 218 (1907).
Bishop, R.L. Newton's telescope revealed. *Sky and Telescope*, **59** (3) p. 207 (1980).
DeVaucouleurs, G. *Astronomical Photography: From the Daguerrotype to the Electron Camera.* New York, NY: MacMillan (1961).
Dick, T. Lord Rosse's great telescope. *Telescope Making* (7), p. 8, (Spring 1980).
Eder, J.M. *History of Photography.* New York, NY: Columbia University Press (also available in Dover edition) (1945).
Gingerich, O. The first photograph of a nebula. *Sky and Telescope*, **60** (5) p. 364 (1980).
Hoffleit, D. The first star photograph. *Sky and Telescope*, **9** (9) p. 207 (1950).
King, E.S. Study of the driving worms of several photographic instruments. *Pub AAS*, **1** p. 212.
– Forms of images in stellar photography. *Annals of the Harvard College Observatory*, **41** (6) p. 153 (1902).
– Photographic photometry on a uniform scale. *Annals of the Harvard College Observatory*, **59** (2) p. 33 (1912).
– Standard tests of photographic plates. *Annals of the Harvard College Observatory*, **59** (1) 1912.
– Lunar photometry and photographic sensitiveness at different temperatures. *Annals of the Harvard College Observatory*, **59** (3) 1912.
– *A Manual of Celestial Photography*, Boston. Eastern Science Supply Co. (1931).
King, H.C. *The History of the Telescope.* New York, NY: Dover (1979).
Mills, D.J. George Willis Ritchey and the development of celestial photography. *American Scientist*, **54** (1) p. 64 (1966).

Miller, Wm C. First color portraits of the heavens. *National Geographic Magazine*, **115** (5) p. 670 (1959).
Norman, D. The development of astronomical photography. *Harvard College Observatory Reprint*, No. 153 (1938) (Reprinted from *Osiris*, **5**, p. 560 (1938).
Roberts, Isaac *Photographs of Stars, Star Clusters and Nebulae*. London Universal Press. (1893).
– *Photographs of Stars, Star Clusters and Nebulae*, Vol. II. London Knowledge Office, (1899).
Russel, H.C. Progress of astronomical photography. *Popular Astronomy*, Part I: **2** (3) p. 101 (1894); Part II: **2** (4) p. 170 (1894); Part III: **2** (7) p. 310 (1895); Part IV: **2** (9) p. 457 (1895).
Stone, R.P.S. The Crossley Reflector: a centennial review. *Sky and Telescope*, Part I: **58** (4) p. 307 (1979); Part II: **58** (5) p. 396 (1979).
Struve, O. E.E. Barnard and Milky Way photography. *Sky and Telescope*, **22** (1) p. 14 (1961).

Chapter 2

Anderson, B. & Sikes, K. The plastic astrocamera. *Sky and Telescope*, **60** (6) p. 530 (1980).
Baker, J.G. A family of flat-field cameras equivalent in performance to the Schmidt camera. *Proc.Am.Phil.Soc.*, **82** (3) p. 339 (1940).
– Planetary telescopes. *Applied Optics*, **2** (2) p. 111 (1963).
Berry, R. Astrographic cameras. *Astronomy*, **4** (11) p. 34 (1976).
– Newtonian telescopes. *Telescope Making* (9), p. 6 (Fall 1980).
Blote, H.W.J. Star images in the presence of aberrations. *Sky and Telescope*, **55** (4) p. 347 (1978).
Bowen, I.S. Telescopes. *AJ*, **69** (10) p. 816 (1964).
Buchroeder, R.A. Cassegrain optical systems. *Telescope Making* (1), p. 1 (Fall 1978).
Cooke, S.R.B. A 35 mm camera for astrophotography. *Sky and Telescope*, **41** (2) p. 109 (1971).
Covitz, F. An automatic guider for astrophotography. *Sky and Telescope*, **41** (3) p. 191 (1974).
– More about an automatic telescope guider. *Sky and Telescope*, **41** (4) p. 259 (1974).
Cox, R.E. Notes on clock-drive speed controls. *Sky and Telescope*, **40** (4) p. 237 (1970).
– Better mirror cells. *Telescope Making* (6), p. 32 (Winter 1979–80).
– Secondary mirrors and spiders. *Telescope Making* (7), p. 4 (Spring 1980).
– The Poncet mount. *Telescope Making* (8), p. 22 (Summer 1980).
DeVany, A.S. A Schmidt–Cassegrain optical system with a flat field. *Sky and Telescope*, Part I: **24** (5) p. 318 (1965); Part II: **24** (6) p. 380 (1965).
Entrop, H.A. Astrophotography with a Poncet mounting. *Sky and Telescope*, **56** (5) p. 462 (1978).
Everhart, E. & Kantorski, J. Diffraction patterns produced by obstructions in reflecting telescopes. *AJ*, **64** (10) p. 455 (1959).
Faix, L.J. Correcting periodic error in a clock drive. *Sky and Telescope*, **55** (5) p. 439 (1978).
Feder, D.P. Optical calculations with automatic computing machinery. *JOSA*, **41** (9) p. 630 (1951).
– Automatic lens design methods. *JOSA*, **47** (10) p. 902 (1957).
– Automatic optical design. *Applied Optics*, **2** (12) p. 1209 (1963).
Gelles, R. A new family of flat field cameras. *Applied Optics*, **2** (10) p. 1081 (1963).
Hawkins, D.G. & Linfoot, E.H. An improved type of Schmidt camera. *MNRAS*, **105** (6) p. 334 (1945).
Herzberger, M. Some remarks on ray tracing. *JOSA*, **41** (11) p. 805 (1951).
– Automatic ray tracing. *JOSA*, **47** (8) p. 736 (1957).
– Some recent ideas in the field of geometrical optics. *JOSA*, **51** (6) p. 661 (1963).
Herzberger, M. & McClure, N.R. The design of superachromatic lenses. *Applied Optics*, **2** (6) p. 553 (1963).
Howell, B.J. Design procedure for Ritchey–Chrétien corrector. *Applied Optics*, **8** (3) p. 685 (1969).
Jensen, N. *Optical and Photographic Reconnaissance Systems*. New York, NY: Wiley (1968).

Kingslake, R. *Lens Design Fundamentals*. New York: Academic Press (1978).
Klaus, G. Maksutov cameras for astronomical photography. *Sky and Telescope*, **28** (3) p. 168 (1964).
Labrecque, J.J. Testing a Schmidt corrector at a finite distance. *Sky and Telescope*, **37** (4) p. 250 (1969).
Larks, L. Optical ray-tracing on a microcomputer. *Sky and Telescope*, **61** (4) p. 356 (1981).
Linfoot, E.H. The Schmidt–Cassegrain systems and their applications to astronomical photography. *MNRAS*, **104** (1) p. 48 (1944).
Lurie, R.J. Anastigmatic catadioptric telescopes. *JOSA*, **65** (3) p. 261 (1975).
Maksutov, D.D. New catadioptric meniscus systems. *JOSA*, **34** (5) p. 270 (1944).
Meinel, A.B. Aspheric field correctors for large telescopes. *ApJ*, **118** p. 335 (1953).
Nash, C. An amateur-built chain drive for a 12.5-inch reflector. *Sky and Telescope*, **42** (6) p. 378 (1971).
Peters, W.T. & Pike, R. The size of the Newtonian diagonal. *Sky and Telescope*, **53** (3) p. 220 (1977).
Plummer, H.C. Star images formed by parabolic mirrors. *AJ*, **19** (3) p. 21 (1898).
– On the images formed by a parabolic mirror. *MNRAS*, **62** (5) p. 352 (1902).
Poncet, A. An equatorial table for astronomical equipment. *Sky and Telescope*, **53** (1) p. 64 (1977).
Price, W.H. The photographic lens. *Scientific American*, **235** (2) p. 72 (1976).
Provin, R.W. & Wallis, B.D. Design considerations for deep-sky photographic Newtonians. *Proceedings Southern California Astrophotography Seminar II*, (3/77) p. 65.
– Telescope design for astrophotography. *Telescope Making* (5), p. 22 (Fall 1979).
Richter, J.L. Rx for the Newtonian telescope. *Sky and Telescope*, **69** (5) p. 456 (1985).
Rosin, S. Optical systems for large telescopes. *JOSA*, **51** (3) p. 331 (1961).
– Corrected Cassegrain systems. *Applied Optics*, **3** (1) p. 151 (1964).
– Ritchley–Chrétien corrector system. *Applied Optics*, **5** (2) p. 675 (1966).
Ross, F.E. Lens systems for correcting coma of mirros. *Astrophysical Journal*, **81** (2) p. 156 (1935).
Rutten, H.G.J. & van Venrooij, M.A.M. The optical performance of eight-inch Schmidt–Cassegrain telescopes. *Telescope Making* (13), p. 12 (Fall 1981).
– Optical performance of astrocameras. *Telescope Making* (18), p. 12 (Winter 1982).
Schulte, D.H. Prime focus correctors involving aspherics. *Applied Optics*, **5** (2) p. 313 (1966).
– Anastigmatic Cassegrain-type telescopes. *Applied Optics*, **5** (2) p. 309 (1966).
Sherman, B. What's wrong with your lens? *Modern Photography*, **32** (10) p. 118 (1968).
Sigler, R.D. Family of compact Schmidt–Cassegrain telescope designs. *Applied Optics*, **13** (8) p. 1765 (1974).
– A high performance Maksutov telescope. *Sky and Telescope*, **50** (3) p. 190 (1975).
Simmons, M. Optical configurations for astronomical photography. *Sky and Telescope*, **60** (1) p. 64 (1980).
Sinnott, R.W. (ed.) Spin-offs of the Poncet mounting. *Sky and Telescope*, **59** (2) p. 163 (1980).
– Further notes on the Poncet platform. *Sky and Telescope*, **59** (3) p. 251 (1980).
Smiley, C. The Schmidt camera. *Popular Astronomy*, **44** (8) p. 415 (1936).
Smith, W.J. *Modern Optical Engineering*. New York, NY: McGraw-Hill (1966).
Sterrenburg, F. Is that lens a star performer? *Astronomy*, **19** (12) p. 43 (1981).
Stoltzmann, D. Newtonian aberrations. *Telescope Making* (12) p. 4 (Summer 1981).
– Star images in the presence of aberrations. *Sky and Telescope*, **59** (4) p. 347 (1978).
– Optical design with a pocket calculator. *Proceedings of Society of Photo-Optical Instrumentation Engineers*, **327** p. 115 (1982).
– Resolution criteria for diffraction limited telescopes. *Sky and Telescope*, **65** (2) p. 176 (1983).
Waineo, T.J. Fabrication of a Wright telescope. *Sky and Telescope*, **38** (2) p. 112 (1969).
Waland, R.L. Flat-fielded Maksutov–Cassegrain optical systems. *JOSA*, **51** (3) p. 359 (1961).
Wallis, B.D. & Provin, R.W. On the road to better astrophotography. *Sky and Telescope*, Part I: **53** (4) p. 314 (1977); Part II: **53** (5) p. 399 (1977); Part III: **53** (6) p. 484 (1977).

– Reality vs. raytrace. *Telescope Making* (19), p. 5 (Spring 1983).
– Raytrace interpretation for astrographic applications. *Proceedings Southern California Astrophotography Seminar VI*, (3/85) p. 40.
Warkoczewski, S.J. A guided camera for a 16.4-inch Newtonian telescope. *Sky and Telescope*, **41** (3) p. 175 (1971).
Werenskiold, C.H. A note on curved spiders. *Sky and Telescope*, **38** (4) p. 262 (1969).
Willey, R.R. Jr Cassegrain-type telescopes. *Sky and Telescope*, **23** (4) p. 191 (1962).
– Characteristics and testing of Cassegrain-type telescopes. *Sky and Telescope*, **23** (4) p. 226 (1962).
Wilson, R.N. Corrector systems for Cassegrain telescopes. *Applied Optics*, **7** (2) p. 253 (1968).
Wright, F.B. An aplanatic reflector with a flat field related to the Schmidt telescope. *PASP*, **47** (280) p. 300 (1935).
Wynne, C.G. Field correctors for parabolic mirrors. *Proceedings of the Physical Society of London*, **62** (12) p. 772 (1949).
– Field correctors for large telescopes. *Applied Optics*, **4** (9) p. 1185 (1965).
– Afocal correctors for paraboloidal mirrors. *Applied Optics*, **6** (7) p. 1227 (1967).
Military Standardization Handbook: Optical Design. (MIL-HDBK-141). Washington, DC: Defense Supply Agency (1962).

Chapter 3

Berry, R. Star test your telescopes. *Astronomy*, **4** (10) p. 27 (1976).
Custer, C.P. Photographic polar alignment of an equatorial mounting. *Sky and Telescope*, **46** (5) p. 329 (1973).
Dickinson, R.C. A simple knife edge focusing attachment. *Sky and Telescope*, **42** (4) p. 235 (1971).
Fleischer, R. Guiding small telescopes. *Sky and Telescope*, **7** (10) p. 243 (1948).
Provin, R.W. & Wallis, B.D. Guiding. *Astronomy*, **8** (12) p. 39 (1980).
Wallis, B.D. Another alternative to the guiding problem. *Griffith Observer*, **38** (3) p. 14 (1974).
Wallis, B.D. & Provin, R.W. On the road to better astrophotography. *Sky and Telescope*, Part I: **53** (4) p. 314 (1977); Part II: **53** (5) p. 399 (1977); Part III: **53** (6) p. 484 (1977).
– Polar alignment for astrophotography. *Astronomy*, **5** (9) p. 48 (1977).
Whipple, F.L. The polar alignment of photographic telescopes. *Popular Astronomy*, **57** (9) p. 419 (1949).
Some notes on polar alignment. *Sky and Telescope*, **58** (3) p. 280 (1979).

Chapter 4

Campbell, W.W. On atmospheric conditions required for astronomical observation. *PASP*, **27** (158) p. 65 (1915).
Evans, D.S. Climatic factors and telescope output. *Astronomy and Astrophysics*, **3** (2) p. 247 (1969).
Everhart, E. Finding your telescope's magnitude limit. *Sky and Telescope*, **63** (1) p. 28 (1984).
Evershed, J. Note on the atmospheric conditions required for astronomical observations, *PASP*, **27** (160) p. 179 (1915).
Fowler, T.B. A monogram for astrophotographers. *Sky and Telescope*, **51** (5) p. 353 (1976).
Gaviola, E. On seeing: fine structure of stellar images and inversion layer spectra. *Astronomical Journal*, **54** (6) p. 155 (1949).
Hubble, E.P. The surface brightness of threshold images. *ApJ*, **76** (2) p. 106 (1932).
Hussey, W.J. Report by W.J. Hussey on certain possible sites for astronomical work in California and Arizona. *Carnegie Institute Yearbook* (2) p. 71 (1903).
Keller, G. Astronomical seeing and its relation to atmospheric turbulence. *Astronomical Journal*, **58** (5) p. 113 (1953).
King, E.S. Forms of images in stellar photography. *Annals of the Harvard College Observatory*, **41** (6) p. 153 (1902).

– Photographic gradation as affected by delay in development. *Pub. AAS*, **3**, p. 271 (1916).
Latham, D.W. & Furenlid, I. The influence of background exposures on the detective performance of photographic plates. *AAS Photo-Bulletin*, (1) p. 11 (1976).
McClure, R.D. A study of the factors affecting seeing at the David Dunlap Observatory. *Astronomical Journal*, **71** (4) p. 273 (1966).
Pease, F.G. Notes on atmospheric effects observed with the 100-inch telescope. *PASP*, **36** (212) p. 191 (1924).
Ross, F.E. Photographic sharpness and resolving power. *ApJ*, **52** (4) p. 201 (1920).
– Physics of the developed photographic image. *Eastman Kodak Monograph*, No. 5 (1924).
– Astrometry with mirrors and lenses. *ApJ*, **77** (4) p. 243 (1933).
– Limiting magnitudes. *ApJ*, **88** (5) p. 548 (1938).
Sinnott, R.W. (ed.) Limiting magnitudes with amateur telescopes. *Sky and Telescope*, **45** (6) p. 401 (1973).
– Further notes on exposure times. *Sky and Telescope*, **51** (5) p. 355 (1976).
Wallis, B.D. & Provin, R.W. Reality vs. raytrace. *Telescope Making* (19), p. 5 (Spring 1983).
– A look into photographic resolution. *AAS Photo-Bulletin*, (2) p. 4 (1983).
Whipple, F. & Rubenstein, P.J. The limiting magnitudes of photographic telescopes. *Popular Astronomy*, **50** (1) p. 24 (1942).

Chapter 5

King, E.S. Absorption of photographic wedges. *Annals of the Harvard College Observatory*, **41** (9) p. 237 (1902).
– Photographic photometry on a uniform scale. *Annals of the Harvard College Observatory*, **59** (2) p. 33 (1912).
Mees, C.E.K. *The Theory of the Photographic Process*. New York, NY: McMillan (1942).
Mees, C.E.K. (ed.) *The Theory of the Photographic Process*. 3rd edn. New York, NY: McMillan (1966).
Miller, Wm C. *Observatory Photographic Manual: Practical and Theoretical Considerations Leading to Optimum Results*. Pasadena, CA: Wm C. Miller, Hale Observatories (1979).
Neblette, C.B. *Photography: Its Materials and Processes*. New York, NY: Van Nostrand (1952).
– *Photography: Its Materials and Processes*. New York, NY: Van Nostrand (1962).
Sanders, N. *Photographing for Publication*. New York, NY: R.R. Bowker (1983).
Shapiro, C. (ed.) *The Lithographers Manual*. Pittsburgh, PA: Graphic Arts Technical Foundation (1980).
Smith, A.G. Measurement of absolute photographic sensitivity. *AAS Photo-Bulletin*, (2) p. 13 (1980).
– Comparison of the absolute sensitivity of Kodak Technical Pan Film 2415 with standard astronomical emulsions. *AAS Photo-Bulletin*, (2) p. 6 (1982).
Spurr, S.H. *Photogrammetry and Photo-Interpretation*. New York, NY: Ronald Press (1960).
Todd, H.N. & Zakia, R.D. *Photographic Sensitometry: The Study of Tone Reproduction*. 2nd edn. New York, NY: Morgan & Morgan (1974).
Wakefield, G.L. *Practical Sensitometry*. London: Fountain Press (1970).
Wentzel, F. *Graphic Arts Photography: Color*. Pittsburgh, PA: Graphic Arts Technical Foundation (1980).
White, M., Zakia, R. & Lorenz, P. *The New Zone System Manual*. New York, NY: Morgan Press (1976).

Chapter 6

Adams, A. *Basic Photo Series*, Vol. 2, *The Negative*. New York, NY: Morgan & Morgan (1968).
Brown, G.P., Keene, G. & Millikan, A. An evaluation of films for astrophotography. *Sky and Telescope*, **59** (5) p. 433 (1980).
Cogoli, J. *Graphic Arts Photography: Black and White*. Pittsburgh, PA: Graphic Arts Technical Foundation (1981).

Conrad, C.M., Smith, A.G. & McCuiston, W.B. Evaluation of nine developers for hypersensitized Kodak Technical Pan Film 2415. *AAS Photo-Bulletin* (1) p. 3 (1985).
Dilfey, J.A. Two developers for use in astronomy. *AJ*, **73** (2) p. 762 (1968).
Eaton, G.T. *Photographic Chemistry*. New York, NY: Morgan & Morgan (1957).
Everhart, E. Adventures in fine-grain astrophotography. *Sky and Telescope*, **61** (2) p. 100 (1981).
Horder, H. (ed.) *Ilford Manual of Photography*. London: Ilford (1958).
Lutnes, J.H. & Davidson, D. A wide latitude developer for use in astronomical photography. *PASP*, **78** (465) p. 511 (1966).
Lyalikov, K.S. *The Chemistry of Photographic Mechanisms*. London: Focal Press (1968).
Marling, J.B. The 2415 revolution. *Astronomy*, **10** (3) p. 59 (1982).
Mees, C.E.K. *The Theory of the Photographic Process*. New York, NY: McMillan (1942).
Mees, C.E.K. (ed.) *The Theory of the Photographic Process*. 3rd edn. New York, NY: McMillan (1966).
Miller, Wm C. *Observatory Photographic Manual: Practical and Theoretical Considerations Leading to Optimum Results*. Pasadena, CA: Wm C. Miller, Hale Observatories (1979).
– Photographic processing for maximum uniformity and efficiency. *AAS Photo-Bulletin*, (2) p. 3 (1971).
Miller, Wm C. & Franz, D.W. Astrophotography: an introduction. *Photomethods*, **18** (12) p. 17 (1975).
Miller, Wm C. & Hoag, A.A. Applications of photographic materials in astronomy. *Applied Optics*, **8** (12) p. 2417 (1969).
Neblette, C.B. *Photography: Its Materials and Processes*. New York, NY: Van Nostrand (1952).
– *Photography: Its Materials and Processes*. New York, NY: Van Nostrand (1962).
Sanders, N. *Photographing for Publication*. New York, NY: R.R. Bowker (1983).
Shapiro, C. (ed.) *The Lithographers Manual*. Pittsburgh, PA: Graphic Arts Technical Foundation (1980).
Spurr, S.H. *Photogrammetry and Photo-Interpretation*. New York, NY: Ronald Press (1960).
White, M., Zakia, R. & Lorenz, P. *The New Zone System Manual*. New York, NY: Morgan Press (1976).

Chapter 7

Ables, H.D. & Christy, J.W. Color photographs of astronomical objects. *PASP*, **78** (465) p. 495 (1966).
Alt, E., Brodkorb, E. & Rushe, J. More about indirect color astrophotography. *Sky and Telescope*, **48** (5) p. 333 (1974).
Clerc, L.P. *Photography*, Vol. 6. *Colour Processes*. New York, NY: *AMPHOTO* (1971).
Evans, R.M. *An Introduction to Color*. New York, NY: Wiley (1948).
Everhart, E. Color negative films for astrophotography. *AAS Photo-Bulletin*, (2) p. 9 (1982).
Horder, H. (ed.) *Ilford Manual of Photography*. London: Ilford (1958).
Hunt, R.W.G. *The Reproduction of Colour*. London: Fountain Press (1975).
Iburg, B. The wet side of color astrophotography. *Astronomy*, **6** (12) p. 42 (1978).
Itten, J. *The Art of Color*. New York, NY: Van Nostrand Reinhold (1973).
Jones, S.E. & Cook, N.D. Color pictures of planets from black-and-white images. *Sky and Telescope*, **47** (1) p. 57 (1974).
Keene, G.T. Color in astrophotography. *Sky and Telescope*, **26** (2) p. 74 (1963).
Lyalikov, K.S. *The Chemistry of Photographic Mechanisms*. London: Focal Press (1960).
Malin, D.F. The colors of deep space. *Astronomy*, **8** (3) p. 6 (1980).
Malin, D. & Murden, P. *Colours of the Stars*. Cambridge University Press (1984).
Marling, J.B. Advances in astrophotography. *Sky and Telescope*, **67** (6) p. 582 (1984).
Miller, Wm C. First color portraits of the heavens. *National Geographic Magazine*, **115** (5) p. 670 (1959).
– Scientific color accuracy. *Ansconian* (GAF House Magazine) (111) p. 1 (1961).
– Color photography in astronomy. *PASP*, **74** (441) p. 457 (1962).
Neblette, C.B. *Photography: Its Materials and Processes*. New York, NY: Van Nostrand (1952).

– *Photography: Its Materials and Processes*. New York, NY: Van Nostrand (1962).
Norton, O.R. Color infrared photography of some astronomical objects. *Sky and Telescope*, **45** (6) p. 396 (1973).
Provin, R.W. & Wallis, B.D. Color emulsions for deep-sky photography, *Proceedings Southern California Astrophotography Seminar I*, (3/75) p. 45.
Rihm, K. Color portraits of deep-sky objects. *Sky and Telescope*, **48** (2) p. 120 (1974).
Sanders, N. *Photographing for Publication*. New York, NY: R.R. Bowker (1983).
Smith, A.G. & Schrader, H.W. Balanced hypersensitization of a fast reversal color film. *AAS Photo-Bulletin*, (2) p. 9 (1979).
Walker, M.F. A two-color dye-transfer photograph of M33. *PASP*, **79** (467) p. 119 (1967).
Wallis, B.D. & Provin, R.W. Is faster better?. *Astronomy*, **4** (12) p. 52 (1976).
Williamson, S.J. & Cummins, H.Z. *Light and Color in Nature and Art*. New York, NY: Wiley (1983).

Chapter 8

Adams, A. *Basic Photo Series*, Vol. 3, *The Print*. New York, NY: Morgan & Morgan (1968).
Clerc, L.P. *Photography*, Vol. 5, *Positive Materials*. New York, NY: Amphoto (1971).
Curtin, D. & DeMaio, J. *The Darkroom Handbook*. New York, NY: Van Nostrand Reinhold (1979).
Horder, H. (ed.) *Ilford Manual of Photography*. London: Ilford (1958).
Jacobson, C.I. & Mannheim, L.A. *Enlarging. 21st edn*. New York, NY: Amphoto (1972).
Kelly, J. (ed.) *Darkroom 2*. New York, NY: Lustrum Press (1978).
Lewis, E. (ed.) *Darkroom*. New York, NY: Lustrum Press (1977).
Miller, Wm C. & Bedke, J. Comparative tests of some commercially available photographic fixing baths. *AAS Photo-Bulletin*, (2) p. 12 (1976).
Neblette, C.B. *Photography: Its Materials and Processes*. New York, NY: Van Nostrand (1952).
– *Photography: Its Materials and Processes*. New York, NY: Van Nostrand (1962).
Poland, A.I. & Gosling, J.T. Reticulation: effects and cure. *AAS Photo-Bulletin*, (2) p. 14 (1975).
White, M., Zakia, R. & Lorenz, P. *The New Zone System Manual*. New York, NY: Morgan Press (1976).

Chapter 9

Benham, F. Color copy requirements. *Technical Photography*, **16** (4) p. 24 (1984).
Carlos, W. The polarized corona in color. *Sky and Telescope*, **63** (2) p. 210 (1982).
Clerc, L.P. *Photography*, Vol. 5, *Positive Materials*. New York, NY: Amphoto (1971).
Croy, O.R. *Retouching*. New York, NY: Amphoto (1964).
Espenak, F. Isophote mapping. *Astronomy*, **6** (3) p. 42 (1978).
Gorski, A.B. Enhancing astronomical photographs. *Sky and Telescope*, **58** (2) p. 184 (1979).
Gotze, H. *Photographic Retouching*. Watford, Herts: Argus Books (1980).
Johnson, H.L., Neville, H.F. & Iriarte, B. A method of increasing the photographic limiting magnitude of an astronomical telescope. *Lowell Observatory Bulletin*, No. 93, **4** (5) p. 83 (1959).
Jones, S.E. Methods, advantages and limitations of compositing photographic images. *AAS Photo-Bulletin*, (1) p. 15 (1976).
Kirby, D.S. Multiple image printing for planetary photography. *PASP*, **71** (421) p. 334 (1959).
Kneller, R.F. Black and white printing for reproduction. *Photomethods*, **27** (6) p. 15 (1984).
Macrie, J. Isophote mapping. *Astronomy*, **12** (8) p. 35 (1984).
Malin, D.F. The colors of deep space. *Astronomy*, **8** (3) p. 6 (1980).
– Photographic enhancement of direct astronomical images. *AAS Photo-Bulletin*, (2) p. 4 (1981).
Malin, D.F. & Zealey, W.J. Astrophotography with unsharp masking. *Sky and Telescope*, **57** (4) p. 355 (1979).
Maron, S.P. The Gum nebula. *Scientific American*, **225** (6) p. 21 (1971).

Matteson, J.L. Two astrophotography projects. *Sky and Telescope*, **44** (3) p. 200 (1972).
Miller, Wm C. *Observatory Photographic Manual: Practical and Theoretical Considerations Leading to Optimum Results*. Pasadena, CA: Wm C. Miller, Hale Observatories (1979).
Miller, Wm C. & Franz, D.W. Astrophotography: an introduction. *Photomethods*, **18** (12) p. 17 (1975).
Miller, Wm C. & Hoag, A.A. Applications of photographic materials in astronomy. *Applied Optics*, **8** (12) p. 2417 (1969).
Reeves, R. Making an astrophotographic mosaic. *Astronomy*, **13** (3) p. 51 (1985).
Sanders, N. *Photographing for Publication*. New York, NY: R.R. Bowker (1983).
Shapiro, C. (ed.) *The Lithographers Manual*. Pittsburgh, PA: Graphic Arts Technical Foundation (1980).
Spurr, S.H. *Photogrammetry and Photo-Interpretation*. New York, NY: Ronald Press (1960).
Walker, M.F. A two-color dye-transfer photograph of M33. *PASP*, **79** (467) p. 119 (1967).
Wallis, B.D. & Provin, R.W. Integration printing as an aid to long exposure astronomical photography. *Astronomy*, **1** (3) p. 35 (1973).
– Highlight masking: a method of detail enhancement. *Astronomy*, **1** (5) p. 19 (1973).
– On the Road to Better Astrophotography. *Sky and Telescope*, Part I: **53** (4) p. 314 (1977); Part II: **53** (5) p. 399 (1977); Part III: **53** (6) p. 484 (1977).
Weber, A.C. Discussions and personal papers.

Chapter 10

Babcock, T.A. A review of methods and mechanisms of hypersensitization. *AAS Photo-Bulletin*, (3) p. 3 (1976).
Babcock, T.A., Ferguson, P., Lewis, W.C. & James, T.H. A novel form of chemical sensitization using hydrogen gas. *PSE*, **19** (1) p. 49 (1975).
– Chemical sensitization using hydrogen gas, part II: with other types of chemical sensitization. *PSE*, **19** (4) p. 211 (1975).
Babcock, T.A., Lewis, W.C. & James, T.H. The effect of temperature upon photographic sensitivity for exposures in room air, vacuum and dry ozygen: I. Undyed emulsion coatings. *PSE*, **15** (4) p. 297 (1971).
Babcock, T.A., Lewis, W.C., McCue, P.A. & James, T.H. The effect of temperature upon photographic sensitivity for exposures in room air, vacuum and dry oxygen: II. Spectrally sensitized emulsion. *PSE*, **16** (2) p. 104 (1972).
Babcock, T.A., Michrina, P.A., McCue, P.A. & James, T.H. Effects of moisture on photographic sensitivity. *PSE*, **17** (4) p. 373 (1973).
Babcock, T.A., Sewell, M.H., Lewis, W.C. & James, T.H. Hypersensitization of spectroscopic films and plates using hydrogen gas. *AJ*, **79** (12) p. 1479 (1974).
Ernest, O. Deep-sky photography with cooled emulsions. *Sky and Telescope*, **45** (3) p. 189 (1973).
Everhart, E. Hypersensitization and astronomical use of Kodak Technical Pan Film 2415. *AAS Photo-Bulletin* (2) p. 3 (1980).
Harlan, E.A. Increasing the sensitivity of commercial films. *Sky and Telescope*, **27** (3) p. 187 (1964).
Heudier, J.L. Astronomical photography: its present status. *AAS Photo-Bulletin*. (1) p. 3 (1981).
Hoag, A.A. Cooled emulsion experiments. *PASP*, **73** (434) p. 301 (1961).
– Experiences with cooled color emulsions. *Sky and Telescope*, **28** (6) p. 332 (1964).
Hollars, D.R. Temperature effects on photographic sensitivity. *AAS Photo-Bulletin*, (2) p. 18 (1971).
King, E.S. Lunar photometry and photographic sensitiveness at different temperatures. *Annals of the Harvard College Observatory*, **59** (3) (1912).
Kreimer, E. Experiments in cooled emulsion photography. *Sky and Telescope*, **30** (6) p. 384 (1965).

– An improved cooled emulsion camera for astrophotography. *Sky and Telescope*, **32** (2) p. 106 (1966).
Lapointer, M. Which films are worth hydrogenating? *Sky and Telescope*, **55** (5) p. 401 (1978).
Lewis, W.C., Babcock, T.A., & James, T.H. The effects of vacuum on the sensitivity and reciprocity characteristics of some Kodak spectroscopic emulsions. *AAS Photo-Bulletin* 1, p. 7 (1971).
Lewis, W.C., & James, T.H. Effects of evacuation on low intensity reciprocity failure and on desensitization by dyes. *PSE*, **13** (2) p. 54 (1969).
Marchant, J.C. & Millikan, A.G. Photographic detection of faint stellar objects. *JOSA*, **55** (8) p. 907 (1965).
Marling, J.B. Gas hypersensitization of Kodak Technical Pan Film 2415. *AAS Photo-Bulletin*. (2) p. 9 (1980).
Miller, Wm C. The application of pre-exposure to astronomical photography. *PASP*, **76** (452) p. 328 (1964).
– A pre-flasher for photographic plates. *PASP*, **76** (453) p. 433 (1964).
– Reduction of low-intensity reciprocity failure in photographic plates by controlled baking. *AAS Photo-Bulletin*, (2) p. 15 (1970).
– Summary of tests with spectroscopic plates baked in nitrogen. *AAS Photo-Bulletin*, (2) p. 3 (1975).
– Preliminary results of vacuum hypersensitization applied to several Kodak spectroscopic emulsions. *AAS Photo-Bulletin*, (2) p. 4 (1972).
– Discussions, personal papers and letters.
Miller, Wm C. & Franz, D.W. Astrophotography: an introduction. *Photomethods*, **18** (12) p. 17 (1975).
Miller, Wm C. & Hoag, A.A. Applications of photographic materials in astronomy. *Applied Optics*, **8** (12) p. 2417 (1969).
Morrison, D. Recent research in hypersensitization. *AAS Photo-Bulletin*, (1) p. 12 (1969).
Newton, J. A cold camera for astrophotography. *Astronomy*, **9** (2) p. 39 (1981).
Provin, R.W. & Wallis, B.D. Pre-exposure of color films in astrophotography. *Proceedings Southern California Astrophotography Seminar III*, (3/79) p. 10.
Santos, J.M. Hypersensitizing films for astrophotography. *Sky and Telescope*, **40** (5) p. 322 (1970).
Sasaki, T. Hypersensitization of Kodak IIIa-J plates by baking in forming gas in a sealed tube. *AAS Photo-Bulletin*, (1) p. 3 (1983).
Scott, R.L. Hypersensitization of Kodak Technical Pan Film with forming gas, dry nitrogen and silver nitrate. *AAS Photo-Bulletin*, (2) p. 14 (1983).
Scott, R.L. & Smith, A.G. Comparison of nitrogen and hydrogen separately and in combination for hypersensitizing Kodak emulsions 103a-O and IIIa-J. *AAS Photo-Bulletin*, (2) p. 6 (1976).
Scott, R.L., Smith, A.G., & Leacock, R.J. The use of forming gas in hypersensitizing Kodak spectroscopic plates. *AAS Photo-Bulletin*, (2) p. 12 (1977).
Sliva, R. Hypersensitizing. *Astronomy*, Part I: **9** (4) p. 39 (1981); Part II: **9** (5) p. 48 (1981).
Smith, A.G. New trends in celestial photography. *Sky and Telescope*, **53** (1) p. 24 (1977).
– Comparison of the absolute sensitivity of Kodak Technical Pan Film 2415 with standard astronomical emulsions. *AAS Photo-Bulletin*, (2) p. 6 (1982).
– Reciprocity failure of hypersensitized and unhypersensitized Kodak Technical Pan Film 2415. *AAS Photo-Bulletin*, (3) p. 9 (1982).
Smith, A.G. & Hoag, A.A. Advances in astronomical photography at low light levels. *Annual Review of Astronomy and Astrophysics*, **17** p. 43 (1979).
Smith, A.G. & Schrader, H.W. Balanced hypersensitization of a fast reversal color film. *AAS Photo-Bulletin*, (2) p. 9 (1979).
Solomon, S.J. Speed gain addition of nitrogen baking and pre-flashing of Kodak IIIa-J emulsion. *AAS Photo-Bulletin*, (2) p. 16 (1977).
Spinrad, H. & Wilder, J. Water hypersensitization of Kodak special plates type 098–02. *AAS Photo-Bulletin*, (1) p. 15 (1972).

Wallis, B.D. & Provin, R.W. Hydrogen hypering: hot stuff or hot air?. *Proceedings Southern California Astrophotography Seminar II*, (3/77) p. 7.
– On the road to better astrophotography. *Sky and Telescope*, Part II: **53** (5) p. 399 (1977); Part III: **53** (6) p. 484 (1977).
Williams, B. A cold camera for astronomical photography. *Scientific American*, **229** (6) p. 122 (1973).
Wiseman, J.D. Jr Cooled emulsion photography without a vacuum. *Sky and Telescope*, **37** (2) p. 109 (1969).
Young, A.T. Optimum ammoniation of Kodak spectroscopic plates, IV–N. *AAS Photo-Bulletin*, (2) p. 3 (1977).

Chapter 11

Armstrong, E.B. Airglow photography with hypersensitized infrared film. *AAS Photo-Bulletin*, (3) p. 3 (1980).
Berry, R. Nebular filters. *Astronomy*, **7** (3) p. 51 (1979).
– More about nebular filters. *Astronomy*, **7** (8) p. 46 (1979).
Davis, J., Tobin, Wm & Eaton, J. Red light sky photography. *Astronomy*, **4** (8) p. 40 (1976).
Everhart, E. Constructing a measuring engine. *Sky and Telescope* **64** (3) p. 279 (1982).
Henzl, L.C. Sky photography without a telescope. *Astronomy*, **2** (11) p. 35 (1974).
– Piggy back astrophotography. *Astronomy*, **3** (1) p. 60 (1975).
Hodge, P.W. Celestial photography with fiber optics image tubes. *Sky and Telescope*, **39** (4) p. 234 (1970).
Marling, J.B. Filters for astrophotography. *Proceedings Southern California Astrophotography Seminar V*, (3/83), p. 49.
– Advances in astrophotography. *Sky and Telescope*, **67** (6) p. 582 (1984).
– Night-sky light pollution: measurement, evaluation and reduction by filter, unpublished.
Marsden, B.G. How to reduce plate measurements. *Sky and Telescope*, **64** (3) p. 284 (1982).
Mayer, B. A self operating meteor camera. *Sky and Telescope*, **48** (1) p. 54 (1974).
Oriti, R.A. Catch a falling star. *Astronomy*, **2** (8) p. 31 (1974).
Patterson, J. & Michaud, P. Photographing stellar spectra. *Astronomy*, **8** (2) p. 39 (1980).
Pike, R. & Berry, R. A bright future for the night sky. *Sky and Telescope*, **55** (2) p. 126 (1978).
Valleli, P. The focal reducer as a telescope accessory. *Sky and Telescope*, **46** (6) p. 405 (1973).
Waber, R. & McPherson, R. Photographing star spectra. *Sky and Telescope*, **33** (5) p. 322 (1967).

Chapter 12

Baker, J.G. Planetary telescopes. *Applied Optics*, **2** (2) p. 111 (1963).
Campbell, W.W. On atmospheric conditions required for astronomical observation. *PASP*, **27** (158) p. 65 (1915).
Dragesco, J. High resolution lunar photography. *Astronomy*, **12** (10) p. 35 (1984).
Evans, D.S. Climatic factors and telescope output. *Astronomy and Astrophysics*, **3** (2) p. 247 (1969).
Everhart, E. Constructing a measuring engine. *Sky and Telescope*, **64** (3) p. 279 (1982).
Evershed, J. Note on the atmospheric conditions required for astronomical observations. *PASP*, **27** (160) p. 179 (1915).
Felbab, J. Photographing the Sun. *Astronomy*, **8** (3) p. 39 (1980).
Fuller, G.C. A new way to photograph the Sun. *Sky and Telescope*, **59** (2) p. 170 (1980).
Gaviola, E. On seeing: fine structure of stellar images and inversion layer spectra. *AJ*, **54** (6) p. 155 (1949).
Goff, J., Hansen, R. & Lacey, L. Uniform dodging of H-alpha filtergrams of the solar disk. *AAS Photo-Bulletin*, (2) p. 3 (1972).
Harvey, J.W. Real-time masking of solar photographs with photochromic glass. *AAS Photo-Bulletin*, (1) p. 5 (1972).

Hicks, J. & Trombino, D. Shooting the Sun in hydrogen-alpha. *Astronomy*, **11** (7) p. 34 (1983).
Hirsch, E. Photographing the Sun in H-alpha. *Astronomy*, **6** (1) p. 37 (1978).
Hosfield, R. Comparisons of stellar scintillation with image motion. *JOSA*, **44** (4) p. 284 (1954).
Hussey, W.J. Report by W.J. Hussey on certain possible sites for astronomical work in California and Arizona. *Carnegie Institute Yearbook*, (2) p. 71 (1903).
Keller, G. Astronomical seeing and its relation to atmospheric turbulence. *AJ*, **58** (5) p. 113 (1953).
Marling, J.B. The 2415 revolution. *Astronomy*, **10** (3) p. 59 (1982).
Marsden, B.G. How to reduce plate measurements. *Sky and Telescope*, **64** (3) p. 284 (1982).
McClure, R.D. A study of the factors affecting seeing at the David Dunlap Observatory. *AJ*, **71** (4) p. 273 (1966).
Minton, R.B. Hints on planetary photography for amateurs. *Sky and Telescope*, Part I: **40** (1) p. 56 (1970); Part II: **40** (2) p. 116 (1970).
Pope, T. & Osypowski, T. High resolution photography. *Sky and Telescope*, **29** (1) p. 52 (1965).
Price, R.S. Planetary photography at high resolution. *Sky and Telescope*, **52** (3) p. 220 (1976).
Pease, F.G. Notes on atmospheric effects observed with the 100-inch telescope. *PASP*, **36** (212) p. 191 (1924).
Rouse, J.K. Tell the planets to say cheese. *Astronomy*, **2** (9) p. 50 (1974).
Smith, H.J. & McCrosky, R.E. Night cloud coverage in the southwest with reference to astronomical observing conditions. *AJ*, **59** (4) p. 156 (1954).
Stock, J. Site survey for the Inter-American Observatory. *Science*, **148** (3673) p. 1054 (1965).
Vehrenberg, H. A homebuilt machine for scanning plates. *Sky and Telescope*, **38** (3) p. 187 (1969).
Walker, M.F. Polar star trail observations of astronomical seeing in Arizona, Baja California, Chile, and Australia. *PASP*, **83** (494) p. 401 (1971).
– California site survey. *PASP*, **82** (487) p. 672 (1970).
Young, A.T. Seeing and scintillation. *Sky and Telescope*, **42** (3) p. 139 (1971).
Wallis, B.D. Photographing a total solar eclipse. *Griffith Observer*, **37** (6) p. 8 (1973).
Wallis, B.D. & Provin, R.W. A look into photographic resolution. *AAS Photo-Bulletin*, (2) p. 4 (1983).
Zeh, H.F. A Toledo amateur's methods in astrophotography. *Sky and Telescope*, **36** (4) p. 256 (1968).
Two adapters for projection photography. *Sky and Telescope*, **43** (2) p. 123 (1972).

Index

In this index, the principal sections dealing with any subject are indicated by the use of page numbers in **bold**. Illustrations are shown by page numbers in *italic*, with '*CP*' signifying that a particular illustration is a color plate. A page number may be followed by 'n' (to show that the subject is mentioned in a footnote), or by 'T' (to indicate that there is a relevant table).